Chemical Contaminants in Canadian Aquatic Ecosystems

EDITORS

Ronald C. Pierce [1]

D.M. Whittle [2]

Jonathan Burton Bramwell [3]

1   Director, Environmental Science Branch, Department of Fisheries and Oceans, 200 Kent Street, Ottawa, Ontario, CANADA
2   Manager, Ecotoxicology Division, Great Lakes Laboratory for Fisheries and Aquatic Sciences, Burlington, Ontario, CANADA
3   President, Polymorph Publications, Inc., P.O. Box 74151, Ottawa, Ontario, K1M 2H9, CANADA

Table of Contents

Acknowledgments

The chapters were prepared by individuals as follows: Introduction: R.C. Pierce; Perspectives: R.F. Addison, and R.C. Pierce; St. Lawrence Marine Ecosystems: D. Caissie, W. Fairchild, M. Gilbert, and L. White; Bay of Fundy Marine Ecosystems: L. White; Northwest Atlantic Marine Ecosystems: J. Kiceniuk; Great Lakes Ecosystems: D.M. Whittle; Inland Freshwater Ecosystems: R. Currie, and D.F. Malley; Arctic Marine Ecosystems: J. La Rue-van Es, and D.F. Malley; Pacific Marine and Freshwater Ecosystems: R.F. Addison; Conclusions: R.C. Pierce and D.M. Whittle. Permission to use text from White and Johns (1997) in the St. Lawrence Marine Ecosystem Chapter is gratefully acknowledged.

We gratefully acknowledge the contribution of information, text, and reviews from the following individuals: R. Anderson, J. Arsenault, J. Banoub, D. Bodaly, J.M. Brewers, K. Brinkley, M. Brylinsky, V. Cairns, C. Chou, C. Couillard, S. Courtney, W. Cretney, G. Daborn, A.J. Evans, J. Gearing, C. Gobeil, G. Harding, B. Hargrave, S. Hawboldt, K. Haya, J. Hellou, P.W. Hicklin, D. Huebert, S. Huestis, M. Ikonomou, R. Kiriluk, J. Klaverkamp, M. Lebeuf, K. Lee, W.L. Lockhart, D. Loring, R. MacDonald, D. Muir, K. Munkittrick, A. Niimi, T. Packard, J. Payne, J.A. Percy, J. Piuze, N. Prouse, S. Ray, J. Rudd, J. Rudd, D. Scruton, M. Servos, D. Shaw, T. Shortt, J. Smith, J. Thompson, J. Uthe, P.G. Wells, D. Willis, P. Yeats, V. Zitko.

Overall project management, and scientific and technical editing were provided by Polymorph Publications Inc. including: J.B. Bramwell (executive editor), P. Bergeron, H. Bramwell, J. Dyer, L. Gauvreau, L. Howard, K. Mahon, G. Moody, S. Perry, L. Reaume, and G. Stewart.

Layout graphics and design were provided by GLS DEZiGN Inc. including: G.L. Sreblowski, C. MacPhail, and R. Robillard.

Support project coordination was provided by D. Williams of Environalysis Consulting. Cover artwork and design were provided by the Multimedia Design Group Inc.

Translation services were provided by D. Campillo and her colleagues at the Science and Technology Section, Translation Bureau, PWGSC, Montreal.

Introduction

Canada has a long history of reliance upon rich and diverse fisheries. These valuable natural resources have sustained generations of fishers, their families, and communities, and have contributed substantially to the nation's economic prosperity and to the health and cultural identity of its citizens.

The dawning of industrialization in the 19th century began to impose serious threats to the health and sustainability of Canada's aquatic ecosystems. These threats now include release of industrial and domestic effluents, changing land use patterns, increasing urbanization, loss of fish habitat, overfishing of some species, and the inappropriate use and disposal of chemicals harmful to fish. The chemicals causing risk include many substances made by humans used to produce a myriad of industrial and consumer products.

Some chemicals can have immediate and long-lasting effects on Canada's aquatic environment and the life it sustains. Chemicals once deemed harmless have been found to accumulate in the aquatic environment, eventually reaching concentrations that can cause serious and unpredictable harm to aquatic life and to its economic value.

In recent years, understanding of how chemicals are transported throughout the environment, and the resulting consequences for fish, other aquatic life, and the habitat has dramatically improved. Scientific studies have shown that some chemicals can travel great distances, remain for long periods in the environment, accumulate and magnify in aquatic food chains, and cause harm to aquatic life, including fish and marine mammals. Some chemicals freely cross the boundaries between air, land, and water and the ultimate sinks for many chemicals are aquatic sediments, irrespective of where the chemicals are used and released. These scientific studies form the basis upon which effective and timely action can be taken to solve chemical problems, prevent future problems, and promote sustainability of fisheries resources.

This report focuses on those chemicals that persist in the Canadian aquatic environment and that are accumulated by aquatic organisms. The terms "toxic chemical," "contaminant," and "pollutant" are often used interchangeably in the literature. In this report, "contaminant" is used to describe a chemical that is foreign to an ecosystem, or a natural chemical present in unnaturally high concentrations. A "pollutant" is defined as a chemical that has an adverse effect on an ecosystem, or on some component of it. From the perspective of determining harm to aquatic life, and also the need for regulatory or other actions, this distinction is very important. A contaminant may be present in a particular aquatic ecosystem but may not be regarded as a pollutant until an adverse effect has occurred. In this report, "fish" is frequently used as it is in the *Fisheries Act* to include marine mammals, finned fish, fish larvae and eggs, and the organisms that live in and comprise the living habitat, including algae, microorganisms, invertebrates and larger plants.

OBJECTIVES OF THIS REPORT

The primary objectives of this report concerning contaminants in Canada's fisheries resources are:

- to examine how contaminants get into and are transported through the aquatic environment;
- to examine the geographic variability and temporal trends of chemical contamination in fish and fish habitats;
- to understand how chemical contaminants affect fish and fish habitat;
- to identify important gaps in our knowledge; and
- to identify chemical contaminant issues that may become future problems.

SCOPE AND CONTENT

The report summarizes scientific information collected on contaminants previously, with emphasis given to Canadian research findings published in the last 10 years. Where appropriate, studies from other countries are included for comparative purposes. Although the amount of scientific information on chemical contaminants has dramatically increased in recent years, it is beyond the scope of this report to provide an exhaustive review of all available information. Instead, it highlights those chemical contaminants of primary concern to Canada's fish and fish habitat by using examples of scientific studies to help explain contaminant fate and effects. The report includes studies in fresh, estuarine and marine waters. Some information on chemical contaminants in air, on land, in vegetation, and in the water column is given to provide a context for explaining the fate and effects of the chemicals discussed in this report.

The report emphasizes those chemical contaminants that persist in the environment, accumulate in aquatic ecosystems, and are toxic to fish and fish habitat. It does not encompass the significant problems of eutrophication and acidification in which the causative agents are not persistent nor accumulated. These subjects are treated extensively in other publications. This report also does not attempt to deal with radionuclides in detail and deals with petroleum hydrocarbons only briefly. Where information is not available on contaminants in fish, studies on other aquatic species (birds and mammals) are reviewed to provide insight on how such substances behave in aquatic foodwebs. Suggested readings of recently published information are identified at the end of each of the regional chapters.

Chemical contaminants can adversely affect Canada's fisheries resources in a number of ways. Some contaminants are directly toxic to fish and other aquatic life. Others cause subtle changes to the structure and functioning of aquatic ecosystems that eventually lead to adverse effects such as reduced fish populations. Still others accumulate to unacceptable concentrations in edible fish. While the matter of safety of fish and shellfish for human consumption is beyond the scope of this report and is the concern of various agencies in Canada that are directed towards the protection of human health, this report does record cases where the human consumption of fish is restricted because of chemical contamination.

There is enormous natural variability and unpredictability in environmental conditions, both within and among aquatic ecosystems. Depending upon the environmental conditions, a chemical contaminant may cause harm to fish in one location, but not in another. Fish and other aquatic life display a vast biological diversity in Canadian waters. Because of such variability and unpredictability, it is often difficult to differentiate human-induced effects of chemicals from natural processes.

Chemical contaminants are not the sole agents responsible for causing harm to Canada's fisheries. A combination of many different stresses, including natural processes, habitat alteration, and the discharge of effluents, can affect fish and fish habitat. Although it is beyond the scope of this report to examine all stress factors in aquatic ecosystems, some information on natural environmental processes and human-induced changes is included to help place the role of chemical contaminants into the overall context of ecosystem health.

STRUCTURE OF THE REPORT

The report is divided into three parts. The first part introduces the general features of aquatic ecosystems and gives basic information on the fate and effects of chemical contaminants.

The second part is divided into broad geographic areas. Contaminant issues are reviewed for the following areas: Atlantic marine ecosystems including the St. Lawrence, Bay of Fundy and general northwest Atlantic; freshwater ecosystems including the Great Lakes and inland waters; the Arctic marine ecosystem; and Pacific aquatic ecosystems including the British Columbia coastal ecosystem, the open coast and some of the major rivers. Although each chapter is written to reflect the unique features of the particular aquatic ecosystems, all chapters have a similar structure in that they discuss the major contaminant issues, important oceanographic or limnological features, biological resources, and the sources of contaminants as well as contaminant distribution, trends, and effects. Current and emerging chemical contaminant issues and associated research topics are summarized at the end of each ecosystem chapter. Detailed information is contained in tables and figures. Throughout the text, all measurements are given in metric units using two significant figures.

The third part is an overall assessment of current knowledge of chemical contaminants and their effects on fish and fish habitat and an identification of major current and emerging contaminant issues confronting Canada's fish and fish habitat.

The report is structured to permit the reader to gain a broad perspective on chemical contaminants as they relate to Canada's fisheries. It is intended to be understandable by the non-specialist, albeit one with a general knowledge of contemporary natural science. At the same time, the report is intended to give the specialist a broad overview of current information on the fate and effects of chemical contaminants. Highlights at the beginning of each chapter capture the main points of interest. Conclusions at the end of each chapter provide overviews of the key findings and their context and importance. Throughout the report, examples of specific scientific investigations are illustrated in Case Studies. Brief natural histories of some fish, birds, and mammals and their importance in the chemical contaminant story appear throughout the text. For those seeking more detailed information, references are listed at the end of each chapter. A glossary of terms provides definitions of technical terms used in the report. A second glossary of chemicals outlines the major physical-chemical and toxicological properties of substances covered in the text. A detailed index permits readers to locate specific topics within the report.

This report relies extensively on data and information in published literature. However, since scientific information is not static, some pre-publication information is included.

All information has been reviewed by both the authors of this report and other scientific experts. The results of Canadian research used in the preparation of this report can be accessed either through the scientific literature or by contacting the Environmental Science Branch, Department of Fisheries and Oceans, Ottawa, Ontario, Canada, K1A 0E6.

Perspectives

Table of Contents

Perspectives

Introduction

List of Figures

Introduction

Threats to the health and sustainability of Canadian fisheries can result from activities ranging from changes in land use patterns to the discharge of chemicals in industrial and domestic effluents. The release of chemicals into air and water and on land result in serious and unpredictable effects on aquatic ecosystems.

This chapter introduces some basic concepts concerning how chemicals behave in Canadian aquatic ecosystems and how they can affect fish and fish habitat. The main features of aquatic ecosystems are described to emphasize that all ecosystem components are interconnected, so that changes to any one component may affect others. The physical-chemical properties that control the behaviour of chemicals in aquatic ecosystems are also summarized, as well as some of their better-studied biological effects. For detailed descriptions of those chemicals discussed in this report, see the Glossary of Chemical Contaminants.

The terms "toxic chemical," "contaminant," "toxic contaminant," and "pollutant" are used interchangeably in the literature. In this report, the term "chemical contaminant" is used to describe a chemical which is foreign to an aquatic ecosystem, or a natural chemical present in unnaturally high concentrations (Government of Canada 1991). A "pollutant" is defined as a substance which has an adverse effect on an ecosystem, or on some component there in (derived from GESAMP 1987). Such adverse effects can include harm to aquatic life, hazards to human health, hindrance to human activities, or reduction of amenity value. The distinction between "contamination" and "pollution" is particularly important. A chemical contaminant that is present in one or more components of an aquatic ecosystem cannot be regarded as a pollutant unless it elicits an adverse effect.

Generally, the main factors that determine whether a chemical contaminant poses a risk of harm are its quantity or concentration and the duration of exposure of the aquatic organism to it. It must be noted however, that there is also a limitation in our ability to relate an observed effect to a specific cause.

AQUATIC ECOSYSTEMS

An ecosystem is a dynamic complex of living organisms including plant, animal, and micro-organism communities (the "biotic component") and their non-living environment of the substrates including sediments, water, and atmosphere, (the "abiotic component"). These two components interact as a functional unit (UNEP 1992). The ecosystems discussed in this report include freshwater streams; rivers and lakes; estuaries and coastal waters; and the open ocean. The division of ecosystems into separate categories is somewhat arbitrary, since these ecosystems interact on local, regional, and global scales. With respect to the environmental dynamics of chemical contaminants, the significant feature of ecosystems is the interdependence of their components — events in one component will affect others (Figure 1). Substances released to the environment may become distributed widely throughout the environment and may reach a variety of aquatic ecosystems, where they may be transported to or incorporated by certain abiotic and biotic components. The concept of a chemical moving and becoming distributed among the different compartments of an ecosystem is fundamental to understanding and eventually predicting how such chemicals behave in the environment. Lindane, an organochlorine insecticide, may be used as an illustration of the mechanisms for the distribution of chemical contaminants in the environment. When sprayed on crops to control insect pests, a certain portion of the lindane will evaporate into the air, some of the lindane will adsorb to soil or other atmospheric particles and may eventually enter water bodies through run-off, and some of the lindane will be accumulated by organisms. The lindane will thus become distributed between air, water, soil and biota, depending on the relative volume of these environmental compartments, and on lindane's physical-chemical properties. A dynamic equilibrium will exist among abiotic and biotic compartments, so that some fraction of lindane will be continuously moving among them.

THE BEHAVIOUR OF CHEMICAL CONTAMINANTS IN AQUATIC ECOSYSTEMS AND THEIR EFFECTS

The behaviour of a chemical in aquatic ecosystems is controlled by factors including the physical-chemical properties of the chemical; its sources, pathways, and sinks; and the physical,

chemical and biological processes that all interact within aquatic ecosystems. The effects of exposure to a chemical can be manifest at the cellular, organ, whole organism, population or community level.

PHYSICAL-CHEMICAL PROPERTIES

A chemical's behaviour in the environment is largely controlled by its physical and chemical properties, all of which can be measured in the laboratory. These properties, when entered into mathematical models, allow reasonably accurate predictions of where a chemical is likely to end up if it is released to the environment (Mackay 1991). The properties of greatest significance for chemical behaviour in the environment include volatility, solubility, partitioning onto solids, and

**Figure 1
The interdependence of ecosystem components.**

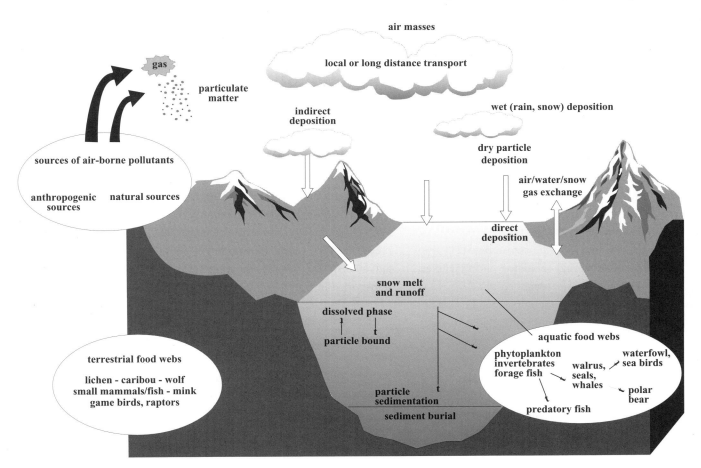

Source: D.A. Williams 1997.

stability or persistence. All of these properties are a function of temperature.

Vapour pressure or volatility describes how a chemical is distributed between the chemical's solid or liquid phase and the air. This property has an important effect on the magnitude of atmospheric transport. Chemicals with appreciable vapour pressures are likely to become widespread, volatilizing where they have been used and moving to locations where they have never been used.

The distribution of a chemical between the air, water and solids controls the movement of the chemical, either through the environment or through the food web.

Movement from liquid or gaseous phases on to particles is a crucial process that controls chemical transport in the environment. Whereas fluids such as air and water are constantly in motion carrying the dissolved components with them, particles tend to settle. Chemicals that attach strongly onto particles, therefore, are prone to removal from the air or water and often end up near the source in soils or sediments. For example, lindane readily separates out of the atmosphere and into water, but only weakly onto solids. Lakes, rivers, ground water and oceans now contain most of the lindane released to the environment. Examples of chemicals that strongly attach to particles include lead, high-molecular-weight polycyclic aromatic hydrocarbons (PAHs), highly chlorinated biphenyls (PCBs) and dibenzo-*p*-dioxins (PCDDs) and dibenzo-furans (PCDFs). Organic carbon strongly enhances the attachment of many chemicals (i.e. organochlorine compounds, PAHs and mercury) to solids, as does a small particle size because of its large surface area per unit mass. Therefore, fine, organic carbon-rich sediments are particularly prone to capturing the particle-reactive chemicals. Chemicals with low water solubility and high lipid solubility or high attraction to particulate material tend to adsorb (attach) to particulates and are therefore said to be hydrophobic.

Chemicals that do not degrade rapidly in the environment are said to be persistent. In principle, because metals are part of the natural geological makeup of the earth's crust, they persist indefinitely in the environment. However, concentrations can become elevated in aquatic ecosystems as a result of human activities. Metals are often predominantly accumulated in their organic form by biota; i.e. mercury as methylmercury.

Perhaps of greatest interest is the uptake of chemicals by biota because the uptake will clearly have the potential to produce toxic effects. Chemicals that have a high affinity for fats (lipids) are particularly prone to accumulate in the fat of aquatic animals. High lipid solubility is the basis for both bioaccumulation and biomagnification of many of the organochlorine compounds through the food chain. For example, PCB concentrations can increase by factors of greater than 10^8 going from water to the top predators.

Some chemicals exhibit more than one form in the environment. Different forms have different physical and chemical properties. The best example is mercury for which transformations between elemental mercury (dissolved or gaseous), dissolved ionic mercury, and methylmercury are possible. Methylmercury, which bioaccumulates and exhibits high toxicity, has a greater biological effect than the other forms which tend to attach to organic and sulphur-containing sediments. Elemental and methylmercury can be transported in both water and atmosphere, whereas the ionic form is transported in water.

The behaviour of all contaminants is determined by a range of physical-chemical properties. The property of greatest influence will depend upon specific circumstances within a particular ecosystem. For example, the insecticide lindane is appreciably lipid soluble, but is also measurably soluble in water, and it is volatile. Therefore, its distribution in an ecosystem will depend on the interrelationship of these three properties, in addition to their relationship to the properties of the ecosystem. Most of the chemical contaminants discussed in this report have a relatively high lipid solubility and low water solubility. These properties tend to lead to elevated concentrations in biota and in sediments, thus much of the information about contaminants relates to their presence in sediments and in aquatic biota.

SOURCES, PATHWAYS AND SINKS

Environmental scientists think of chemical contaminants in terms of their sources (where they come from), pathways (how they are distributed) and sinks (where they accumulate). As indicated above, chemicals cycle continually through the environment; although sinks may sometimes be considered as the eventual repositories for some chemicals, they can also act as sources. Recycling of contaminants from sinks is becoming more common as anthropogenic sources to the environment are increasingly limited. For the chemicals discussed in this report, sinks are of importance since they tend to represent the primary media through which many aquatic biota and their consumers are exposed to chemical contaminants. To understand how chemicals arrive in these sinks, it is helpful to know something about their sources and pathways.

Chemicals are released to the environment through natural processes or by human activity. Human activities can introduce new chemicals or can cause a change in distribution patterns of naturally occurring chemicals, as in the case of the redistribution of metals. New chemicals that are introduced by human activities are termed anthropogenic chemicals, synthetic chemicals or xenobiotics. Examples include organochlorine pesticides such as DDT and industrial compounds such as PCBs. Although some information on natural sources is included to provide context, the focus of this report is on those chemicals that enter aquatic ecosystems as a result of human activities.

Chemicals reach aquatic ecosystems from several sources. Point sources include industrial and municipal effluents that discharge water-borne waste material; industrial and municipal incinerators that emit substances into the air that eventually reach water bodies; solid waste discharges such as the disposal of sewage sludge and dredged materials; and intentional direct discharges or accidental spills. The significance of point sources is that the distribution of a chemical emanating from them can often be traced back to its starting point or elimination. Diffuse sources, or non-point sources, include general watershed run-off, urban run-off, or long range atmospheric transport. Although noteworthy from a regulatory or control perspective, the distinction between point and diffuse sources is probably less important than the medium into which the chemical is released, as the latter will have a direct influence on the pathways by which the chemical will become dispersed within an aquatic ecosystem.

Chemicals discharged into the atmosphere can remain in the vapour phase or can become adsorbed (attached) to atmospheric particles, or both. The distribution between vapour and adsorbed forms depends on factors such as the properties of the chemical, environmental temperature, and the size distribution and composition of particles in the atmosphere. For example, trichloroethylene, a compound used in dry cleaning and industrial cleaning, is quite volatile, and, once in the atmosphere, remains predominantly in the vapour phase. It therefore tends to be widely distributed throughout the world's atmosphere. In contrast, compounds such as PCDDs and PCDFs, which can be discharged into the atmosphere from municipal incinerators, are much less volatile and adsorb readily to atmospheric particles. Hence they tend to remain closer to their sources as atmospheric particles settle out or are stripped out by precipitation.

Chemicals discharged to water behave similarly. Some remain mostly dissolved, and move with water masses, while others become adsorbed to particulate material and eventually settle out as sediment. For the chemical contaminants discussed in this report, particulate matter occurring either in water (suspended sediments, suspended particulate matter and particulate organic matter) or present on the bottom of water bodies — is important in determining their fate and eventual sinks in aquatic ecosystems. An equilibrium will exist between dissolved and adsorbed phases, which will determine the actual distribution of the chemical. Chemicals that adsorb to particulate matter will be carried along in the water until the particles settle to the bottom through sedimentation. Examples of such chemicals include PCBs, some organochlorine compound pesticides, high-molecular-weight PAHs, and some metals. These chemicals are usually found in appreciably higher concentrations in sediments than in water. Depending on environmental conditions, particulate material will be settle out and uncontaminated sediments may build up over time and eventually bury these chemicals. While shorter-lived or less stable chemicals may break down, those which persist can remain in sediments for decades. If these sediments remain undisturbed, the chemicals are unlikely to be released into the water. However, if sediments are disturbed, larger amounts of the chemicals may be once again introduced into the water. Sediments can be disturbed by dynamic natural events, such as tides, storms, or the movement of organisms living on the bottom (bioturbation), as well as by activities such as dredging, bottom trawling and ship passage.

The size, amount and nature of organic matter found in particulate material are important properties relating to contaminant dynamics. Particle size governs the surface area available to adsorb a chemical. Smaller particles generally have more surface area per unit weight or volume and thus possess a greater adsorptive capacity. The amount of organic matter also affects the amount of contaminants adsorbed to particles since higher organic matter content tends to increase adsorptive capacity.

Suspended particulates and sediments in aquatic ecosystems are highly complex, variable chemical and biological mixtures, consisting of inorganic (clay and other minerals), living and non-living organic components. The non-living organic components may arise from biological activity outside the aquatic ecosystem (allochthonous origin) or from aquatic primary and secondary production within the aquatic ecosystem (autochthonous origin). Thus, the organic matter may include breakdown products of woody and herbaceous portions of terrestrial

Figure 2
Accumulation of contaminants by predators.

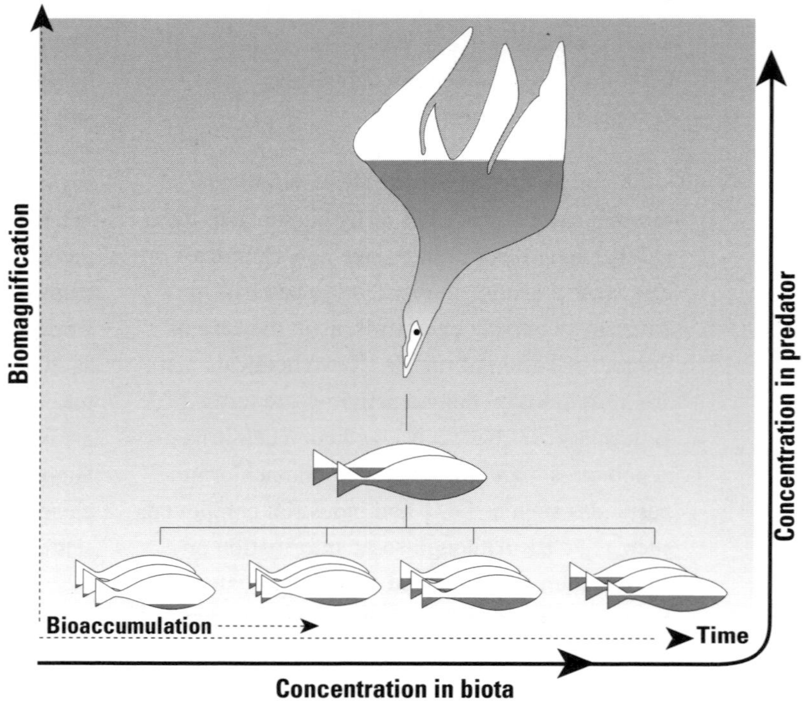

Source: D.A. Williams 1997.

plants and pollen, as well as breakdown products, excretions, exuviae (cast exoskeletons) and wastes (fecal pellets) from algae, zooplankton, and other aquatic organisms. Some of the chemical components of the organic matter are simple (amino acids, sugars, fatty acids) and readily absorbed and assimilated by bacteria and some algae. Other components are complex (proteins, cellulose, chitin, lipids, some alcohols, some esters, humic acids, fulvic acids, lignin) and are less easily assimilated. Bacterial assemblages which comprise the living portion of the organic matter contained in particulate material are responsible for consuming and chemically transforming non-living organic matter. They also synthesize new compounds as the particulates sink through the water column or become buried deeper in the sediment.

The amount and character of the organic matter of sediments, including the characteristics and amount of bacteria present, are important in that they control the chemical properties of sediment. These properties include buffering capacity, and metal binding capacity for and the availability of hydrophobic organic chemicals. Therefore, if contaminant concentrations in sediments and their effects are to be compared over time and space, it is important to compare sediments that are similar in grain size, have the same organic matter content and organic matter quality, and other chemical parameters such as oxygen content. Depending upon the circumstances, the organic matter of sediments can either increase or decrease the bioavailability of some chemical contaminants in aquatic ecosystems.

All species have evolved the ability to selectively extract and use essential chemical elements and natural compounds from their environment. The uptake, distribution and accumulation of chemical contaminants in biota depend on their physical-chemical properties, as well as on the behaviour of the organism. The higher the lipid solubility of a chemical, the more readily is it bioaccumulated. Persistent chemicals tend to biomagnify as they move up the food

chain. Predators retain contaminants in the prey and accumulate them to increasing concentrations (Figure 2). Uptake of chemicals by aquatic organisms may be passive or active. For example, phytoplankton can accumulate chemical contaminants such as PCBs by adsorption to their surface. It seems that at higher trophic levels, ingestion gradually replaces absorption as a route by which contaminants are accumulated. For example, seals accumulate virtually all of their chlorinated hydrocarbon burden through feeding, whereas clams may accumulate significant amounts by absorption through gills.

Once taken up by complex organisms, chemical contaminants may undergo several processes. Very lipid-soluble (lipophilic) compounds (e.g., PCBs) become deposited in fatty tissue. The compound's behaviour in the organism generally follows the way in which fat is used. A dynamic equilibrium between lipid and aqueous phases will exist within organisms. Some fraction of the overall amount of PCBs that are circulating in the bloodstream are presumably carried to metabolically active tissue. That tissue may be able to convert some compounds to metabolites which could then be excreted. However, in general, complex compounds such as PCBs are quite resistant to biological degradation. As a result, as animals grow older, they generally accumulate increasing concentrations of these compounds. Similarly, mercury becomes bound to proteins in the form of methylmercury and may not be easily excreted, so that mercury concentrations in fish tend to increase as fish get older. Since contaminant concentrations may vary with age, comparisons in biota over time and space must account for age differences.

The distribution, metabolism, and excretion of chemical contaminants is affected by many factors. Contaminants present in media such as sediments and water may not be readily detectable in fish. PAHs are a class of chemicals detected primarily as the unmetabolized compounds at low concentrations in fish. However, metabolites of PAHs can accumulate in liver tissues of fish where they may produce toxic effects. Metals, when present in their organic form (for example, methylmercury, alkyllead, tributyltin) are more toxic than their inorganic counterparts. The organic form of some metals is preferentially accumulated by biota. More than 90% of the mercury found in fish consists of the organic form. Female fish and marine mammals can export large amounts of fat into eggs or milk, respectively, with their accompanying burdens of chemical contaminants, which are then passed along to their offspring. Consequently, females often contain lower contaminant burdens than males. Often, the early life stages of species are more susceptible than adults to the adverse effects of chemical contaminants.

EFFECTS

The presence of a chemical in an aquatic ecosystem or in an organism does not necessarily imply an adverse effect. A chemical contaminant may elicit an adverse effect if the organism is exposed to and accumulates a sufficiently high concentration. Some chemicals may adversely affect an organism at very low concentrations such as those chemicals responsible for genetic alterations. Some aquatic organisms may be killed rapidly upon exposure for a very short time to a particular chemical. Alternatively, a chemical may be present at a very low concentration in water but may gradually bioaccumulate or biomagnify in an aquatic ecosystem. Lifetime or multigenerational exposure to low concentrations of some chemicals may cause chronic effects. Once a chemical reaches a certain concentration within an organism or a sensitive tissue, an adverse effect may be detectable.

The effects of exposure to a contaminant may be observed at the cellular, organ, whole organism, population or community level (Government of Canada 1991). Since all components of an aquatic ecosystem are linked, the effects of a chemical contaminant on a particular organism can have far reaching impact on other organisms and ultimately affect both the overall structure and functioning of ecosystems and the sustainability of the resources they contain.

Two biochemical processes involving chemical contaminants that can occur within organisms are of special interest, as they are used increasingly as indicators of exposure to chemicals and their biological effects. These biochemical indicators are the induction of mono-oxygenases, or mixed-function oxidases (MFO), and the induction of metallothionein.

Mono-oxygenases are a group of enzymes (biological catalysts) found in vertebrate liver. They appear to contribute to detoxification by promoting reactions that convert lipid-soluble compounds to more water-soluble products which

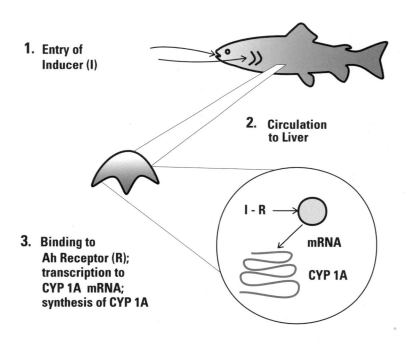

Figure 3
The enzyme induction process.

1. **Entry of Inducer (I)**

2. **Circulation to Liver**

3. **Binding to Ah Receptor (R); transcription to CYP 1A mRNA; synthesis of CYP 1A**

I - R → mRNA

CYP 1A

4. **Increased mono-oxygenase activity:**

$$\text{Substrate} \xrightarrow[\text{O}_2\,,\ \text{NADPH}]{\text{CYP 1A}} \text{Product}$$

Source: R.F. Addison 1997.

can then be more easily excreted. The distinctive feature of mono-oxygenases is that they can be induced. When an organism is exposed to certain chemicals, its mono-oxygenase activity increases, apparently assisting in its metabolism and eventual excretion of the target chemical. Mono-oxygenases can be thought of as a kind of latent defence mechanism. Under normal circumstances, their activity remains low, but if the organism is exposed to certain chemicals, their activity is induced to help eliminate the chemical.

Over the past few years, measurements of mono-oxygenase activity have been used increasingly to diagnose exposure to chemicals and their effects on organisms. These measurements are usually expressed as activities of indicator enzymes, such as EROD (ethoxyresorufin-*O*-deethylase), AHH (aryl hydrocarbon hydroxylase) or, more recently, as concentrations of the catalytic protein, cytochrome P-450 1A (CYP 1A). Figure 3 illustrates the induction process for these enzymes. Inducers include contaminants such as PAHs, PCBs and PCDDs.

Induced MFO activity can precede the onset of more serious cellular and physiological changes, such as liver damage, tumour formation, reproductive toxicity and other toxic effects. The persistent induction of MFO may lead to increased oxidative stress to cells and organs, increasing elimination of endogenous substances such as vitamins and hormones, and increased formation of carcinogenic and cytotoxic compounds (J. Payne personal communication 1996).

Metallothioneins function much like the mono-oxygenases. They are a group of low molecular weight proteins rich in sulphydryl groups which bind strongly to certain metals, such as mercury, copper, zinc and cadmium. Metallothioneins can also be induced. When an organism (either vertebrate or invertebrate) is exposed to a metal, metallothionein concentrations will increase so

that metals bond to the protein and toxicity is thereby reduced.

Measurements of mono-oxygenase and metallothionein activities are increasingly being used along with supporting chemical analyses of contaminants to investigate relationships between cause and effect in aquatic ecosystems. The chapters in this report dealing with specific aquatic ecosystems describe some of these data.

ADVANCES IN THE SAMPLING AND ANALYSIS OF CHEMICAL CONTAMINANTS

Our ability to compare concentrations of chemical contaminants over space and time, and our ability to relate those concentrations to biological effects, are limited by the sampling and analysis techniques used. Several abiotic and biotic variables must be measured during sampling. These include abiotic factors such as grain size, organic matter content, and particulate matter composition; and biotic factors such as age, size, sex, reproductive status and lipid content of organisms. Even with the elimination of these variables, there will still be unexplainable variations in chemical concentrations between individual organisms in a sample. These variations may reflect differences in biological factors such as diet, growth rate or reproductive patterns. It is often not possible to control all variables and, as a result, strictly comparable samples may not be available at different places or times. In this report, attempts are made to report and interpret all data taking into account as many of the confounding variables as possible. Where information is available, ranges of data are given, as well as mean values.

Most information about concentrations of contaminants in aquatic ecosystems have been gathered during the last three decades. Since the early 1970s, many of the common analytical methods used in environmental studies have evolved considerably. The sensitivity of most analytical techniques has increased by about one thousand-fold per decade for the past two decades. Thus, in the early 1970s, PCBs could be measured with some difficulty in the microgram-per-gram range (i.e. 10^{-6} g.g^{-1} range). Today, individual compounds (congeners) in a complex mixture of PCBs can be routinely analyzed in the picogram-per-gram range (i.e. 10^{-12} g.g^{-1} range). The fact that modern analytical methods can now measure individual components of complex mixtures and at very low concentrations makes it difficult to compare modern results with older data. However, where possible in this report, data collected over long time periods are compared in an attempt to determine whether concentrations of certain chemical contaminants are increasing or decreasing in aquatic ecosystems.

For the analysis of metals, advances have been made in analytical capability, particularly with respect to the use of clean techniques. In the past decade, samples of water, sediment and biota were often unknowingly contaminated with trace amounts of metals during sampling and analysis. Hence, reported concentrations were often high. Since using clean techniques, more precise and accurate measurements have been possible. As a result, the concentrations of metals reported in the recent literature more accurately reflect the actual concentrations occurring in the environment.

CHEMICAL CONTAMINANTS IN THE CONTEXT OF GENERAL CONTAMINATION AND POLLUTION

Although this report focuses on a limited suite of chemical contaminants, their presence is not the only factor to threaten aquatic ecosystems and the resources that they contain. Other chemical, physical and biological stressors can have a dramatic and adverse effect on aquatic ecosystems.

Sewage usually comprises a mix of domestic and industrial waste. The ratio will depend on the community from which it is collected and discharged. Most sewage which contains a considerable amount of organic matter consists primarily of feces, paper products, food waste, and household chemicals, along with related micro-organisms. Historically, industrial effluents containing a high concentration of organic matter, such as pulp mill wastes, have also been discharged into both freshwater and marine ecosystems. Many Canadian coastal communities discharge such wastes directly into estuaries or into the sea, sometimes with only minimal treatment. When such material enters the aquatic environment, natural chemical and biological processes begin to break these organic materials down to simpler products. These degradation processes use oxygen dissolved in the water, and depending upon the process, the discharged waste creates a chemical oxygen demand (COD) or a biochemical oxygen demand (BOD) in the water receiving the wastes. BOD and COD can appreciably reduce oxygen concentrations in water, making aquatic habitat unsuitable for fish or other organisms.

The discharge of nutrients such as phosphate, nitrate and silicate, into water can lead to eutrophication. In this process aquatic plants (usually microscopic algae) take up these nutrients, undergo significant increases in growth (algal blooms), and then die. Both living and dead algae create a large BOD and COD, which in turn lead to a reduction in the amount of oxygen in water thereby creating aquatic habitats unsuitable for fish or other aquatic life. Eutrophication can result from either natural or human activities, such as the introduction of fertilizers into aquatic ecosystems or the release of nutrients in sewage. Eutrophication caused by human activities is termed cultural eutrophication. A well-publicized example of cultural eutrophication occurred in the Great Lakes during the 1970s, when discharges of domestic detergents containing high amounts of phosphates stimulated massive algal growth, lowered oxygen levels and resulted in widespread fish kills (Vallentyne 1974). Although controls on the use and discharge of phosphates have significantly reduced cultural eutrophication in the Great Lakes, the discharge of nutrient-rich material into marine coastal areas may be of growing concern, particularly in areas of high population density and limited flushing rates to disperse discharges. Eutrophication of marine coastal areas may be related to the incidence of algal blooms which release natural toxins, such as the compounds responsible for paralytic shellfish poisoning.

One of the most significant and pervasive impacts on fish and fish habitat in eastern Canada is the acidification of freshwater aquatic ecosystems resulting from emissions of sulphur dioxide and oxides of nitrogen into the atmosphere. Approximately 38% of Canada's surface fresh water is in regions susceptible to acidic atmospheric deposition. This includes some 390 000 lakes on the Precambrian Shield of eastern Canada with its thin soils and rocks that resist weathering. These lakes generally have low productivity, clear water, and are poorly buffered (Kelso *et al.* 1990). Lakes in the Sudbury metal-smelting region and Nova Scotian rivers were acidified in the 1950s, 1960s and early 1970s and have well-documented losses of fish species and fish productivity (Kelso *et al.* 1990). More than 15 000 of the susceptible lakes are estimated to be very acidic (pH less than 4.7) and to have lost fish populations, resulting in an economic loss of $53 million per year (Kelso *et al.* 1990).

Acidification has affected more than fish populations. Using mathematical models that combined geological, chemical and biological data, scientists have predicted that at least 20% of all lakes (i.e. about 55 000 lakes) in 15 of 38 secondary watersheds in eastern Canada had lost at least 20% of their species by 1980. The highest loss of species is believed to have been in mollusc populations. Rotifer and crustacean populations were determined to have the fewest species losses (Minns *et al.* 1990). Acidification of water bodies can mobilize chemicals, such as aluminum and other metals, from soils and sediments (Schindler 1988). Acidification and elevated concentrations

of metals in soft fresh waters often occur together because non-ferrous metal smelters — the major source of sulphur dioxide — also emit metals. In response to legislation by the governments of Canada and the United States to control smelter emissions, there is evidence of chemical and biological recovery in lakes in Ontario's Sudbury region, one of the longest affected areas in Canada (Keller *et al.* 1992). The historical contamination of these lakes by metals is believed to slow the rate of biological recovery. With time however, the metal concentrations in the water are decreasing (Yan *et al.* 1996).

Physical alteration of fish habitat has also had a significant impact on aquatic ecosystems (Environment Canada/U.S. EPA 1995). Such changes have included the filling of lakes and wetlands in urban and agricultural areas, the dredging of channels and harbours, the construction of docks, ports, breakwaters, and buildings on shorelines, and the erection of dams and other water control devices. Similarly, the use of water bodies as areas for human recreation, water supply and waste disposal has also had a significant effect on the physical character and functioning of some ecosystems. Such changes have not only eliminated and altered habitats, but have also fundamentally modified the structure and functioning of ecosystems. The removal of river bank vegetation, for example, can change water temperature and increase sedimentation. These sediments, in turn, can blanket valuable aquatic habitats including nursery areas for many fish species as well as alter the cycling of chemical contaminants.

The introduction of new or exotic species can have a significant impact on aquatic ecosystems. For example, zebra mussels, which were accidentally introduced through the discharge of ballast waters into the Great Lakes in 1986, have not only altered biological patterns through increased competition for food and space, but have also effectively changed the flow of energy, nutrients and chemical contaminants in these ecosystems through their feeding on plankton and detritus (Kidd 1994).

Alterations in fish community composition occur over time. These changes may be the result of invasion of exotic species, over-harvesting of stocks including the selective removal of top predators, loss of habitat, purposeful introduction of desired species, or potential effects of climate change. Resulting changes in ecosystem structure and function serve to alter the pathways and fate of contaminants in aquatic systems. Thus, information on trends over time may reflect shifts in contaminant accumulation patterns caused by community alterations as much as changes in chemical loadings to the system (Egerton 1985).

References

ADDISON, R.F. Department of Fisheries and Oceans. Sidney, BC. 1997.

GESAMP (Joint Group of Experts on the Scientific Aspects of Marine Pollution) - O/FAO/UNESCO/WMO/IAEA/UN/UNEP. 1987. Land/Sea Bound. Flux Contaminants: Contributions from Rivers. Rep. Stud. GESAMP No. 32. UNESCO. Paris.

GOVERNMENT of Canada. 1991. Toxic Chem. G. Lakes Assoc. Effects. 3 Volumes, 755 p.

EGERTON, F. G. 1985. Overfishing or Pollution? Case History of a Controversy on the Great Lakes. Great Lakes Fishery Commission. Tech. Rep. No. 41. 28 p.

ENVIRONMENT Canada, and U.S. Environmental Protection Agency. 1995. SOLEC 1994 State Lakes Ecosys. Conf. Backgr. Paper: Toxic Contam.

KELLER, W., J.M. Gunn, and N.D. Yan. 1992. Evidence of biological recovery in acid-stress lakes near Sudbury, Canada. Environ. Poll. 78: 79-85.

KELSO, J.R.M., M.A. Shaw, C.K. Minns, and K.H. Mills. 1990. An evaluation of the effects of atmospheric acidic deposition on fish and the fishery resource of Canada. Can. J. Fish. Aquat. Sci. 47: 644-655.

KIDD, Joanna. 1994. State of the Lakes Ecosystem Conference: Integration Paper. Environment Canada, Burlington.

MACKAY, D. 1991. Multimedia Environmental Models: The Fugacity Approach. Lewis, Chelsea, MI.

MINNS, C.K., J.E. Moore, D.W. Schindler, and M.L. Jones. 1990. Assessing the potential extent of damage to inland lakes in eastern Canada due to acidic deposition. III. Predicted impacts on species richness in seven groups of aquatic biota. Can. J. Fish. Aquat. Sci. 47: 821-830.

SCHINDLER, D.W. 1988. Effects of acid rain on freshwater ecosystems. Sci. 239: 149-157.

UNEP (United Nations Environment Programme). 1992. Conv. Biolog. Divers. Na. 92-7807.

VALLENTYNE, J.R. 1974. The Algal Bowl. Department of the Environment, Fisheries and Marine Services. Misc. Public. No. 22: 1-185.

WILLIAMS, D.A. Department of Fisheries and Oceans. 1997.

YAN, N.D., P.G. Welch, H. Lin, D.J. Taylor, and J.-M. Filion. 1996. Demographic and genetic evidence of the long-term recovery of (*Daphina galeata mendotae*) (*Crustacea: Daphniidae*) in Sudbury lakes following additions of base: the role of metal toxicity. Can. J. Fish. Aquat. Sci. 53: 1328-1344.

Personal Communications

PAYNE, J. Department of Fisheries and Oceans. 1996.

ST. LAWRENCE
Marine
Ecosystem

Highlights

- The St. Lawrence marine ecosystem links the Great Lakes and St. Lawrence River to the Atlantic Ocean, and comprises the Estuary and Gulf of St. Lawrence as well as the Saguenay Fjord.

- More than 25 species of marine fish and shellfish in this region are fished commercially. These generated a market value of $450 million in 1994. The St. Lawrence marine ecosystem also supports a growing ecotourism industry mainly related to the observation of marine mammals in their natural habitat.

- Contaminants accumulated by aquatic biota in the St. Lawrence marine ecosystem include the metals mercury, cadmium, and lead and organic compounds including polycyclic aromatic hydrocarbons (PAHs), polychlorinated biphenyls (PCBs), polychlorodibenzo-*p*-dioxins (PCDDs), polychlorodibenzo-furans (PCDFs), dichlorodiphenyltrichloroethane (DDT), mirex and other pesticides. These compounds originate from the numerous industrial, municipal and agricultural activities occurring in the Great Lakes and the St. Lawrence drainage basin.

- Contaminants enter the St. Lawrence marine system via freshwater runoff, especially from the St. Lawrence River, and via atmospheric deposition, and marine currents originating from the Atlantic Ocean.

- The main area for the deposition of contaminants associated with suspended particles is the Laurentian Channel, where significant amounts of mercury, lead, PCBs, PCDDs and PCDFs have accumulated in bottom sediments since the beginning of industrialization. Industrial activity in the Saguenay region has also resulted in the accumulation of mercury and PAHs in sediments of the Saguenay Fjord's Inner Basin.

Highlights

■ There are several small estuaries, embayments and nearshore areas that are affected by local industrial and municipal effluents. These include the Baie des Anglais on the north shore of the St. Lawrence Estuary and Belledune Harbour in Chaleur Bay. In both of these areas, contaminants have accumulated in sediments.

■ Commercially valuable fisheries in the St. Lawrence marine ecosystem affected by chemical contamination are the northern shrimp of the Saguenay Fjord (mercury), and the American lobster of the Belledune Harbour area in Chaleur Bay (cadmium). The beluga population in the St. Lawrence Estuary (mercury, PCBs and various pesticides) is also affected.

■ Contamination has resulted in the closure of some commercial fisheries (i.e. the northern shrimp fishery in the Saguenay Fjord). Species living in the ecosystem have also shown symptoms of stress. Health and reproductive impairment likely related to contaminants have been observed in belugas of the St. Lawrence. These stresses may affect the long term recovery of the population.

■ In general, chemical contamination has significantly decreased in the St. Lawrence marine ecosystem over the past 20 years. However, this general trend does not apply to all contaminants or to all habitats and resources found in the region.

Table of Contents

The St. Lawrence Marine Ecosystem

List of Figures

List of Sidebars

Introduction

The St. Lawrence marine ecosystem is an integral part of the Great Lakes–St. Lawrence drainage basin. The area measures 1.6 million km^2 (St. Lawrence Center 1996). The marine ecosystem, together with the St. Lawrence River, has played an important role in the cultural and economic development of Canada, especially as it is the navigational link between the Great Lakes and the Atlantic Ocean.

For the purposes of this chapter, the St. Lawrence marine ecosystem comprises the Estuary and Gulf of St. Lawrence as well as the Saguenay Fjord (Figure 1). The freshwater portion of the St. Lawrence, which includes the St. Lawrence River upstream of Quebec City and other coastal rivers, is discussed in the Inland Freshwater Ecosystems Chapter.

MAJOR CONTAMINANT ISSUES

The St. Lawrence marine ecosystem is of major economic importance in Atlantic Canada and eastern Quebec. Several invertebrate and fish species, representing 25% of Canadian fisheries landings are harvested commercially in the Estuary and Gulf of St. Lawrence (DFO 1995). The area also has a growing recreational and

Figure 1

The St. Lawrence marine system showing the various subdivisions and bathymetry.

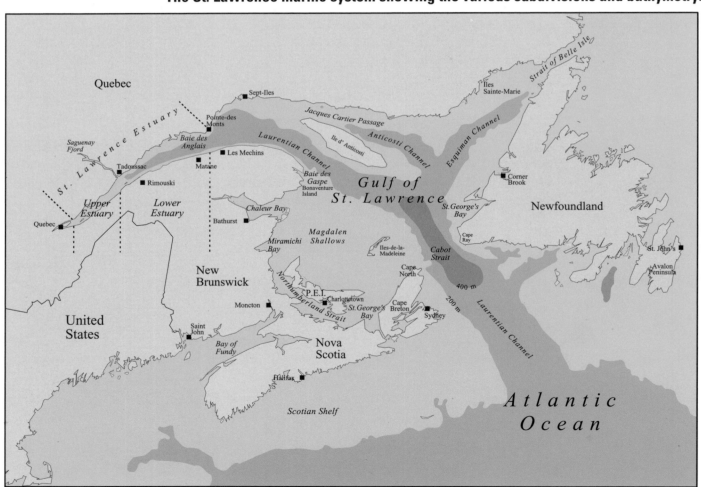

Source: White and Johns 1997.

tourism industry, particularly for the observation of marine mammals in their natural habitat. However, the Estuary and Gulf of St. Lawrence form the downstream part of a large drainage basin which is the site of intense human activities, including several large cities, numerous industries, and large scale agricultural activities. The St. Lawrence is a major shipping route to the interior of the North American continent. Thus, a number of contaminants are introduced into the system either directly via local municipal and industrial effluents discharges or, more importantly, indirectly through freshwater run-off and atmospheric deposition.

Chemical contamination in the St. Lawrence marine ecosystem has led to the closure of some fisheries, such as those of the northern shrimp in the Saguenay Fjord and the American lobster in the Belledune Harbour area in Chaleur Bay. The health of marine species may also suffer, shown by symptoms of stress in aquatic biota including those observed in the St. Lawrence beluga population and in American eels migrating in the

St. Lawrence Estuary. These consequences of chemical contamination affect the fishing and ecotourism industries.

OCEANOGRAPHIC FEATURES OF IMPORTANCE TO CONTAMINANT DYNAMICS

The St. Lawrence marine ecosystem is very heterogeneous and nearly every known oceanographic process occurs within it. Fresh water for the Estuary and Gulf of St. Lawrence originates from a very large drainage basin that includes the Great Lakes System. The St. Lawrence River and other coastal rivers along the Estuary and Gulf discharge a considerable volume of freshwater into the ecosystem and their influence may be felt as far downstream as the Scotian Shelf, in the Atlantic Ocean. The freshwater outflow of the St. Lawrence River alone ($424 \text{ km}^3.\text{yr}^{-1}$) exceeds the run-off ($353 \text{ km}^3.\text{yr}^{-1}$) from the entire Atlantic coast of the U.S. (Sutcliffe *et al.* 1976).

Figure 2
Summer water circulation in the Laurentian Channel.

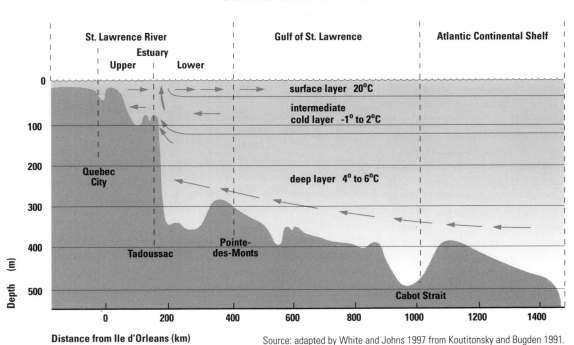

Source: adapted by White and Johns 1997 from Koutitonsky and Bugden 1991.

The Laurentian Channel (Figure 1) is the dominant feature of the bottom topography in the Estuary and Gulf of St. Lawrence. This 300 to 500 m deep trench in the continental shelf extends a distance of over 1 300 km from Cabot Strait to the mouth of the Saguenay Fjord in the St. Lawrence Estuary (Koutitonsky and Bugden 1991). The circulation along the Laurentian Channel in the Estuary and Gulf of St. Lawrence is typical of estuaries with inflowing deep waters and outflowing surface waters (Figure 2). Deep oceanic waters from the Atlantic Ocean are thus brought into the Gulf of St. Lawrence; and their influence on temperature, salinity, and density is felt as far inland as Île d'Orléans, near Quebec City.

American Lobster (*Homarus americanus*)

Lobster is the most important fishery in Atlantic Canada accounting for a third of the total landed value of all Atlantic fisheries. Lobsters are crustaceans and grow through periodic moulting, which decreases in frequency with age. Young lobsters moult 3 to 4 times a year, mature lobsters moult every year, while older lobsters moult every 3 to 4 years. Lobsters mature between 65 and 110 mm in carapace length. At that length, their average weight is 0.7 kg. The female mates after moulting in mid summer and the following summer produces eggs which hatch in early June. The larvae spend 30 to 60 days feeding and growing near the surface before settling on the bottom of inshore waters including harbours and bays. Lobster eat dead fish and immobile or slow moving invertebrates such as mussels, sea urchins and worms on the sea bottom. Lobsters in commercial catches range from 18 to 30 cm in length and weigh from 0.75 to 1 kg. Giant lobsters can exceed 20 kg and live from 40 to 65 years. Lobsters make seasonal migrations moving to shallower waters in summer and deeper waters in winter. Although the range of most lobsters is less than 30 km, some lobsters are known to migrate long distances from tens to hundreds of km. Tagging studies have shown that at least some lobsters return to the same area each year.

From a toxic contaminants perspective, lobsters are important because the commercial fisheries are near the coastline. Lobster fisheries are found within many harbours. There, exposure to toxic chemicals is likely to be highest. People eat various lobster tissues, one of which, tomalley, is known to store a number of chemical contaminants. Surprisingly high concentrations of various natural (e.g., mercury, and B[a]P) and synthetic chemicals (e.g., PCBs) may be found in tomalley because lobsters are bottom-dwellers, in contact with sediment. Lobsters also lack enzymes needed to remove many contaminants from their tissues.

More than 220 000 kg are taken annually by the fishery from Halifax Harbour, and substantial lobster fishing occurs in many other harbours in eastern Canada. However, harbours may receive untreated industrial and municipal effluent, and there is concern as to whether harbour lobsters are safe to eat. Except for a few instances, such as the elevated cadmium concentrations in lobsters captured in Belledune Harbour, New Brunswick in 1980 and high concentrations of PAHs in lobsters from the south arm of Sydney Harbour, Nova Scotia in the 1980s, chemical contaminant concentrations have not been considered by Health Canada to be high enough to be a human health problem.

The St. Lawrence Estuary has two main sub-regions, the upper and lower estuary (Figure 1). The upper estuary is the portion of the St. Lawrence that extends from Île d'Orléans to the mouth of the Saguenay Fjord. Its width varies markedly from a few to 20 km and its depth reaches 100 m in some areas. The upper estuary comprises a Maximum Turbidity Zone (MTZ), which is located between Île d'Orléans and Île aux Coudres. Highly turbid waters are caused by the local estuarine circulation, strong tidal currents, and turbulence, which entrap particles and prevent their permanent deposition (d'Anglejan 1990).

The lower estuary extends from the mouth of the Saguenay Fjord to Pointe des Monts, Quebec. Its width varies from 24 to 50 km, and its depth exceeds 300 m in the Laurentian Channel. The lower estuary is characterized by a cold intermediate layer, located between the surface and deep layers (Figure 2). Waters of the cold intermediate layer originate from the Labrador Current. This current enters the Gulf through the Strait of Belle Isle, and from the local cooling of surface waters in the Gulf during the winter (Gilbert and Pettigrew 1997). The head of the Laurentian Channel, at the mouth of the Saguenay Fjord, is a site of strong upwelling of waters from the deep and cold intermediate layers, an activity resulting from the change in topography which forces the upward movement of inflowing waters.

The Saguenay Fjord extends 170 km inland in a narrow (1 to 6 km) steep sided-glacial valley, and comprises three distinct basins. The most important of these is the inner basin, located in the upstream part of the Fjord and covering at least two-thirds of its length. The Saguenay Fjord opens into the St. Lawrence Estuary at the head of the Laurentian Channel, but is partly isolated from the Estuary by a shallow 20 m deep sill at its mouth. Nevertheless, salt water exchanges occur between the deep waters of the Saguenay Fjord and those of the lower estuary through inflowing and outflowing currents associated with tides.

In the Gulf of St. Lawrence, the surface circulation follows a counter clockwise pattern, and is dominated by the Gaspé Current, a strong coastal current that originates from the outflow of the Estuary (Koutitonsky and Bugden 1991; Figure 3). It has been suggested that a portion of

Figure 3
Summer water circulation in the Gulf of St. Lawrence.

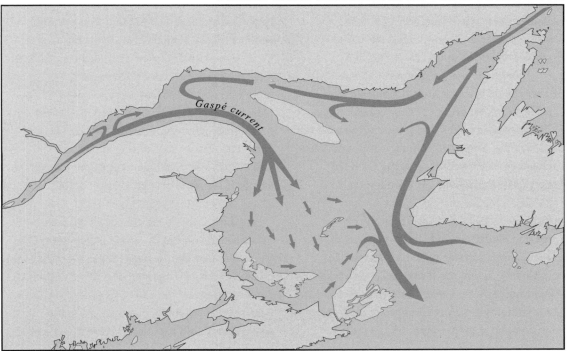

Source: adapted by White and Johns 1997 from P. Smith, B. Petrie and G. Bugden personal communication; Ardisson and Bourget 1992.

the surface waters recirculate around the Gulf, rather than leaving through Cabot Strait (Boudreault and Héritier 1971; Messieh 1974).

BIOLOGICAL RESOURCES

More than 25 different species of marine fish are harvested commercially in the Estuary and Gulf of St. Lawrence (Messieh and El Sabh 1988). In 1994, total landings for the St. Lawrence marine system reached 252 000 t with a landed value of $450 million. Since the establishment of the 200 mile limit in 1976, landings (total catch) of commercial fish from the Gulf represent 20% of the fishery in the Northwest Atlantic (Chadwick and Sinclair 1991).

Since 1960, cod, herring and redfish comprised 75% of the total catch in the Gulf of St. Lawrence. This proportion fell to 65% in the 1980s, largely because of an increase in invertebrate landings, including crab, lobster and shrimp (Chadwick and Sinclair 1991). Following a collapse in groundfish stocks in the early 1990s, the cod, redfish and silver hake fisheries were gradually closed from 1994 to 1996 throughout the entire Gulf. The limited fisheries for American plaice and winter flounder have remained unchanged from previous years. Lobster, shrimp and snow crab fisheries have expanded steadily through the 1980s and currently provide the greatest economic return and employment in the Gulf (Chouinard and Fréchet 1994). At present, the largest landings of herring, mackerel, crab and lobster originate from the Magdalen Shallows, while most groundfish, capelin and shrimp landings come from the Laurentian Channel and the northern Gulf.

Commercially landed species in the St. Lawrence Estuary include the American eel (dominant diadromous species) and Atlantic herring and capelin (dominant pelagic species) (Gagné and Sinclair 1990). Between 1968 and 1981, landings of demersal fishes in the Estuary were dominated by Atlantic cod and redfish.

Greenland halibut, first harvested commercially in the Estuary in 1977, represented between 20 and 70% of the total catch of demersal fishes between 1978 and 1986 (Gagné and Sinclair 1990). Invertebrate species constituted between 20 and 40% of landings in Quebec between 1968 and 1981. Soft shell clams and northern shrimp were of major importance. The snow crab fishery has expanded significantly in recent years and this species now dominates invertebrate landings in the Estuary.

Commercial fisheries in the St. Lawrence Estuary account for less than 10% of the total landings for all fishing activities in Quebec (Gagné and Sinclair 1990). For most species, the available biomass is low and often distributed over areas difficult to access with commercial fishing gear (Andersen and Gagnon 1980). In addition, several commercial fish species move into the Estuary for only a few months in the summer and are therefore available only for short periods of time. Furthermore, large areas of the St. Lawrence Estuary are ice-covered for several months during the winter.

At least three species of seals and five species of whales are observed in the Estuary during summer. These animals provide the basis of a local ecotourism industry. The most common seals are the grey, harp and harbour seals. The most common whales are beluga, minke, fin and blue whales and the harbour porpoise. There are occasional sightings of humpback and sperm whales as well as dolphins. Commercial harvesting in marine mammals in the Estuary and Gulf is now restricted to seal hunting.

Breeding seabirds are concentrated most heavily in the northern, central and western Gulf. One quarter of Gulf seabirds breed off the eastern Gaspé Peninsula; large numbers are also found on the north shore of the Gulf and at the Magdalen Islands. Seabird numbers are lowest in the southeastern Gulf and along the west coast of Newfoundland.

Assessment of Chemical Contamination

SOURCES

There are many contaminant sources to the Estuary and Gulf of St. Lawrence. In addition to the industries found in the Great Lakes Basin (See the Great Lakes Ecosystems Chapter), 50 major industrial sites along the St. Lawrence River and Estuary were identified as significant contributors to contamination of the Estuary (Bouchard and Gingras 1992; Figure 4). These sites include four pulp and paper facilities, three aluminum smelters, and one mining operation in the Saguenay Fjord region. In the Gulf, industrial contaminant sources include pulp and paper mills, power generating stations and petroleum refineries, and mining and smelting facilities (Figure 5).

Six million people live in the Quebec portion of the St. Lawrence River and Estuary discharge basin. Municipal sewage contributes the full spectrum of industrial and human wastes including metals and organic compounds. In 1984, only 8% of total municipal waste discharge into the St. Lawrence underwent treatment (Harding 1992). However, the proportion of municipalities along the River, Estuary and Gulf that treated their waste water has increased since 1986, reaching 30% in 1992. This new proportion represented 65% of the population (St. Lawrence Center 1996). In the Saguenay Fjord region, releases of municipal wastewater are the second major source of contaminants such as trace metals, pesticides, oils and greases, and various organic compounds. Logimer (1987) estimated that 187 000 $m^3.d^{-1}$ of municipal wastewater are discharged into the Saguenay/Lake Saint Jean hydrographic basin.

Agricultural activities also contaminate the Estuary and Gulf of St. Lawrence. The St. Lawrence drainage basin includes a large amount of agricultural land. Run-off from these areas contributes nutrients, mainly phosphates and nitrates, as well as pesticides and herbicides to the St. Lawrence marine ecosystem (Wells and Rolston 1991).

Figure 4
Industries in the St. Lawrence Estuary.

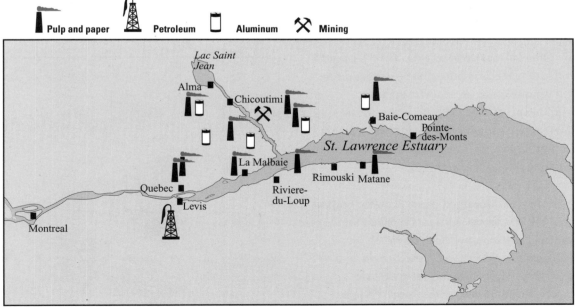

Source: modified by White and Johns 1997 from Bouchard and Gingras 1992.

Other potential sources of contaminants to the St. Lawrence marine ecosystem include maritime traffic and dredging activities. The St. Lawrence-Great Lakes system is an international shipping route that averages 2 000 vessel movements per year. There is a continuing risk of the accidental release of hazardous materials or petroleum products, which comprise almost 20% of the total volume of merchandise handled in the ports of the St. Lawrence (Environment Canada 1991). Fortunately very few accidents have occurred in the Estuary and Gulf. The most well-known case is that of the *Irving Whale*, which sank on September 7, 1970, in the southern Gulf of St. Lawrence. Its cargo included 4 100 t of Bunker C fuel oil, as well as 7.5 t of PCBs and 1.8 t of chlorobenzenes. The PCBs and chlorobenzenes were contained in the cargo heating system (Gilbert and Walsh 1996). Some 400 to 600 t of oil and unknown quantities of PCBs and chlorobenzenes were released at the time of the sinking. The barge was successfully recovered in July 1996 with much of its oil cargo, but more than 70% of the initial quantity of PCBs was lost into the environment during the 26 years that the barge remained in the Gulf (Government of Canada 1997).

Routine dredging is carried out periodically in harbours and shipping lanes to increase the depth and width of navigation channels. One potential major impact of the dredging is the resuspension of sediments into water column. The resuspension will lead to the remobilization of contaminants. It is estimated that an annual average of 5 million m^3 of dredged material are removed from fishing harbours and ports in the southern Gulf of St. Lawrence, and disposed of by ocean dumping (Messieh *et al.* 1991).

Freshwater discharges, particularly the outflow from the St. Lawrence River, are one of the principal pathways by which chemical contaminants enter the St. Lawrence marine ecosystem. Most contaminants which are exported with freshwater discharges from the Great Lakes Basin and the rivers feeding into the St. Lawrence River become bound to suspended particles that are transported downstream towards the Estuary and Gulf of St. Lawrence. However, most contaminated particles do not settle to bottom sediments until they reach the Laurentian Channel, in the lower estuary, because strong currents in the upper estuary prevent their local deposition. For example, mirex was only discharged into Lake Ontario (Holdrinet *et al.* 1978). Its presence can be traced far downstream in the bottom sediments of the lower estuary (Comba *et al.* 1993). This indicates that contaminants originating upstream in the Great Lakes Basin can reach the marine environment via the St. Lawrence River.

Coastal rivers contribute to the contamination of the St. Lawrence marine ecosystem, although these rivers contribute less than the St. Lawrence. For example, aluminum smelting industries have released large amounts of mercury and polycyclic aromatic hydrocarbons (PAHs) into the Saguenay River and Fjord (Pelletier *et al.* 1989; Smith and Levy 1990). The Miramichi River provides another example; here contaminants such as polycyclic aromatic hydrocarbons (PAHs), polychlorinated biphenyls (PCBs), polychlorodibenzo-*p*-dioxins (PCDDs),

Figure 5
Industries in the Gulf of St. Lawrence.

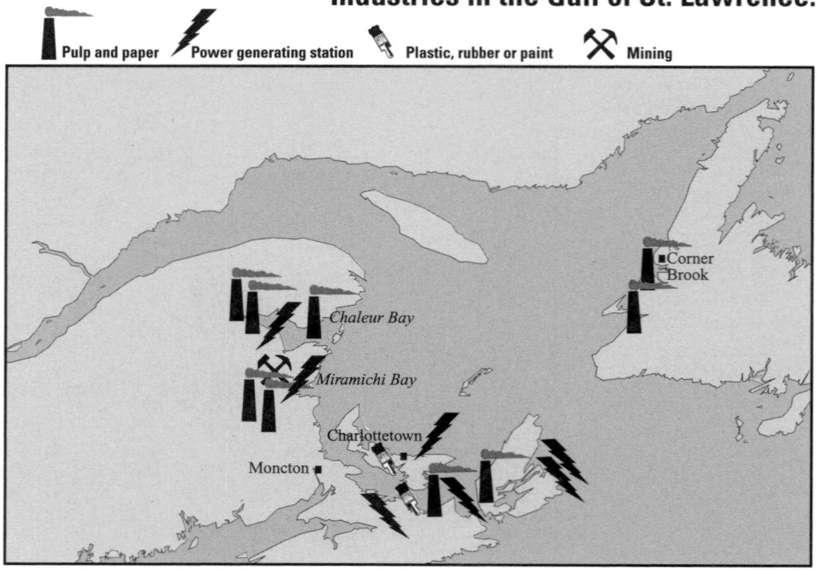

Pulp and paper   Power generating station   Plastic, rubber or paint   Mining

Source: adapted by White and Johns 1997 from Eaton *et al.* 1994.

polychlorodibenzo-furans (PCDFs), originate mainly from the local pulp and paper mills (Courtenay *et al.* 1995).

The atmosphere is a significant pathway for the transport and deposition of anthropogenic chemicals. This is particularly important for Atlantic Canada which is in the downwind path of many weather patterns that first pass through highly industrialized regions of North America. Several contaminants that are released into the atmosphere in these regions may be subsequently deposited in eastern Canada (Eaton *et al.* 1986). Brun *et al.* (1991) have shown that in the early 1980s, PCBs, chlorinated pesticides, and PAHs were deposited with precipitation in Atlantic Canada.

The atmosphere is an important source of contaminants found in the Estuary and Gulf of St. Lawrence. For example, a significant proportion of the anthropogenic mercury in sediments of the Laurentian Channel in the Gulf of St. Lawrence was likely brought to the system via the atmosphere (Gobeil and Cossa 1993). The isotopic composition of anthropogenic lead in sediments of the St. Lawrence Estuary is similar to that of atmospheric lead in urban regions in Canada (Gobeil *et al.* 1995). This suggests that atmospheric transport and deposition of lead is an important loading mechanism. As well, for organochlorine compound contaminants, the recent accumulation of PCDDs and PCDFs in sediments of the St. Lawrence Estuary has been attributed mostly to atmospheric inputs (Lebeuf *et al.* 1995a, 1996a).

DISTRIBUTION, TRENDS AND EFFECTS

Covering a total area of about 230 000 km^2, the Estuary and Gulf of St. Lawrence constitute a large ecosystem in which contaminants adsorbed to suspended particles are not uniformly distributed. A large proportion of incoming contaminants accumulate in sediment depositional areas. A smaller proportion is incorporated into the food chain or remains dissolved in the water column. Thus, sediments in these regions are important reservoirs where large amounts of contaminants have accumulated since the beginning of industrialization. The major area of sediment deposition in the St. Lawrence marine ecosystem is the Laurentian Channel, in the lower estuary and the Gulf of St. Lawrence. Another important area for the deposition of sediments and associated contaminants is the inner basin of the Saguenay Fjord. The accumulation of contaminants in these areas decreases downstream because the highest rates of sedimentation occur upstream where abrupt changes in topography and circulation favour the deposition of outflowing suspended particles (Loring and Nota 1973; Silverberg *et al.* 1986; Smith and Walton 1980).

METALS AND ORGANOMETALLIC COMPOUNDS

Mercury has accumulated in sediments of both the inner basin of the Saguenay Fjord and the Laurentian Channel. Before the 1971 government regulations on effluents from chlor-alkali plants limited the discharge of industrial mercury into the St. Lawrence drainage basin, mercury concentrations up to 50 and 10 times higher than natural concentrations were measured in sediments of the inner basin and the Laurentian Channel respectively (Loring 1975). Although anthropogenic inputs of mercury into the Saguenay Fjord and the St. Lawrence Estuary have since decreased significantly, concentrations of mercury in recently deposited sediments of the Saguenay's inner basin and the Laurentian Channel are still much higher than pre-industrial concentrations (Smith and Loring 1981; Gobeil and Cossa 1993). Overall, it has been estimated that approximately 120 t of industrial mercury have been deposited and remain stored in sediments of the Saguenay Fjord (Loring 1978a), and 170 t in the lower estuary (Gobeil and Cossa 1993).

Mercury concentrations in some fish of the St. Lawrence marine ecosystem parallel spatial variations of this contaminant in sediments (Gobeil *et al.* 1997). The muscle tissue of northern shrimp, snow crab, Greenland halibut, and Atlantic cod in the Saguenay Fjord are more contaminated by mercury than those of the same species in the Estuary and Gulf of St. Lawrence (Figure 6). However, only shrimp and snow crab exhibit mercury concentrations close to the Health Canada guideline of 0.5 $\mu g.g^{-1}$ in edible portions of commercial fishery products. In particular, northern shrimp from the Saguenay Fjord were heavily contaminated by mercury in the early 1970s when muscle concentrations reached values up to 10 $\mu g.g^{-1}$ (Cossa and Desjardins 1984; Pelletier *et al*. 1989). This led to the closure of the fishery, which is still closed despite a significant decrease in contamination in the late 1970s (See section on A Recent Decrease in Contaminant Concentrations).

Of all metals, lead appears to have had the highest inputs into the St. Lawrence marine ecosystem. In the main depositional areas of the Saguenay Fjord and the lower estuary, lead occurs up to a depth of 30 cm in sediments, with maximum concentrations reaching values up to 3 or 4 times higher than natural concentrations. There are approximately 13 000 t of anthropogenic lead in the sediments of the estuarine part of the Laurentian Channel (Gobeil *et al*. 1995).

Other metals of anthropogenic origin have also been detected in sediments of the Saguenay Fjord and the Laurentian Channel, including zinc, copper and chromium (Loring 1976a, 1976b, 1978b, 1979; Barbeau *et al*. 1981; Pelletier and

Figure 6
Mercury concentrations in muscle tissue of shrimp, crab, halibut and cod in the Saguenay Fjord, the St. Lawrence Estuary and the Gulf of St. Lawrence.

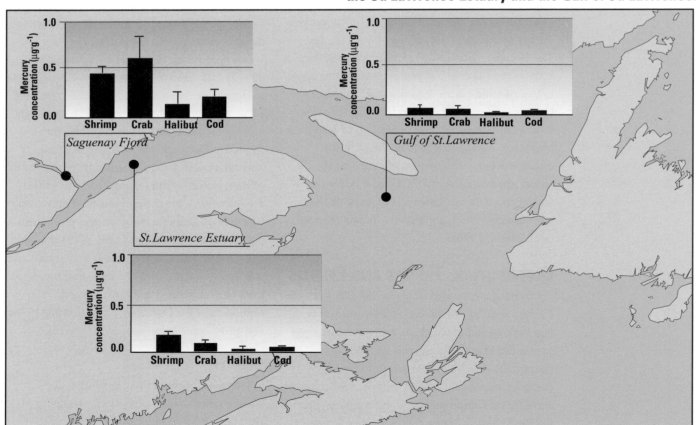

Source: Gobeil *et al*. 1997.

Canuel 1988). Cadmium does not appear to be a significant contaminant in sediments of the St. Lawrence Estuary, in which concentrations are typical of the natural abundance of this metal (Gobeil *et al.* 1987).

However, cadmium is a significant contaminant in fisheries of the St. Lawrence marine system, although problems related to this contaminant are localized. Cadmium concentrations in the muscle tissue of fisheries resources from most areas of the Estuary and Gulf of St. Lawrence are below detection limits (Gobeil *et al.* 1997), except in Belledune Harbour on the south coast of Chaleur Bay. Cadmium concentrations in the muscle tissue of lobsters from this area reached 2.7 $\mu g.g^{-1}$ in 1979 and were high enough to close the fishery in the harbour in 1980. Following a decrease in contamination in the early 1980s (See section on a Recent Decrease in Contaminant Concentrations), the lobster fishery was reopened under harvesting restrictions.

ORGANIC CHEMICALS

Several organic contaminants are found in habitats and biota of the St. Lawrence marine ecosystem. There are concerns regarding organochlorine compound and other organic contaminants, given that some fish of the St. Lawrence are highly contaminated and suffer from health problems (See Case Studies on the beluga and the American eel at the end of this chapter).

ORGANOCHLORINE COMPOUNDS

Organochlorine compound contaminants have been detected in sediments of the Laurentian Channel in the Lower St. Lawrence Estuary, including PCBs (Cossa 1990; Gobeil and Lebeuf 1992; Lebeuf *et al.* 1995b), dichlorodiphenyltrichloroethane (DDT) (Cossa 1990), mirex (Comba *et al.* 1993), as well as PCDDs and PCDFs (Lebeuf *et al.* 1995a). They are not naturally found in the environment, and the accumulation history of these contaminants in sediments follows the evolution of their industrial and commercial use since the 1930s and 1940s.

PCBs, PCDDs, and PCDFs in sediments of the Laurentian Channel and Saguenay Fjord include specific congeners which are highly toxic but their concentrations are rather low (Brochu *et al.* 1995; Lebeuf *et al.* 1995a, 1995b). Current concentrations of most organochlorine compound contaminants in sediments of the Saguenay Fjord and Laurentian Channel are approximately 5 to 10 times lower than highest values reported for Lake Ontario sediments (Czuczwa and Hites 1986; Oliver *et al.* 1989; Cossa 1990; Gobeil and Lebeuf 1992; Comba *et al.* 1993).

High organochlorine compound contamination occurs, however, in sediments of Baie des Anglais, in the lower estuary (See Case Study Contamination of Near-Shore Sediments in Baie des Anglais), and at the immediate site in the southern Gulf where the sunken barge *Irving Whale* rested for 26 years with its cargo and load of fuel oil, PCBs and chlorobenzenes. After the recovery of the barge in July 1996, PCB concentrations ranging from 10 to 10 000 $\mu g.g^{-1}$ were found in sediments within a small area of less than 0.02 km^2 (Government of Canada 1997). The total amount of PCBs remaining in sediments around the *Irving Whale* site was estimated at 150 to 350 kg, which represents less than 0.8% of the 38 to 44 t of PCBs that are believed to be present in sediments of the Gulf of St. Lawrence (Gilbert and Walsh 1996).

The first evidence of organochlorine compound contamination in biota of the Estuary and Gulf of St. Lawrence was provided by studies conducted in the 1960s and 1970s. By the early 1970s, plankton from the southern Gulf of St. Lawrence was found to be heavily contaminated by PCBs (Ware and Addison 1973; Harding *et al.* 1997); (See Case Study Biomagnification of Contaminants in Marine Food Webs: PCBs). Concentrations reached 20 $\mu g.g^{-1}$ off the northern coast of Prince Edward Island in 1972 and were reported as the highest concentrations ever found in natural plankton (Ware and Addison 1973). PCBs and other organochlorine compound contaminants were also found to accumulate in other biota of the Estuary and Gulf in the 1970s. PCBs and DDT were detected in the muscle tissue of at least 11 species of fish and invertebrates. The most contaminated of these were herring and mackerel (Couillard 1982). In the late 1960s, oil extracts from herring and seals in the Gulf were contaminated by PCBs and DDT, but dieldrin was detected only in herring oil (Addison *et al.* 1972).

Recent studies conducted with more sophisticated analytical techniques than those used in the 1970s show that some organisms of the Estuary and Gulf St. Lawrence marine ecosystem are still contaminated by organochlorine compounds, particularly the American eel and the beluga (See Case Studies in this chapter). In the Maximum Turbidity Zone of the upper estuary, the PCB contamination of the local food chain was found to be significant, with total PCB concentrations in smelt and tomcod exceeding $0.1 \ \mu g.g^{-1}$ (Gagnon *et al.* 1990). Further downstream, concentrations of PCDDs and PCDFs do not exceed $0.06 \ ng.g^{-1}$ in snow crab, whelk, northern shrimp, Atlantic cod, Greenland halibut, and American plaice from the Saguenay Fjord and the Lower St. Lawrence Estuary (Brochu *et al.* 1995). In the Estuary, some highly toxic PCB congeners are also found to accumulate at low concentrations in the lipid-rich liver of Atlantic cod and in the muscle tissue of Greenland halibut (Lebeuf *et al.* 1995a).

A monitoring program of organochlorine compound contamination in seabird eggs has been conducted since 1972 by the Canadian Wildlife Service of Environment Canada (Noble and Elliot 1986; Elliot *et al.* 1988; Pearce *et al.* 1989). This program, which involves several species from different marine habitats, has shown that PCBs and DDT-group chemicals are the dominant organochlorine compounds found in seabird eggs. In the early 1980s, concentrations were higher in seabird eggs from the Gulf, and particularly the Estuary, than in those from all other Canadian coastal sites. In the early 1990s, however, concentrations of organochlorine

compounds in the Estuary and Gulf were similar to those found in the Bay of Fundy but 2 to 4 times lower than in similar species from the Great Lakes. Nevertheless, seabird eggs remain more contaminated by organochlorine compounds in the Estuary than in Newfoundland, the west coast, and Arctic regions of Canada (N. Burgess personal communication).

OTHER AROMATIC COMPOUNDS

PAHs have accumulated in considerable amounts in sediments of the Saguenay Fjord (Cossa 1990; Martel *et al.* 1986; Schafer *et al.* 1990). Until 1964, the principal source of PAHs in the Saguenay Fjord was atmospheric transport. However, between 1964 and 1976, direct discharges of industrial effluents far exceeded atmospheric sources (Smith and Levy 1990). Around 1968, the rate of PAH inputs into the sediments of the Fjord reached about $1400 \ kg.yr^{-1}$ (Martel *et al.* 1987). Smith and Levy (1990) estimated that, by 1976, more than 1100 t of PAHs had entered the Saguenay River Fjord system. The highly carcinogenic benzo[a]pyrene (B[a]P) represents 3% of the total PAHs generated by smelters and its loading in the Saguenay Fjord is estimated at approximately 500 kg, 22% of which would be buried in sediments (Lun and Mackay in press).

PAHs were also detected at relatively low concentrations in some fish of the St. Lawrence marine ecosystem (Hellou *et al.* 1994, 1995). Concentrations of PAHs in the muscle tissue of Atlantic cod caught at contaminated sites of the Saguenay Fjord and the St. Lawrence Estuary (Baie des Anglais) do not exceed $0.05 \ \mu g.g^{-1}$ on average (Hellou *et al.* 1994). American plaice and Greenland halibut from the same areas have higher concentrations of PAHs than cod, with average concentrations reaching 0.14 and $1.5 \ \mu g.g^{-1}$ respectively (Hellou *et al.* 1995). PAH concentrations in the muscle tissue of Greenland halibut decrease with increasing distance from sources of this contaminant, which are located in the upper reaches of the Saguenay Fjord (Hellou *et al.* 1995).

ECOLOGICAL CONSEQUENCES

Studies of the biological effects of organic contamination in the St. Lawrence marine ecosystem are rather limited. Most fisheries resources in the system are exposed to chemicals such as PAHs, PCBs, chlorinated pesticides, PCDDs and PCDFs, but concentrations of these compounds are generally below those for which biological effects have been reported. Nevertheless, health problems have been observed in the St. Lawrence beluga population (See Case Study: Contamination of the St. Lawrence beluga population) and in American eels migrating in the St. Lawrence Estuary (See Case Study: Contamination of the St. Lawrence American eel population). Although these disorders were thought to be linked to chemical contamination, their relation to a specific contaminant could not be made, and the effects may be the result of exposure to a variety of chemicals, or have some other non-chemical cause.

An array of biological effects of organic contamination have also been observed in biota from locally contaminated areas that receive industrial and municipal effluents (See Case Study: Contamination of Near-Shore Sediments in Baie des Anglais). Disorders of the liver and other organs have been found in winter flounder collected near pulp and paper mill discharges in western Newfoundland (Khan *et al.* 1992, 1994). In the Miramichi Estuary, fin erosions, vertebral lesions and modifications of the reproductive cycle were observed in mummichogs living downstream of a bleached kraft pulp and paper mill (Leblanc *et al.* 1997; Couillard and Légaré 1994). Also, exposure indicators related to enzymes such as cytochrome P450 are found at higher concentrations in tomcod from the Miramichi Estuary than in those from undeveloped areas of the Gulf (Courtenay *et al.* 1995); but concentrations are lower than in tomcod from the highly industrialized Hudson River, on the east coast of the United States (Wirgin *et al.* 1994).

The history of the Northern Gannet community is the best documented case of chemical contamination causing the population decline of a Gulf seabird species. Approximately 70% of the North American population of Northern Gannet breed at three colonies in the Gulf (Nettleship and Chapdelaine 1988). Nesting gannets substantially increased over the last century, reaching a peak in 1966 (Nettleship 1975). The population then decreased by 23% between 1966 and 1972 to 1973 due to very low hatching success between 1966 and 1970. The data strongly implicate dichlorodiphenyldichloroethylene-(DDE) induced shell thinning and embryonic mortality as the main factors in reduced productivity (30% net productivity in 1966 to 1967 versus 77% in 1979; Chapdelaine *et al.* 1987). Improved breeding success, with increased eggshell thickness and increased population have coincided with a significant drop in DDE and dieldrin levels in eggs (Elliot *et al.* 1988; Figure 7). On Bonaventure Island, where 75% of gannets in the Gulf breed (Nettleship and Chapdelaine 1988), a 1984 survey indicated that the population had recovered to pre-1966 levels.

A RECENT DECREASE IN CONTAMINANT CONCENTRATIONS

Sediment studies have shown a significant decrease in contaminant inputs into marine environment of the St. Lawrence over the past two decades. Signs of improvement were first obtained from the inner basin in the Saguenay Fjord where mercury concentrations decreased in surface sediments during the 1970s (Loring and Bewers 1978; Barbeau *et al.* 1981; Smith and Loring 1981). Contaminant accumulation profiles in sediments of the Laurentian Channel in the Lower St. Lawrence Estuary confirmed a similar trend for mercury (Gobeil and Cossa 1993) and other contaminants, including lead (Gobeil *et al.* 1995), PCBs (Cossa 1990; Gobeil and Lebeuf 1992; Lebeuf *et al.* 1995b), DDT (Cossa 1990), PCDDs and PCDFs (Lebeuf *et al.* 1995a). PAH contamination also decreased in sediments of the

Saguenay Fjord (Martel *et al.* 1987; Smith and Levy 1990). In some cases, concentrations in recently deposited sediments were 2 to 3 times less than maximum concentrations measured in deeper sediments (Figure 8). However, present concentrations for most contaminants which also occur naturally in sediments of the Saguenay Fjord and the Laurentian Channel still remain higher than natural concentrations.

There has been a recent decrease in the contamination of aquatic biota in the St. Lawrence marine ecosystem, particularly in fisheries that were closed due to metal contamination. In the late 1970s, mercury concentrations in edible portions of northern shrimp from the Saguenay Fjord decreased by a factor of 20 following drastic reductions of mercury inputs into the Fjord with the closure of a chlor-alkali plant in Arvida in 1976 (Cossa and Desjardins 1984; Pelletier *et al.* 1989; Cossa 1990; Figure 9). However, concentrations have stabilized and remained at values close to consumption limits since the early 1980s (Cossa and Desjardins 1984; Pelletier *et al.* 1989; Cossa 1990).

In Belledune Harbour, cadmium concentrations in lobsters decreased between 1981 and 1985, by 63 to 69% in the muscle tissue and by 56 to 64% in the digestive gland (Uthe *et al.* 1986). Improved effluent treatment by the local lead smelter industry may have been the main cause of the decrease.

Decreases in organochlorine compound contamination around the late 1970s and early 1980s were reported in the southern Gulf, in particular for PCBs in plankton and pelagic fish (Harding *et al.* 1997; See Case Study: Biomagnification of Contaminants in Marine Food Webs: PCBs in St. Georges Bay, Nova Scotia). Concentrations of PCBs and hexachlorobenzenes (HCBs), also decreased in cod liver from the southern Gulf between 1977 and 1985 (Misra and Nicholson 1994).

Contaminant concentrations have significantly decreased in the American eel in recent years (See Case Study: Contamination of the St. Lawrence American eel population). However, most of the contaminated eels migrating in the St. Lawrence Estuary accumulate their burdens upstream in the Great Lakes and in tributaries of the St. Lawrence River (Castonguay *et al.* 1989). Thus, concentration in this species reflects improving conditions in the upstream part of the St. Lawrence drainage basin rather than in the Estuary itself (Hodson *et al.* 1994). Nevertheless, eels represent potential vectors through which contaminants may be transferred to the beluga population of the St. Lawrence Estuary (Béland *et al.* 1993). This risk is declining with time as the overall levels of contamination in migrating eels decrease (Hodson *et al.* 1994).

The monitoring program established by the Canadian Wildlife

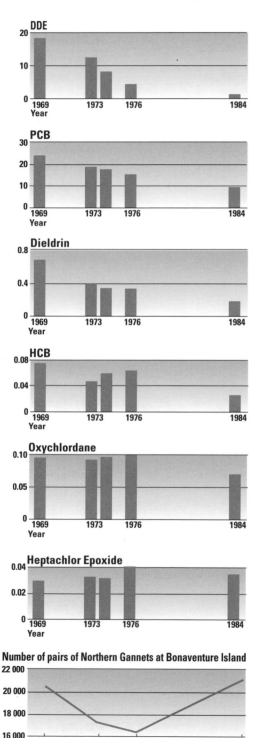

Figure 7

Trends in contaminant concentrations in Northern Gannet eggs and changes in the number of pairs of Northern Gannets at Bonaventure Island.

Source: Noble and Burns 1990.

Service to measure organochlorine compounds in seabird eggs provides insight into recent time trends of this type of contamination in the Estuary and Gulf of St. Lawrence. Between 1972 and 1992, Double-crested Cormorant eggs showed significant declines in concentrations of PCBs, DDE, dieldrin and hexachlorocyclohexane

(HCH) (Figure 10). These time trends are similar to those reported in gannet eggs from the Gulf of St. Lawrence (Elliot *et al.* 1988 ; Figure 7) and have been attributed to a cessation of PCB use in 1977, severe restrictions on the use of dieldrin and lindane in the early 1970s, and to the ending of the extensive spraying of DDT in forests of

Figure 8

Metal and organochlorine compound concentrations in sediment cores collected in the Saguenay Fjord and the St. Lawrence Estuary in the early 1990s.

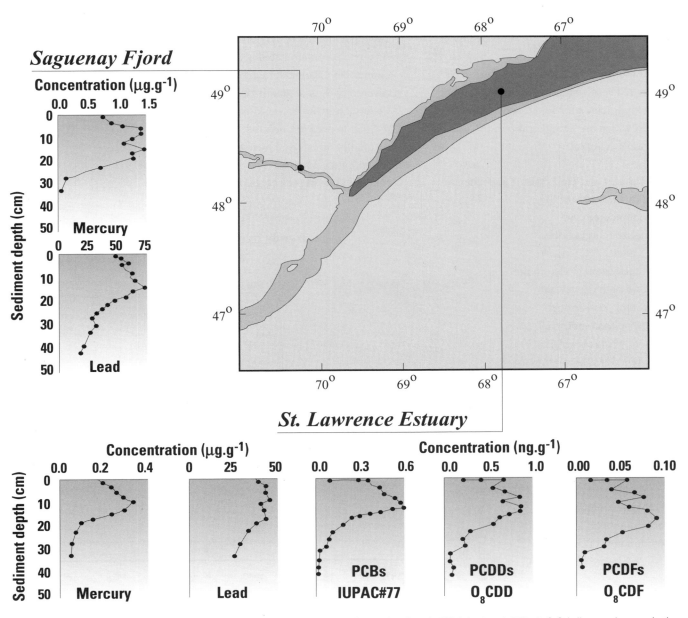

Source: Gobeil *et al.* 1995; Lebeuf *et al.* 1995a, b; C. Gobeil personal communication.

New Brunswick to control the spruce budworm (Pearce *et al.* 1989). However, concentrations of HCBs, oxychlordane and mirex continue to fluctuate and now appear to be as high as in the mid 1970s (Figure 10).

Conclusions

The St. Lawrence marine ecosystem ranks among Canada's most diverse aquatic environments. It includes one of the world's largest semi-enclosed seas, the Gulf of St. Lawrence, with a drainage basin whose freshwater outflow exceeds the run-off from the entire east coast of the United States. Several species of invertebrates as well as diadromous and marine fish are found in the Estuary and Gulf of St. Lawrence. These fish provide the support for commercial, recreational and subsistence fisheries which are an integral part of the region's economy.

The Estuary and Gulf of St. Lawrence are located downstream of the heavily populated and industrialized Great Lakes Basin. The St. Lawrence marine ecosystem is also in the downwind path of weather systems that pass through highly populated industrial centres of North America. There are several local sources of contaminants within the St. Lawrence marine ecosystem, including a number of municipalities and industries, and agricultural and shipping activities.

Over time, contaminants have accumulated in sediment deposition areas of the

ecosystem. Mercury, lead, PCBs, PCDDs and PCDFs are found in sediments of the Laurentian Channel, particularly in the Lower St. Lawrence Estuary. In the Saguenay Fjord, sediments of the inner basin contain high concentrations of mercury and PAHs. Much of this came from a chlor-alkali plant in the Saguenay River. The plant was closed in 1976. Contaminated sediments also occur in small estuaries, embayments, and nearshore areas in the Estuary and Gulf of St. Lawrence. These include the Baie des Anglais, Belledune Harbour, and the Miramichi Estuary.

Since the 1970s contaminant inputs into the St. Lawrence marine ecosystem have been reduced. As a result contaminated sediments in the Laurentian Channel and inner basin are gradually covered by less-contaminated materials. Nevertheless the burial of contaminants in sediments does not preclude them from being eventually transferred to fish. Benthic communities may increase the incorporation of contaminants in sediments into the food chain.

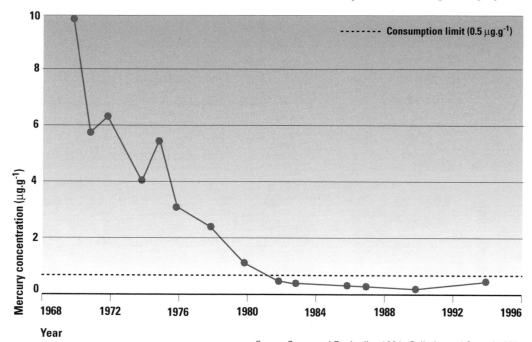

Figure 9
Trends in mercury concentrations in edible parts of northern shrimp from the Saguenay Fjord.

Source: Cossa and Desjardins 1984; Pelletier and Canuel 1988; Pelletier *et al.* 1989; Hodson *et al.* 1994; Cossa 1990.

This process would aid in their eventual uptake by commercially valuable fish. In addition, human activities may result in the resuspension of bottom sediments and their associated contaminants. The resuspension would enhance bioavailability of the contaminants. For example, the mercury contamination of northern shrimp in the Saguenay Fjord has remained stable since the early 1980s even though during that time, concentrations of mercury were decreasing in sediments of the inner basin.

In addition to the northern shrimp, several other aquatic biota in the St. Lawrence marine ecosystem accumulate chemical contaminants. The cadmium contamination of the lobster fishery in Belledune Harbour is local. However, other organisms have symptoms of stress that could affect the long term survival of entire populations. This is the case for the St. Lawrence beluga, the American eel, and was the case for the Northern Gannet populations of the St. Lawrence ecosystem. Although the latter has recovered from its pesticide-associated reproduction impairments during the 1970s, beluga and eel numbers are currently at low levels. These low levels make the beluga and eel populations even more vulnerable to the effects of contamination.

Despite the recent decrease in contaminant inputs, a number of contaminant-related issues remain. Although control of industrial and municipal effluents has improved, distant atmospheric sources may continue to be important. As a result, the long range transport of atmospheric pollutants (LRTAP) may play a relatively increasing role in bringing contaminants into the system. Continued use of the Estuary and Gulf of St. Lawrence as a major shipping route to the interior of the continent leads to the risk of accidental spills. Although the *Irving Whale* was recovered from the southern Gulf of St. Lawrence in July 1996, some PCB contamination of sediments persists at the site where the barge sank in 1970 (Government of Canada 1997).

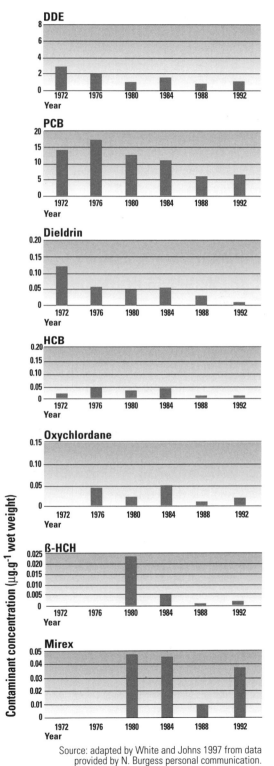

Figure 10

Trends in contaminant concentrations in Double-crested Cormorant eggs.

Contaminant concentration (μg.g^{-1} wet weight)

Source: adapted by White and Johns 1997 from data provided by N. Burgess personal communication.

Current shifts in usage of the St. Lawrence marine ecosystem and its commercially valuable fisheries also affect contaminant dynamics. Hitherto unexploited species with increasing commercial value are now being caught. This has possible effects on the structure of food chains and therefore, on the contaminant transfers that occur within the altered food chains. As these changes are taking place the aquaculture industry is expanding in some areas of the Gulf of St. Lawrence. This expansion means increasing the use of therapeutic chemicals. The potential for contamination by these threrapeutic chemicals is presently not known.

Case Study:

BIOMAGNIFICATION OF CONTAMINANTS IN MARINE FOOD WEBS: PCBS IN ST. GEORGES BAY, NOVA SCOTIA

St. Georges Bay is a shallow marine embayment open to the southern Gulf of St. Lawrence and relatively remote from local sources of industrial and domestic pollution. For these reasons, the Bay was chosen for a study of the biomagnification of PCBs coming from atmospheric sources. Because of its remoteness, the Bay receives most of its PCB inputs from LRTAP (Ware and Addison 1973; Harding *et al.* 1997). The majority (98%) of the PCBs detected in the Bay are found in seawater, associated with suspended particulate matter. Only 2% of the PCBs

are found in marine biota, including plankton, fish and marine mammals (G.C. Harding personal communication).

Small amounts of PCBs can be biomagnified through the marine food web. Studies have shown that PCB concentrations were 38 times higher in fish than in plankton and were 2 times higher in marine mammals than in fish, providing evidence of PCB biomagnification in the marine food web (Harding *et al.* 1997).

Although biomagnification occurs, PCB concentrations in marine organisms in the Bay are declining. For example, PCB concentrations in plankton dropped exponentially by a factor of 6 000 from the early to the late 1970s, although concentrations have remained relatively constant since 1977 (Figure 11). PCBs in pelagic fish, such as herring, show a consistent downward trend from the 1970s to the 1990s (Figure 12; Harding *et al.* 1997).

**Figure 11
Trends in average PCB concentrations in plankton in the southern Gulf of St Lawrence.**

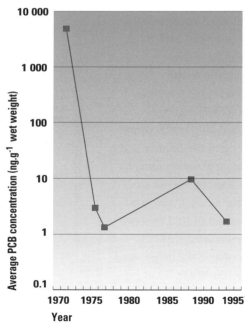

Source: Adapted from Harding *et al.* 1997.

**Figure 12
Trends in average PCB concentrations in fish in the southern Gulf of St. Lawrence.**

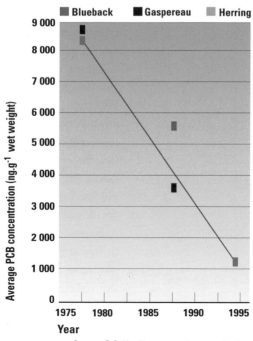

Source: G.C. Harding personal communication.

Case Study:

CONTAMINATION OF THE ST. LAWRENCE AMERICAN EEL POPULATION

American eels spawn in the Sargasso Sea; larvae and juveniles migrate to coastal streams in North America. They spend the majority of their life growing to maturity in freshwater before migrating back again to the Sargasso Sea for spawning (Schmidt 1925). In the St. Lawrence drainage basin, eels ascend as far as Lake Ontario and may spend as much as 13 to 15 years growing to a size of approximately 4 kg (Castonguay *et al.* 1994). Their diet ranges from large invertebrates to small fish, and they may accumulate high concentrations of contaminants depending on local chemical contamination.

Large-scale surveys in 1982 provided the first evidence of chemical contamination of American eels migrating in the St. Lawrence Estuary (Desjardins *et al.* 1983a, 1983b; Dutil *et al.* 1985; Castonguay *et al.* 1989). Most adult eels caught in the St. Lawrence River and Estuary were found to be heavily contaminated by PCBs and chlorinated pesticides, especially mirex (Desjardins *et al.* 1983a, 1983b; Castonguay *et al.* 1989). Eels contaminated by mirex were first presumed to have migrated from the Great Lakes. This is because mirex sources were initially restricted to Lake Ontario and two of its tributaries (Holdrinet *et al.* 1978; Dutil *et al.* 1985). However, further study revealed that mirex contamination was not restricted to eels migrating from Lake Ontario (Castonguay *et al.* 1989). Nevertheless, eels migrating from the upstream St. Lawrence River and Lake Ontario area were more contaminated than those migrating from tributaries of the St. Lawrence River and Estuary (Desjardins *et al.* 1983b; Castonguay *et al.* 1989).

In 1990, another large-scale survey was conducted to assess changes in contamination of American eels since 1982 (Hodson *et al.* 1992, 1994). In eight years, PCB, mirex and DDT concentrations in the muscle of adult eels migrating in the St. Lawrence Estuary had declined by 68, 56 and 77% respectively (Castonguay *et al.* 1989; Hodson *et al.* 1992; Figure 13). Concentrations of mercury and other pesticides had also declined except for dieldrin which remained unchanged since 1982 (Hodson *et al.* 1992). This overall decrease is believed to reflect the decline in Lake Ontario contamination between the 1970s and 1980s (Hodson *et al.* 1994).

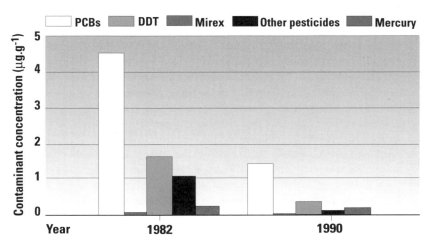

Figure 13

PCB, mirex, DDT, other pesticide and mercury concentrations in adult eels in the Lower Estuary, 1982 and 1990.

Source: Hodson *et al.* 1992, 1994; Desjardins unpublished data.

Adult eels migrating down the St. Lawrence River in 1990 were still over 100 times more contaminated than those originating from tributaries of the St. Lawrence Estuary (Hodson *et al.* 1994). In addition, vertebral malformations and lesions in the liver were more frequent when eels were more heavily contaminated with organochlorine compounds (Couillard *et al.* 1997). However, vertebral malformations may also have been caused by other natural or anthropogenic factors such as the passage through the turbines of two large hydroelectric dams upstream in the St. Lawrence River (Couillard *et al.* 1997).

Given their high contaminant concentrations compared to resident fish species of the Estuary, migrating eels originating upstream from the St. Lawrence River and Lake Ontario may represent significant vectors for the transfer of contaminants to the beluga population of the St. Lawrence Estuary, especially for mirex and PCBs (Béland *et al.* 1993; See Case Study: Contamination of the St. Lawrence beluga population).

Case Study:

CONTAMINATION OF THE ST. LAWRENCE BELUGA POPULATION

Belugas are top predators in the food chain of the St. Lawrence Estuary. They have a maximum life span of 30 years or more in the wild (Sergeant 1973). The population of approximately 700 individuals (Kingsley 1996) summers within a 200 km segment centred at the confluence of the Saguenay Fjord and the upper estuary, both of which are affected by upstream contaminant inputs. As a result, the beluga population is highly vulnerable to contamination in the St. Lawrence.

Studies conducted on stranded carcasses of belugas found along the St. Lawrence Estuary between 1982 and 1990 demonstrated that this population has been exposed to chemical products of both industrial and agricultural origin for a long time (Muir *et al.* 1990; Wagemann *et al.* 1990; Béland *et al.* 1993). Several organochlorine

Beluga Whale
(Delphinapterus leucas)

Along with narwhals, beluga belong to the family Monodontidae. Found in Arctic and subarctic waters along the northern coast of Canada and the St. Lawrence river estuary, belugas are distinguished by the absence of dorsal fins and their white colour. Adult males (3.7 to 4.3 m and 450 to 1000 kg) are larger than females (3.1 to 3.7 m, 250 to 700 kg). Females produce young about every three years with a gestation period of 14.5 months and a lactation period of 18 months. The fat-rich milk results in the rapid growth of the young. The beluga is an opportunistic feeder with a diverse diet that varies seasonally and consists of many fish species such as capelin, Arctic cod and herring, as well as shrimp, squid and marine worms. Most feeding and storage of blubber reserves takes place during their autumn migration. This fat, at times making up to 50% of the total body weight, serves as an insulator and as an energy reserve. In summer belugas gather in estuaries and adjacent waters characterized by shallow, brackish and relatively warm waters and sandy or muddy substrates. Circumpolar winter distribution varies with shifting ice and open water.

Beluga whales are important from a toxic chemicals point of view because they are top predators and consequently occupy the top of the upper level of contaminant bioaccumulation in an aquatic food chains. The beluga population of the St. Lawrence Estuary is particularly important in this context because of its status as an endangered population and the well-known high degree of contamination, including DDT and PCBs, that may affect its recovery. In the Arctic, belugas have always been an important subsistence resource for the Inuit and other northern native communities, providing meat, fat, oil, leather, tools and materials for use in the fabrication of arts and crafts. Despite evidence of significant decreases in both DDT and PCB contamination in some marine mammals, there still exists a concern about the consumption of their tissues on human health (Dewailly et al. 1993).

Figure 14
Organochlorine compound concentrations in
the blubber of belugas from areas of the
Canadian Arctic and the St. Lawrence Estuary, 1983 to 1987.

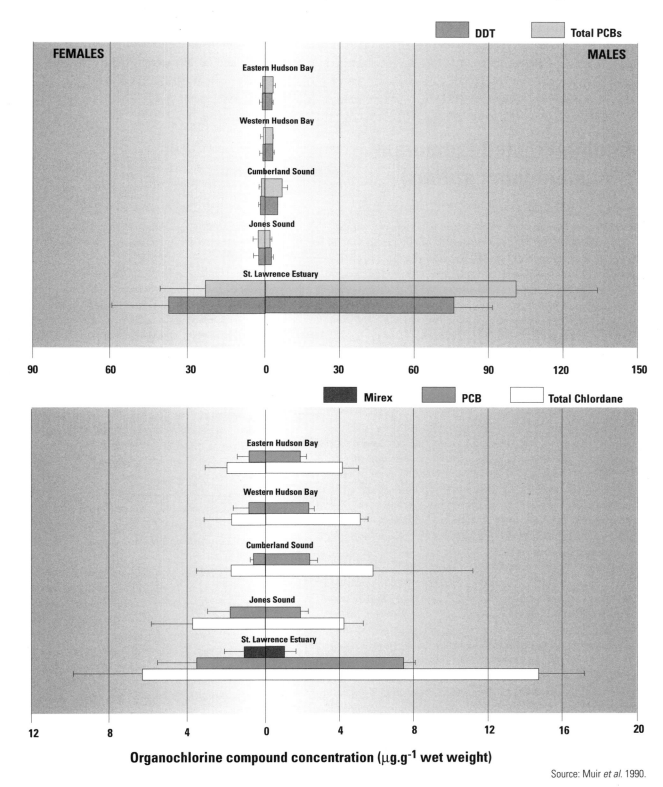

Organochlorine compound concentration (μg.g⁻¹ wet weight)

Source: Muir *et al.* 1990.

compound contaminants such as PCBs, DDT and mirex have been found in the blubber of St. Lawrence belugas at concentrations that were at least 10 times higher than those found in belugas from other populations in the Canadian Arctic (Figure 14). Metals such as mercury and lead were also found in higher concentrations in belugas of the St. Lawrence than in Arctic populations (Wagemann *et al*. 1990). As seen in Figure 14, female belugas are usually less contaminated than males because contaminants can be transferred from mothers to their neonates through the placenta during pregnancy and through maternal milk during nursing (Wagemann and Muir 1984; Stern *et al*. 1994).

Concentrations of other major contaminants such as PAHs, PCDDs and PCDFs were found to be very low or not detectable in St. Lawrence belugas (Béland *et al*. 1992; Muir *et al*. 1996a).

Double-crested Cormorant (*Phalacrocorax auritus*)

The Double-crested Cormorant is one of six species of cormorants breeding in North America. It has the widest distribution, being the only species to occur extensively in the interior as well as on both Atlantic and Pacific coasts. The birds migrate to winter in the southern U.S., chiefly coastal areas from Texas to North Carolina. The Double-crested Cormorant is a fish-eating bird, feeding opportunistically on a wide variety of benthic and pelagic fish in relatively shallow water. Cormorants nest in trees, on ground and on cliffs depending on habitat availability. They tend to be adaptable and are attracted to concentrated food sources. This has resulted in a long history of conflict with fishers and more recently with fish farmers, who typically view them as a competitor for fish stocks.

The cormorant was selected as a monitoring species because of its wide distribution, inshore feeding habits and the its sensitivity to eggshell thinning from DDT. Currently, Great Lakes and Atlantic coast cormorant populations are only slightly affected by contaminants. However, the cormorant is not an ideal monitor since it winters along the U.S. coast, hence its contaminant load results from exposure in both its wintering and breeding sites.

Reproductive failure caused by DDT-induced eggshell thinning and subsequent egg breakage led to almost virtual extinction on the Great Lakes by the early 1970s. With regulatory controls on pesticide use, the populations in the Great Lakes and prairie provinces have recovered to a current level of 220 000 pairs and are increasing at an average rate of 29% per year. This recent increase is faster than in most other parts of its range and is attributed to reduced egg harvesting, and declining concentrations of organochlorine compound contaminants. It is further due to an abundance of forage fish, following decimation of Great Lakes stocks of predatory fish

from the 1940s to the 1960s. By the early 1990s, the Great Lakes cormorant population had grown to approximately 38 000 pairs, much larger than at any time in recorded history.

On the Atlantic coast the population ceased growing as a result of exposure to pesticides. In Atlantic Canada, while individuals were affected by high contaminant concentrations, populations remained relatively unaffected compared to the Great Lakes. Mean concentrations in cormorant eggs from 1970 to 1973 from the Bay of Fundy and Chaleur Bay were in the 4 to 8 $\mu g.g^{-1}$ range, but only 20 percent of eggs contained residues greater than 15 $\mu g.g^{-1}$, the concentration at which eggshell thinning is known to occur. Other factors, such as egg harvesting, high reproductive capability and food availability have been more important for population growth. Populations on the Atlantic coast have recovered from low levels in the early 1970s. The current population is 96 000 pairs.

In the early 1970s, contaminant concentrations in Pacific coast Double-crested Cormorants were lower than in the Great Lakes or Atlantic coasts, averaging 4 $\mu g.g^{-1}$ which was well below concentrations known to cause deleterious effects.

Planar PCB contaminants are now the most important toxic chemicals faced by cormorants and other Great Lake species. In the Great Lakes, colonies in some more contaminated areas are likely to be less productive than those from cleaner areas. An effective metabolic detoxification capability for planar PCBs with fewer than four chlorinations and ability to exploit forage fish has nevertheless provided double-crested cormorants with a significant competitive advantage over their Great Lakes ecosystem competitors.

Based on information contained in Noble and Elliott (1986) and Nettleship and Duffy (1995).

These contaminants appear to be metabolized by belugas so that they do not accumulate in their tissues (Ray *et al.* 1991; Muir *et al.* 1996a). However, potential toxic effects may result from exposure to these contaminants. The degradation products of PAHs have been detected in tissues of both Arctic and St. Lawrence belugas (Ray *et al.* 1991). However, those of B[a]P, which was released in great quantities in the Saguenay Basin as a byproduct of aluminum production (Martel *et al.* 1986), were found only in the St. Lawrence beluga (Shugart *et al.* 1990).

Several contaminants which were found in the tissues of St. Lawrence belugas are toxic. These included mercury, lead, PCBs, DDT and mirex. The stranded belugas in which these compounds were measured had rarely-observed chronic lesions, numerous tumors, and presented signs of an affected reproductive system (Martineau *et al.* 1988; DeGuise *et al.* 1994; Béland *et al.* 1993). In contrast, none of these lesions and diseases were found in much less contaminated Arctic belugas (Béland *et al.* 1993). Contaminants are thus believed to be the cause of health and reproductive impairments observed in St. Lawrence belugas. However, it is not possible to link a specific contaminant to a given observed effect when several toxic compounds are present and act simultaneously in a number of tissues of the same animal (Béland *et al.* 1993).

Although signs of reproductive impairment in St. Lawrence belugas were linked to the presence of contaminants in their tissues (Massé *et al.* 1986; Martineau *et al.* 1987; Béland *et al.* 1993), there is no evidence that contamination has affected the growth of the St. Lawrence beluga population in recent years. Contamination does not appear to affect the longevity of St. Lawrence belugas since more contaminated males die at an age similar to that of females (Lesage and Kingsley, 1995). In addition, a recent study indicates that the beluga population may have increased since the official end of its hunt in 1979, and that the proportion of juveniles in the population is indicative of a normal reproduction rate (Kingsley 1996). Nevertheless, the contaminant burden of the St. Lawrence beluga population raised sufficient concerns to help bring about remediation strategies and regulations for reducing contaminant inputs into the St. Lawrence marine ecosystem. These actions are now showing signs of effectiveness with the recent decrease in contamination of some aquatic biota, including the beluga, where declines in PCB and DDT concentrations were observed in the blubber of males between 1982 and 1994 (Muir *et al.* 1996b).

In 1995, the St. Lawrence Beluga Recovery Plan was developed by representatives of Department of Fisheries and Oceans and the World Wildlife Fund (WWF) in collaboration with independent specialists. The plan sought to improve population numbers and conditions so that natural events and human activities would not threaten the survival of the St. Lawrence beluga whale population. Several complementary strategies were proposed to achieve this goal. One of these strategies involves actions to supplement existing efforts already in place to reduce pollution, and to prevent the introduction of potentially deleterious substances. More specifically, the St. Lawrence Beluga Recovery Plan recommends that: 1) industries discharging PAHs, mercury and lead should develop pollution prevention plans in order to eliminate inefficiencies and adopt the best available non-polluting technologies; 2) non-point sources for these contaminants and banned substances (PCBs, DDT and mirex), such as sewage treatment plants, leachate from dump sites, and LRTAP be identified and reduced; and 3) contaminated sediment sites which pose a threat to belugas should be characterized and decontaminated. The St. Lawrence Beluga Recovery Plan is currently being implemented.

Case Study:

CONTAMINATION OF NEAR-SHORE SEDIMENTS IN THE BAIE DES ANGLAIS

Baie des Anglais, located on the north shore of the Lower St. Lawrence Estuary, provides an example of the contamination of nearshore sediments by local human activities. In addition to the nearby city of Baie Comeau, three major companies are located on the shores of Baie des Anglais: an aluminum smelter, a pulp and paper mill, and a cereal company.

Elevated concentrations of PAHs in bottom sediments have previously been reported for Baie des Anglais and have historically been the result of discharge from the aluminum smelter (Laliberté, 1991; MENVIQ 1993). In general, concentrations of total PAHs in sediments of the Laurentian Channel in the St. Lawrence Estuary range from 0.5 to 1.0 $\mu g.g^{-1}$, but concentrations up to 40 $\mu g.g^{-1}$ are found in sediments of Baie des Anglais, near the smelter (Gearing *et al.* 1994).

PCBs, PCDDs and PCDFs have also been detected in Baie des Anglais sediments (Brochu *et al.* 1995; Lebeuf *et al.* 1996b). Concentrations of PCBs are particularly high, exceeding 200 $ng.g^{-1}$ compared to less than

Figure 15
Spatial variations of total PCB concentrations in sediments of the Baie des Anglais and the St. Lawrence Estuary.

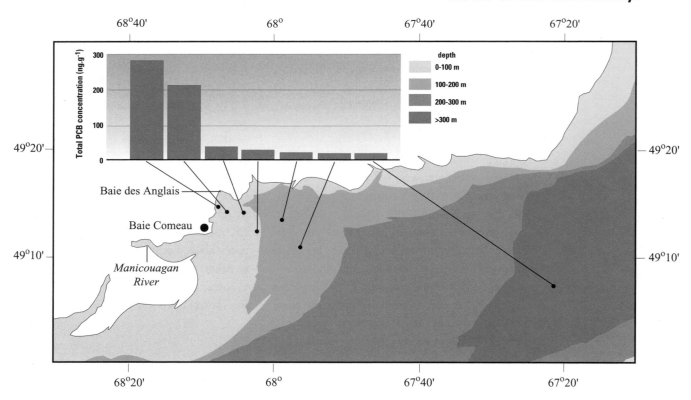

Source: Lebeuf *et al.* 1996a.

30 ng.g^{-1} in sediments of the Laurentian Channel (Lebeuf *et al.* 1996b; Figure 15). However, the elevated PCB contamination of sediments is confined to the immediate vicinity of Baie des Anglais as concentrations decrease rapidly within a few kilometers outside the Bay to those usually found in sediments of the Laurentian Channel (Figure 15). Concentrations of PCDFs are also higher in sediments of Baie des Anglais while those of PCDDs are lower than in sediments of the St. Lawrence Estuary (Lebeuf *et al.* 1996b).

PCDDs and PCDFs in Baie des Anglais sediments apparently originated from the PCB contamination, which is also related to the presence of the aluminum smelter (Brochu *et al.* 1995). Indeed, the congener composition of

PCBs, PCDDs and PCDFs in the Bay is typical of a commercial PCB mixture. However, Baie des Anglais is not a major source of PCBs, PCDDs and PCDFs to the St. Lawrence Estuary (Lebeuf *et al.* 1996b). In the case of PCDDs and PCDFs, most of the contamination in sediments of the Laurentian Channel originated from the atmosphere (Lebeuf *et al.* 1996a,b).

Chronic toxicity tests with American plaice have shown that exposure of males to the contaminated sediments of Baie des Anglais reduces both their capacity for fertilization and the hatching success of fertilized eggs (Nagler and Cyr 1997). However, the long-term impacts of sediment contamination in Baie des Anglais on local fisheries resources are unknown.

References

ADDISON, R.F., M.E. Zinck, and R.G. Ackman. 1972. Residues of organochlorine pesticides and polychlorinated biphenyls in some commercially produced Canadian marine oils. J. Fish. Res. Board Can. 29: 349-355.

ANDERSEN, A., and M. Gagnon. 1980. Fisheries resources of the St. Lawrence Estuary, Quebec, Canada. Can. Ind. Rep. Fish. Aquat. Sci. No. 119.

ARDISSON, P.L., and E. Bourget. 1992. Large-scale ecological patterns: discontinuous distribution of marine benthic epifauna. Marine Ecology Progress Series 83: 15-34.

BARBEAU, C., R. Bougie, and J. E. Côté. 1981. Temporal and spatial variations of mercury, lead, zinc, and copper in sediments of the Saguenay Fjord. Can. J. Fish. Aquat. Sci. 18: 1065-1074.

BÉLAND, P., S. De Guise, and R. Plante. 1992. Toxicology and pathology of St. Lawrence marine mammals: final report. National Institute for Ecotoxicology in the St. Lawrence, World Wildlife Fund, Toronto, Ont. 95 p.

BÉLAND, P., S. De Guise, C. Girard, A. Lagacé, D. Martineau, R. Michaud, D. C. G. Muir, R. J. Norstrom, É. Pelletier, S. Ray, and L. R. Shugart. 1993. Toxic compounds and health and reproductive effects in St. Lawrence beluga whales. J. Great Lakes Res. 19: 766-775.

BOUCHARD, H., and D. Gingras. 1992. Capsules éclair sur le Saint-Laurent. Centre Saint-Laurent, Division Coordination et bilan environmental, Montréal, Que.

BOUDREAULT, F. R., and F. Héritier. 1971. Dériveurs de surface dans le Golfe du Saint-Laurent. Direction Générale des Pêches Maritimes, Québec. Cahier d'information industrie et commerce. No. 57.

BROCHU, C., S. Moore, and É. Pelletier. 1995. Polychlorinated dibenzo-*p*-dioxins and dibenzofurans in sediments and biota of the Saguenay Fjord and the St. Lawrence Estuary. Mar. Poll. Bull. 30: 515-523.

BRUN, G.L., G.D. Howell, and H.J. O'Neill. 1991. Spatial and temporal patterns of organic contaminants in wet precipitation in Atlantic Canada. Environ. Sci. Technol. 25: 1249-1261.

CASTONGUAY, M., J. D. Dutil, and C. Desjardins. 1989. Distinction between American eels (*Anguilla rostrata*) of different geographic origins on the basis of their organochlorine contaminant levels. Can. J. Fish. Aquat. Sci. 46: 836-843.

CASTONGUAY, M., P. V. Hodson, C. M. Couillard, M. J. Eckersley, J. D. Dutil, and G. Verreault. 1994. Why is recruitment of the American eel, *Anguilla rostrata*, declining in the St. Lawrence River and Gulf? Can. J. Fish. Aquat. Sci. 51: 479-488.

CHADWICK, M., and A. Sinclair. 1991. Fisheries production in the Gulf of St. Lawrence. *In* J.C.Therriault [ed.] The Gulf of St. Lawrence: small ocean or big estuary? Can. Spec. Publ. Fish. Aquat. Sci. Vol. 113.

CHAPDELAINE, G., P. Laporte, and D. N. Nettleship. 1987. Population, productivity and DDT contamination trends of northern gannets (*Sula bassanus*) at Bonaventure Island, Quebec, 1967-1984. Can. J. Zool. 65: 2922-2926.

CHOUINARD, G. A., and A. Fréchet. 1994. Fluctuations in the cod fisheries of the Gulf of St. Lawrence. International Council of Exploration of the Sea symposium on cod and climate change, 1993 in Reykjavik, Iceland. 28 p.

COMBA, M. E., R. J. Norstrom, C. R. Macdonald, and L. E. Kaiser. 1993. A Lake Ontario-Gulf of St. Lawrence dynamic mass budget for mirex. Environ. Sci. Techn. 27: 2198-2206.

COSSA, D. 1990. Chemical contaminants in the St. Lawrence Estuary and Saguenay Fjord, p. 239-262. *In* M. El Sabh, and N. Silverberg [eds.] Oceanography of a large scale system, the St. Lawrence. Springer-Verlag, New York, N. Y. Coast. Estuar. Studies. Vol. 39.

COSSA, D., and C. Desjardins. 1984. Évolution de la concentration en mercure dans les crevettes du fjord du Saguenay (Québec) au cours de la période 1970-83. Rapp. techn. hydrogr. sci. océan. No. 32.

COUILLARD, D. 1982. Évaluation des teneurs en composés organochlorés dans le fleuve, l'estuaire et le golfe Saint-Laurent, Canada. Environ. Pollut. (Série B) 3: 239-270.

COUILLARD, C.M., and B. Légaré. 1994. Temporal variations of fin erosions and vertebral malformations in mummichogs living downstream from a bleached kraft pulp mill undergoing chlorine dioxide substitution. Proceedings of the 2nd International Conference on environmental fate and effects of bleached kraft pulp mill effluents, Vancouver, B.C. p.83.

COUILLARD, C.M., P.V. Hodson, and M. Castonguay. 1997. Correlations between pathological changes and chemical contamination in American eels, *Anguilla rostrata*, from the St. Lawrence River. Can. J. Fish. Aquat. Sci. 54: 1916-1927.

COURTENAY, S. C., P. J. Williams, C. Vardy, and I. Wirgin. 1995. Atlantic tomcod (*Microgadus tomcod*) and smooth flounder (*Pleuronectes putnami*) as indicators of organic pollution in the Miramichi Estuary (Chapter 13), p. 211-227. *In* E. M. P. Chadwick [ed.] Water, science, and the public: the Miramichi ecosystem. Can. Spec. Publ. Fish. Aquat. Sci. Vol. 123.

CZUCZWA, J.M., and R.A. Hites. 1986. Airborne dioxins and dibenzofurans: sources and fates. Environ. Sci. Technol. 20: 195-200.

D'ANGLEJAN, B. 1990. Recent sediments and sediment transport processes in the St. Lawrence Estuary. *In* M. I. El Sabh, and N. Silverberg [eds.] Oceanography of a large scale estuarine system, the St. Lawrence. Springer-Verlag, New York, N.Y. Coast. Estuar. Studies. Vol. 39.

DE GUISE, S., A. Lagacé, and P. Béland. 1994. Tumors in St. Lawrence beluga whales (*Delphinapterus leucas*). Vet. Pathol. 31: 444-449.

DESJARDINS, C., J. D. Dutil, and R. Gélinas. 1983a. Contamination d'anguille (*Anguilla rostrata*) du bassin du fleuve de St-Laurent par les biphényles polychlorés. Can. Tech. Rep. Fish. Aquat. Sci. 144.

DESJARDINS, C., J .D. Dutil, and R. Gélinas. 1983b. Contamination par le mirex de l'anguille (*Anguilla rostrata*) du bassin du fleuve de St-Laurent. Can. Tech. Rep. Fish. Aquat. Sci. No. 141.

DEWAILLY, E., P. Ayotte, S. Bruneau, C. Laliberté, D.C.G. Muir, and R.J. Norstrom. 1993. Inuit exposure to organochlorines through the aquatic food chain in Arctic Quebec. Environ. Health Perspect. 101: 618-620.

DFO. 1995. 1994 Statistical review. Department of Fisheries and Oceans, Maritimes Region, Moncton, New Brunswick.

DUTIL, J. D., B. Légaré, and C. Desjardins. 1985. Discrimination d'un stock de poisson, l'anguille (*Anguilla rostrata*), basée sur la présence d'un produit chimique de synthèse, le mirex. Can. J. Fish. Aquat. Sci. 42: 455-458.

EATON, P. B., L. P. Hildebrand, and A. A. d'Entremont. 1986. Environmental quality in the Atlantic region 1985. Environment Canada, Environmental Protection Service, Dartmouth, N.S. 241 p.

EATON, P. B., A. G. Gray, P. W. Johnson, and E. Hundert [eds.] 1994. State of the environment in the Atlantic Region. Environment Canada, Atlantic Region, Dartmouth, N.S.

ELLIOTT, J. E., R. J. Norstrom, and J. A. Keith. 1988. Organochlorines and eggshell thinning in Northern Gannets (*Sula bassanus*) from eastern Canada, 1968-1984. Environ. Pollu. 52: 81-102.

ENVIRONMENT Canada. 1991. St. Lawrence Updates June 1990-91. Environment Canada, St. Lawrence Centre, Montréal, Que. Factsheets on the state of the St. Lawrence River.

GAGNÉ, J., and M. Sinclair. 1990. Marine fisheries resources and oceanography of the St. Lawrence Estuary, p. 358-377. *In* M.I. El Sabh, and N. Silverberg [eds.] Oceanography of a large scale estuarine system, the St. Lawrence. Springer-Verlag, New York, N.Y. Coast. Estuar. Studies. Vol. 39.

GAGNON, M. M., J. J. Dodson, M. E. Comba, and K. L. E. Kaiser. 1990. Congener-specific analysis of the accumulation of polychlorinated biphenyls (PCBs) by aquatic organisms in the maximum Turbidity Zone of the St. Lawrence Estuary, Québec, Canada. Sci. Total Environ. 97/98: 739-759.

GEARING, J. N., P. J. Gearing, M. Noël, and J. N. Smith. 1994. Polycyclic aromatic hydrocarbons in sediments of the St. Lawrence Estuary, p. 58-64. *In* R. V. Collie, Y. Roy, Y. Bois, P. G. C. Campbell, P. Lundahl, L. Martel, M. Michaud, P. Riebel, and C. Thellen [eds.] Proceedings of the 20th annual aquatic toxicity workshop, held October 17, 1993, in Quebec City, P.Q. Can. Tech. Rep. Fish. Aquat. Sci. 1989.

GILBERT, D., and B. Pettigrew. 1997. Interannual variability (1948-1994) of the CIL core temperature in the Gulf of St. Lawrence. Can. J. Fish. Aquat. Sci. 54: 57-67.

GILBERT, M., and G. Walsh [eds.]. 1996. Potential consequences of a PCB spill from the barge Irving Whale on the marine environment of the Gulf of St. Lawrence. Can Tech. Rep. Fish. Aquat. Sci. 2113: xiii + 59 p.

GOBEIL, C., and D. Cossa. 1993. Mercury in sediments and sediment pore water in the Laurentian Trough. Can. J. Fish. Aquat. Sci. 50: 1794-1800.

GOBEIL, C., and M. Lebeuf. 1992. Inventaire de la contamination des sédiments du chenal Laurentien: les biphényles polychlorés. Rapp. Tech. Can. Sci. Halieut. Aquat. 1851.

GOBEIL, C., N. Silverberg, B. Sundby, and D. Cossa. 1987. Cadmium diagenesis in Laurentian Trough sediments. Geoch. Cosm. A. 51: 589-596.

GOBEIL, C., W. K. Johnson, R. W. Macdonald, and C. S. Wong. 1995. Sources and burden of lead in St. Lawrence Estuary sediments: isotopic evidence. Environ. Sci. Techn. 29: 193-201.

GOBEIL, C., Y. Clermont, and G. Paquette. 1997. Concentrations en mercure, plomb et cadmium chez diverses espèces de poissons de fond, de poissons pélagiques et de crustacés de l'estuaire et du golfe du Saint-Laurent et du fjord du Saguenay. Rapp. stat. can. sci. halieut. aquat. 1011: v + 83 p.

GOVERNMENT of Canada. 1997. Scientific assessment of the PCB contamination in sediments and biota around the site of the sinking of the barge Irving Whale. Status Report, Fisheries and Oceans Canada and Environment Canada, March 1997, xii + 48 p.

HARDING, G. C. 1992. A review of the major marine environmental concerns off the Canadian East Coast in the 1980s. Can. Tech. Rep. Fish. Aquat. Sci. 1885.

HARDING, G.C., R.J. Leblanc, W.P. Vass, R.F. Addison, B.T. Hargrave, S. Pearre, Jr., A. Dupuis, and P.F. Brodie. 1997. Bioaccumulation of polychlorinated biphenyls (PCBs) in the marine pelagic food web, based on a seasonal study in the southern Gulf of St. Lawrence, 1976-1977. Mar. Chem. 56: 145-179.

HELLOU, J., C. Upshall, J. F. Payne, and P. V. Hodson. 1994. Polycyclic aromatic compounds in cod (*Gadus morhua*) from the Northwest Atlantic and St. Lawrence Estuary. Sci. Total Environ. 145: 71-79.

HELLOU, J., P. V. Hodson, and C. Upshall. 1995. Contaminants in muscle of plaice and halibut collected from the St. Lawrence Estuary and Northwest Atlantic. Chem. Ecol. 11: 11-24.

HODSON, P. V., C. Desjardins, É. Pelletier, M. Castonguay, and C. M. Couillard. 1992. Decrease in chemical contamination of American eels (*Anguilla rostrata*) captured in the estuary of the St. Lawrence River. Can. Tech. Rep. Fish. Aquat. Sci. 1876. 60 p.

HODSON, P. V., M. Castonguay, C. M. Couillard, C. Desjardins, E. Pelletier, and R. McLeod. 1994. Spatial and temporal variations in chemical contamination of American eels, *Anguilla rostrata*, captured in the estuary of the St. Lawrence River. Can. J. Fish. Aquat. Sci. 51: 464-478.

HOLDRINET, M. V. H., R. Frank, R. L. Thomas, and L. J. Hething. 1978. Mirex in the sediments of Lake Ontario. J. Great Lakes Res. 4: 69-74.

KHAN, R.A., D.E. Barker, R. Hooper, and E.M. Lee. 1992. Effect of pulp and paper effluent on a marine fish, (*Pseudopleuronectes americanus*). Bull. Environ. Contam. Toxicol. 48: 449-456.

KHAN, R.A., D.E. Barker, R. Hooper, E.M. Lee, K. Ryan, and K. Nag. 1994. Histopathology in winter flounder (*Pseudopleuronectes americanus*) living adjacent to a pulp and paper mill. Arch. Environ. Contamin. Toxicol. 26: 95-102.

KINGSLEY, M.C.S. 1996. Population index estimate for the belugas of the St. Lawrence in 1995. Can. Tech. Rep. Fish. Aquat. Sci. 2117. 38 p.

KOUTITONSKY, V. G., and G. L. Bugden. 1991. The physical oceanography of the Gulf of St. Lawrence: a review with emphasis on the synoptic variability of the motion. *In* J. C. Therriault [ed.] The Gulf of St. Lawrence: small ocean or big estuary? Can. Spec. Publ. Fish. Aquat. Sci. Vol. 113.

LALIBERTÉ, D. 1991. Teneurs en HAP dans les sédiments près de cinq alumineries du Québec en 1988. Ministère de l'environnement du Québec. #4QE–91–10.

LEBEUF, M., C. Gobeil, C. Brochu, and S. Moore. 1995a. Polychlorinated dibenzo-*p*-dioxins and dibenzofurans in Laurentian Channel sediments, Lower St. Lawrence Estuary. Organohal. Comp. 24: 293-298.

LEBEUF, M., C. Gobeil, Y. Clermont, C. Brochu, and S. Moore. 1995b. Non-ortho chlorobiphenyls in fish and sediments of the estuary and gulf of St. Lawrence (PCB). Organohal. Comp. 26: 421-426.

LEBEUF, M., S. Moore, and C. Brochu. 1996a. The influence of PCBs and PCDFs in the sediments of Baie-des-Anglais on the Lower St. Lawrence Estuary. Organohal. Comp. 28: 243-247.

LEBEUF, M., C. Gobeil, C. Brochu, and S. Moore. 1996b. Direct atmospheric deposition versus fluvial inputs of PCDD/Fs to the sediments of the Lower St. Lawrence Estuary. Organohal. Comp. 28: 20-24.

LEBLANC, J., C.M. Couillard, and J.-C. Brêthes. 1997. Modifications of the reproductive period in mummichog *(Fundulus heteroclitus)* living downstream from a bleached kraft pulp mill in the Miramichi Estuary, New Brunswick, Canada. Can. J. Fish. Aquat. Sci. 54: 2564-2573.

LESAGE, V., and M. C. S. Kingsley. 1995. Bilan des connaissances de la population de bélugas (*Delphinapterus leucas*) du Saint-Laurent. Rapp. Tech. Can. Sci. Halieut. Aquat. 2041.

LOGIMER. 1987. Évaluation préliminaire de l'évolution à long terme de la qualité de l'écosystème du fjord du Saguenay. Rapport pour Parcs Canada. Logimer Inc.

LORING, D. H. 1975. Mercury in sediments of Gulf of St. Lawrence. Can. J. Earth Sci. 12: 1219-1237.

LORING, D. H. 1976a. The distribution and the partition of zinc, copper and lead in sediments of the Saguenay Fjord. Can. J. Earth Sci. 13: 960-971.

LORING, D. H. 1976b. The distribution and the partition of cobalt, nickel, chromium and vanadium in the sediments of the Saguenay Fjord. Can. J. Earth Sci. 13: 1706-1718.

LORING, D. H. 1978a. Industrial and natural inputs, levels, behavior, and dynamics of biologically toxic heavy metals in the Saguenay Fjord, Gulf of St. Lawrence, Canada, p. 1025-1040. *In* W. E. Krumbein [ed.] Proceedings of the third International symposium environ. Ann Arbor Science, Ann Arbor, Michigan.

LORING, D. H. 1978b. Geochemistry of zinc, copper and lead in the sediments of the estuary and Gulf of St. Lawrence. Can. J. Earth Sci. 15: 757-772.

LORING, D. H. 1979. Geochemistry of cobalt, nickel, chromium, and vanadium in the sediments of the estuary and open Gulf of St. Lawrence. Can. J. Earth Sci. 16: 1196-1209.

LORING, D.H., and J.M. Bewers. 1978. Geochemical mass balances for mercury in a Canadian Fjord. Chem. Geol. 22:309-330.

LORING, D.H., and D.J.G. Nota. 1973. Morphology and sediments of the Gulf of St. Lawrence. Bull. Fish. Res. Board Can. 182: 147 p.

LUN, R., and D. Mackay. In press. Modelling the fate of benzo(a)pyrene in the Saguenay Fjord. Environ. Toxicol. Chem.

MARTEL, L., M. J. Gagnon, R. Massé, A. Leclerc, and L. Tremblay. 1986. Polycyclic aromatic hydrocarbons in sediments from the Saguenay Fjord, Canada. Bull. Envir. Contam. Toxic. 37: 133-140.

MARTEL, L., M. J. Gagnon, R. Massé, and A. Leclerc. 1987. The spatio-temporal variations and fluxes of polycyclic aromatic hydrocarbons in the sediments of the Saguenay Fjord, Quebec, Canada. Wat. Res. 21: 699-707.

MARTINEAU, D., P. Béland, C. Desjardins, and A. Lagacé. 1987. Levels of organochlorine chemicals in tissues of beluga whales, *Delphinapterus leucas*, from the St. Lawrence Estuary, Québec, Canada. Arch. Environ. Contam. Toxicol. 16: 137-147.

MARTINEAU, D., A. Lagacé, P. Béland, R. Higgins, D. Armstrong, and L. R. Shugart. 1988. Pathology of stranded beluga whales (*Delphinapterus leucas*) from the St. Lawrence Estuary, Quebec, Canada. J. Comp. Pathol. 98: 287-311.

MASSÉ, R., D. Martineau, L. Tremblay, and P. Béland. 1986. Concentrations and chromatographic profile of DDT metabolites and polychlorobiphenyl (PCB) residues in stranded Beluga whales (*Delphinapterus leucas*) from the St. Lawrence Estuary, Canada. Arch. Environ. Contam. Toxicol. 15: 567-580.

MENVIQ. 1993. Baie des Anglais: Rapport du groupe de travail sur la contamination. Ministère de l'environnement du Québec.

MESSIEH, S. N. 1974. Surface and bottom currents in the Gulf of St. Lawrence as inferred from recoveries of drift bottles and sea bed drifters released in 1967-68. Fish. Res. Board Can. Manusc. Rep. Ser. 1287.

MESSIEH, S. N., and M. I. El-Sabh. 1988. Man–made environmental changes in the southern Gulf of St. Lawrence, and their possible impact on inshore fisheries, p. 499-523. *In* M.I. El–Sabh, and T. S. Murty [eds.] Natural and man-made hazards: Proceedings of the international symposium, held August 3, 1986, in Rimouski, Quebec, Canada. D Reidel Publishing Company, Dordrecht, NL.

MESSIEH, S. N., T. W. Rowell, D. L. Per, and P. J. Cranford. 1991. The effects of trawling, dredging and ocean dumping on the eastern Canadian continental shelf seabed. Continent. Shelf Res. 11: 1237-1263.

MISRA, R. K., and M. D. Nicholson. 1994. Univariate and multivariate analyses for time trends. International Council for the Exploration of the Sea. ICES C.M. 1994/ENV: 6.

MUIR, D. C. G., C. A. Ford, R. E. A. Stewart, T. G. Smith, R. F. Addison, M. E. Zinck, and P. Béland. 1990. Organochlorine contaminants in Belugas, *Delphinapterus leucas*, from Canadian waters, p. 165-190. *In* T. G. Smith, D. J. St. Aubin, and J. R. Geraci [eds.] Advances in research on the beluga whale, *Delphinapterus leucas*. Can. Bull. Fish. Aquat. Sci. Vol. 224.

MUIR, D.C.G., C.A.Ford, D. Rosenberg, R.J. Nostrom, M. Simon, and P. Béland. 1996a. Persistent organochlorines in beluga whales (*Delphinapterus leucas*) from the St. Lawrence River Estuary - I. Concentrations and patterns of specific PCBs, chlorinated pesticides, and polychlorinated dibenzo-p-dioxins and dibenzofurans. Environ. Poll. 93: 219-234.

MUIR, D.C.G., K. Koczanski, B. Rosenberg, and P. Béland. 1996b. Persistent organochlorines in beluga whales (*Delphinapterus leucas*) from the St. Lawrence River Estuary - II. Temporal trends, 1982-1994. Environ. Poll. 93 :235-245.

NAGLER, J.J., and D.G. Cyr. 1997. Exposure of male American plaice (*Hippoglossoides platessoides*) to contaminated marine sediments decreases the hatching success of their progeny. Environ. Toxicol. Chem. 16: 1733-1738.

NETTLESHIP, D. N. 1975. A recent decline of gannets, *Morus bassanus*, on Bonaventure Island, Québec. Can. Field-Nat. 89: 125-133.

NETTLESHIP, D.N., and D.C. Duffy. 1995. Cormorant and human interactions: An introduction. Colonial Waterbirds 18 (Special Publication 1): 3-6.

NETTLESHIP, D. N., and G. Chapdelaine. 1988. Population size and status of the northern gannet, *Sula bassanus,* in North America, 1984. J. Field Ornithol. 59: 120-127.

NOBLE, D.G., and J.E. Elliot. 1986. Environmental contaminants in Canadian seabirds 1968-1985: trends and effects. Can. Wildl. Serv. Tech. Rep. Ser. 13.

NOBLE, D.G., and S.P. Burns. 1990 Contaminants in Canadian seabirds. State of the Environment Fact Sheet 90-1, Environment Canada, Ottawa, Ont.

OLIVER, B.G., M. Charlton, and R.W. Durham. 1989. Distribution, redistribution, and geochronology of biphenyl congeners and other chlorinated hydrocarbons in Lake Ontario sediments. Environ. Sci. Technol. 23: 200-208.

PEARCE, P.A., J.E. Elliot, D.B. Peakall, and R.J. Norstrom. 1989. Organochlorine contaminants in eggs of seabirds in the Northwest Atlantic, 1968-84. Environ. Pollut. 56: 217-235.

PELLETIER, É., and G. Canuel. 1988. Trace metals in surface sediment of the Saguenay Fjord, Canada. Mar. Poll. Bull. 19: 336-338.

PELLETIER, É., C. Rouleau, and G. Canuel. 1989. Niveau de contamination par le mercure des sédiments de surface et des crevettes du fjord du Saguenay. Rev. Sci. Eau 2: 14-27.

RAY, S., B. P. Dunn, J. F. Payne, L. Fancey, R. Helbig, and P. Béland. 1991. Aromatic DNA–carcinogen adducts in Beluga whales from the Canadian Arctic and Gulf of St. Lawrence. Mar. Poll. Bull. 22: 392-396.

SCHAFER, C. T., J. N. Smith, and R. Cété. 1990. The Saguenay Fjord: a major tributary to the St. Lawrence Estuary. *In* M. El Sabh, and N. Silverberg [eds.] Oceanography of a large scale estuarine system, the St. Lawrence. Springer-Verlag, New York, N.Y. Coast. Estuar. Studies. Vol. 39.

SCHMIDT, J. 1925. The breeding places of the eel. Smithson. Inst. Annu. Rep. 1924: 279-316.

SERGEANT, D. E. 1973. Biology of white whales (*Delphinapterus leucas*) in western Hudson Bay. J. Fish. Res. Board Can. 30: 1065-1090.

SHUGART, L. R., D. Martineau, and P. Béland. 1990. Detection and quantitation of benzo(a)pyrene adducts in brain and liver tissues of beluga whales (*Delphinapterus leucas*) from the St. Lawrence and Mackenzie estuaries, p. 219-223. *In* J. Prescott, and J. Gauquelin [eds.] Compte-rendu du Forum international pour l'avenir du béluga, held 1988, in Tadoussac, Quebec. Université du Québec Press, Sillery, Que.

SILVERBERG, N., H.V. Nguyen, G. Delibiras, M. Koide, B. Sundby, Y. Yokoyama, and R. Chesselet. 1986. Radionuclides profiles, sedimentation rates and bioturbation in modern sediments of the Laurentian Trough, Gulf of St. Lawrence. Oceanol. Acta 9: 285-290.

SMITH, J. N., and E. M. Levy. 1990. Geochronology for polycyclic aromatic hydrocarbon contamination in sediments of the Saguenay Fjord. Environ. Sci. Techn. 24: 874-879.

SMITH, J.N., and D.H. Loring. 1981. Geochronology for mercury pollution in the sediments of the Saguenay Fjord, Québec. Environ. Sci. Technol. 15: 944-951.

SMITH, J.N., and A. Walton. 1980. Sediment accumulation rates and geochronologies measured in the Saguenay Fjord using the Pb-210 dating method. Geochim. Cosmochim. Acta 44: 225-240.

ST. LAWRENCE Center. 1996. State of the environment report on the St. Lawrence River. Volume 2: The state of the St. Lawrence. Environment Canada - Québec Region, Environmental conservation and Éditions Multimondes, Montréal, 153 p.

STERN, G. A., D. C. G. Muir, M. Segstro, R. Dietz, and M. P. Heide Jorgensen. 1994. PCBs and other organochlorine contaminants in white whales (*Delphinapterus leucas*) from West Greenland: variations with age and sex. BioScience 39: 243-257.

SUTCLIFFE, W. H., Jr., R. H. Loucks, and K. Drinkwater. 1976. Coastal circulation and physical oceanography of the Scotian Shelf and Gulf of Maine. J. Fish. Res. Board Can. 33: 98-115.

UTHE, J. F., C. L. Chou, and D. P. Scott. 1986. Management of the cadmium contaminated lobster fishery at Belledune, New Brunswick, Canada. International Council for the Exploration of the Sea. ICES C.M. 1986/E: 27.

WAGEMANN, R., and D. G. C. Muir. 1984. Concentrations of heavy metals and organochlorines in marine mammals of northern waters: overview and evaluation. Can. Tech. Rep. Fish. Aquat. Sci. 1279.

WAGEMANN, R., R. E. A. Stewart, P. Béland, and C. Desjardins. 1990. Heavy metals and selenium in tissues of beluga whales, *Delphinapterus leucas*, from the Canadian arctic and the St. Lawrence Estuary. *In* T.G. Smith, D. J. St. Aubin, and J. R. Geraci [eds.] Advances in research on the beluga whales, *Delphinapterus leucas*. Can. Bull. Fish. Aquat. Sci. Vol. 224.

WARE, D. M., and R. F. Addison. 1973. PCB residues in plankton from the Gulf of St. Lawrence. Nature 246: 519-521.

WELLS, P. G., and S. J. Rolston, and Marine Environmental Quality Advisory Group. 1991. Health of our oceans, a status report on Canadian marine environmental quality. Environment Canada, Dartmouth, N.S. 166 p.

WHITE, L., and F. Johns. 1997. Marine Environmental Assessment of the Gulf of St. Lawrence and Estuary. Fisheries and Oceans Canada. 128 pp.

WIRGIN, I. I., C. Grunwald, S. Courtenay, G-L. Kreamer, W. L. Reichert, and J. Stein. 1994. A biomarker approach in assessing xenobiotic exposure in cancer-prone Atlantic tomcod from the North American Atlantic Coast. Environ. Health Perspect. 102: 764-770.

Additional reading

DEPARTMENT of Fisheries and Oceans. 1996. Mercury contamination of northern shrimp in the Saguenay Fjord. Fact Sheet on the State of the Environment of the St. Lawrence, Maurice Lamontagne Institute. 6 p.

DEPARTMENT of Fisheries and Oceans. 1997. Contamination of Baie des Anglais. Fact Sheet on the State of the Environment of the St. Lawrence, Maurice Lamontagne Institute. 8 p.

DEPARTMENT of Fisheries and Oceans. 1997. PAHs in the Saguenay Fjord. Fact Sheet on the State of the Environment of the St. Lawrence, Maurice Lamontagne Institute. 8 p.

DUSEN, K.V, and A.C. Johnson Hayden. 1989. The Gulf of Maine: Sustaining our common heritage. Maine State Planning Office. 63 pp.

KENNISH, M.J. 1992. Ecology of estuaries: anthropogenic effects. CRC Press, Boca Raton, FL. 494 p.

NORTH Sea Task Force. 1993. North Sea Quality Status Report 1993. Oslo and Paris Commissions, London. Olsen & Olsen, Fredensborg, Denmark. 132+vi pp.

Personal Communications

BURGESS, N. Canadian Wildlife Service, Environment Canada, Sackville, N.B., n.d.

DESJARDINS, C. n.d.

GOBEIL, C. n.d.

HARDING, G.C. Department of Fisheries and Oceans, n.d.

SMITH, P., B. Petrie, and G. Bugden. Department of Fisheries and Oceans. 1997.

The BAY OF FUNDY Marine Ecosystem

Highlights

- The Bay of Fundy is a highly dynamic hydrographic system. It is driven by very strong tidal currents and tidal amplitudes which are among the highest in the world. These are extremely effective at dispersing contaminants and generally prevent contaminant accumulation near point sources.

- The Bay of Fundy has extremely productive fisheries. In the outer bay alone, total landed value in 1993 for all fisheries species was $18 million, with an additional $100 million generated in cultured salmon.

- Major point sources of contamination in the Bay of Fundy include wastewater treatment plants and some industrial facilities concentrated largely within the Saint John River Basin, and, to a lesser extent, river basins around the Minas Basin.

- Metal contamination does not appear to be a significant problem in the Bay of Fundy. The local geology of the area (i.e. weathering of rocks) is the most important contributor to elevated background concentrations in sediments and blue mussels.

Highlights

■ In the 1970s, polychlorinated biphenyls (PCBs) and dichlorodiphenyltrichloroethanes (DDTs) were the most prominent organochlorine compounds found in biological organisms such as seabirds and porpoises. In response to later regulatory measures which banned their use in Canada and the United States, there have been significant decreases in PCBs and DDT-related compounds in seabirds and porpoises in the Bay of Fundy since the 1970s. However, these contaminants still make up the majority of organochlorine compounds found in biological organisms today, demonstrating the extreme persistence of PCBs and DDTs in the environment.

■ Recent studies on seabirds and porpoises indicate that for many organochlorine compound contaminants there appears to be a shift from local sources to remote sources such as the long-range transport via the atmosphere (LRTAP) or ocean currents.

■ There are a number of current and emerging contaminant issues in the Bay of Fundy. These include impacts associated with continued industrial and municipal growth in the region, especially in the Saint John River Basin, and those associated with the intense aquaculture activities in the outer bay.

Table of Contents

The Bay of Fundy Marine Ecosystem

List of Figures

List of Tables

List of Sidebars

Introduction

The Bay of Fundy marine ecosystem is known for its diverse habitats, productive fishery grounds, and unique oceanographic conditions. The Bay is part of the Gulf of Maine ecosystem (Figure 1) that twice daily receives a tidal ebb and flow 2 000 times the volume of the St. Lawrence River. The turbid inner regions of the Bay, where tide heights are among the highest in the world, are as unique biologically as they are oceanographically.

Research on the Bay of Fundy marine ecosystem has been episodic, reaching peaks of intensity when practical questions have arisen. For example, proposals for development of tidal power on a large scale precipitated basic research on the Bay in the 1920s and 1930s, the 1950s, and again in the 1970s (Charlier 1982). Recent information on contaminant issues is generally lacking except for some long-term monitoring programs.

Major Contaminant Issues

The Bay is a high-energy dynamic system. Contaminants tend to become dispersed throughout the system and do not readily accumulate near their point sources. Little information is available on the extent to which near shore areas and embayments are contaminated.

Information on long-term trends in organochlorine compound contaminants is available only for seabirds and porpoises, primarily for the outer bay and the waters of the nearby North Atlantic. Long-term monitoring indicates that polychlorinated biphenyls (PCBs) and dichlorodiphenyltrichloroethanes (DDTs) made up the majority of organochlorine compounds in both seabirds and porpoises in the 1970s and 1980s and continue to do so today, despite declines from high concentrations in the 1970s. This demonstrates the persistence of these chemicals in the environment. It is thought that the decline in PCBs and DDTs can be attributed to government regulations which have banned their use in Canada and the U.S. At present, there are very small inputs of PCBs and DDTs into the Bay of Fundy system. These originate from leaching of landfill sites and long-range transport and deposition from remote sources, respectively.

Long-range transport of contaminants via the atmosphere (LRTAP) or the ocean is an increasingly important contaminant source as regulations have decreased the input from local (riverine) sources. For example, Leach's Storm Petrel, a seabird which feeds on organisms found on the ocean's surface have higher organochlorine compound concentration (except PCBs) than species which feed inshore or in deeper waters. After PCBs, chlorinated bornanes (CHBs) were the most common organochlorine compound found in harbour porpoises. Canadian use of CHBs is very limited and therefore they are thought to originate from LRTAP and ocean currents, from the southern U.S.

Oceanographic Features of Importance to Contaminant Dynamics

The physical oceanography of the Bay of Fundy is briefly summarized below and relies on recent reviews by Greenberg *et al.* (1997) and Daborn (1997). Discussion of sediment dynamics was contributed by T. Milligan (personal communication).

The Bay of Fundy is funnel shaped, 150 km in length, 100 km wide at its mouth and 45 km wide at the head of the main bay. The average depth of the Bay is 75 m (Figure 1). Oceanographic features differ between the upper and lower bay. The boundary between these two areas can be defined by a line joining the Saint John Harbour to the Annapolis Basin and corresponding roughly to the 100 m depth contour (Daborn 1986).

Figure 1
Bathymetry and watershed of the Bay of Fundy/Gulf of Maine.

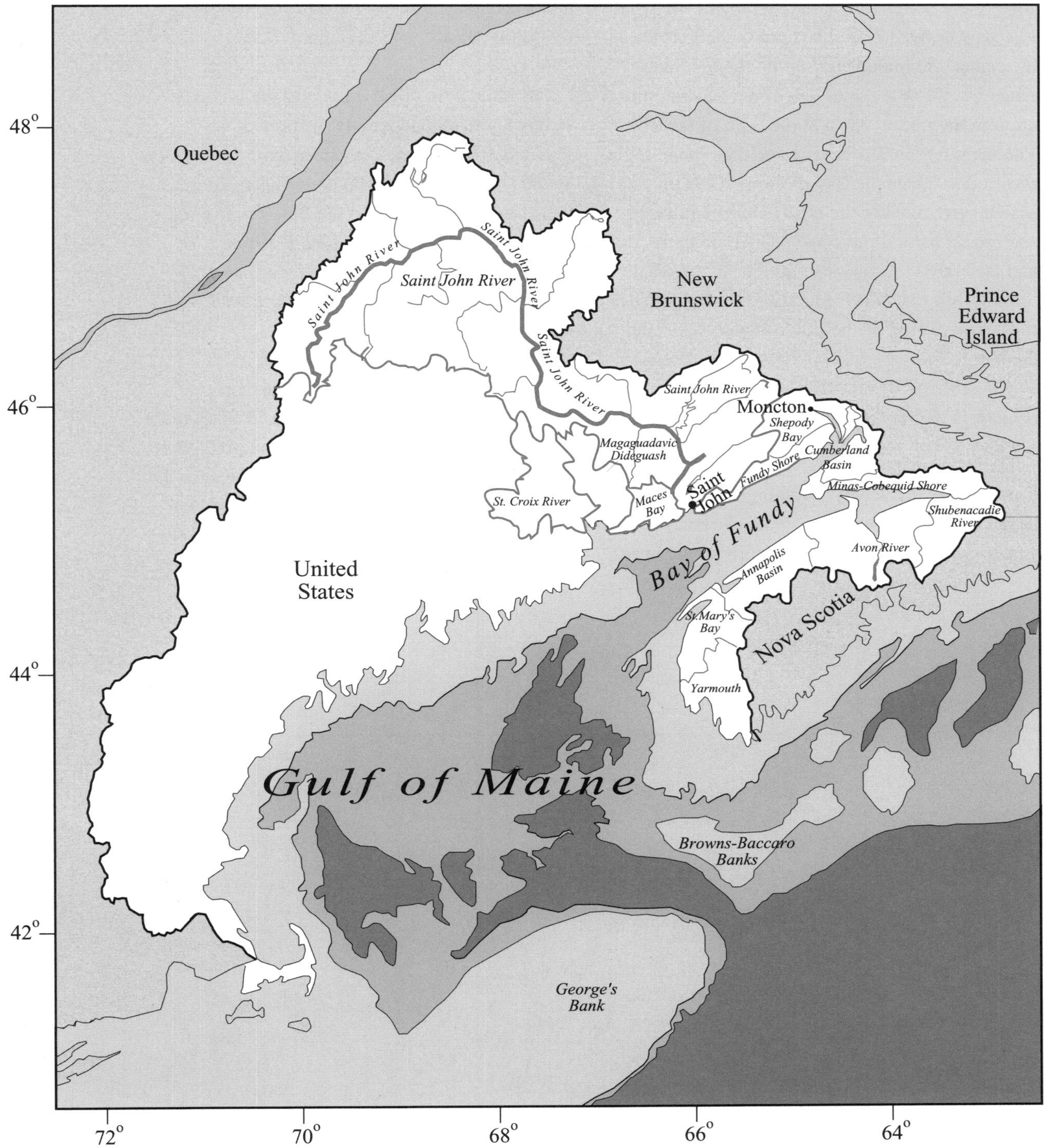

Bathymetry:

0 - 100 m
100 - 200 m
> 200 m

Watershed:

Gulf of Maine – watershed
Bay of Fundy river basins
Saint John River
Secondary Rivers

Source: adapted from Maine State Planning Office
Augusta, Maine, USA.

In contrast to most marine systems, tides completely dominate the functioning of the Fundy system (Garrett 1972). The tidal heights (between low water and high water) in the Bay of Fundy range from 4 m at the mouth of the Bay to 14 to 16 m in the Minas Basin at the head of the Bay. The main reason for the extreme tidal range in the Bay is that the whole Bay of Fundy/Gulf of Maine system out to the edge of the continental shelf has a natural period of oscillation of 13.3 h and is nearly in resonance with the 12.4 h tidal forcing from the North Atlantic. The resonance is further amplified by the Bay's funnel shape and the steady decrease in depth toward the Chignecto Bay and Minas Basin at the head of the Bay (Figures 1 and 2; Greenberg 1984).

The Bay's intertidal zone — the area of land exposed between low tide and high tide — is very large (30 000 ha) because of tidal extremes and a gently sloping shoreline, particularly in the upper

bay (Daborn 1986). Two-thirds of the intertidal area is contained within the Chignecto Bay and the Minas Basin.

The Saint John River emptying into the Bay of Fundy is the second largest river (after the St. Lawrence) in Atlantic Canada (Figure 1). It drains an area of 55 000 km^2 and contributes 70% of the total freshwater input into the Bay. The average discharge rate of the Saint John River is 882 m^{-3}.s^{-1}, although half of the discharge occurs between April and May at the time of spring run-off.

Tidal movements cause strong currents throughout the Bay ranging from 0.75 m.s^{-1} at the mouth of the Bay up to 4 m.s^{-1} in the narrows at Cape Split (Figure 2). Currents at the head of the Bay tend to be reversing rather than rotary, producing marked tidal excursion. Tidal currents may be so strong that the bottom is either scoured

Figure 2
Bay of Fundy.

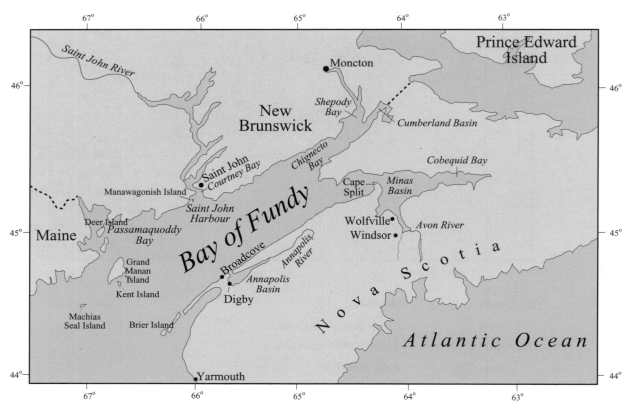

Source: L. White n.d.

to bedrock or covered by mobile sand waves (Amos and Joice 1977). A residual, counter-clockwise circulation is found in the main bay (Figure 3). This results in part from the entry of Atlantic tidal water primarily on the Nova Scotia side and the freshwater flow of the Saint John River on the New Brunswick shore. Water exchanges between the inner and outer bay are minimal despite large tidal influences (Greenberg 1984). In winter, inflow from the Atlantic along the Nova Scotia shore and outflow to the Gulf of Maine is minimal resulting in a relatively "closed" circulation in winter and a relatively "open" circulation in the summer months.

In the outer bay, water is often stratified during the summer so that different depths have different mean values for salinity and temperature. During the fall and winter, stratification is generally eliminated by meteorological conditions such as high winds (Garrett *et al.* 1978). Due to strong

tidal currents, water in the inner bay is vertically mixed all year round. Substantial vertical mixing also takes place in shallower areas of the outer bay such as Grand Manan Island, Deer Island, Brier Island, and areas off southwestern Nova Scotia (Figure 2; Greenberg 1984).

Strong tidal currents produce high concentrations of suspended sediment. The transport and settling of suspended particles ultimately determines much of the distribution of contaminants because most contaminants are associated with the fine particle fraction of suspended and bottom sediments. Suspended sediment load increases dramatically from open ocean values of less than 1 mg.L^{-1} at the mouth of the Bay to concentrations in excess of 2 500 mg.L^{-1} in the Cumberland and Minas Basins at the head of the Bay (Amos and Long 1980).

Figure 3
Surface circulation in the Bay of Fundy.

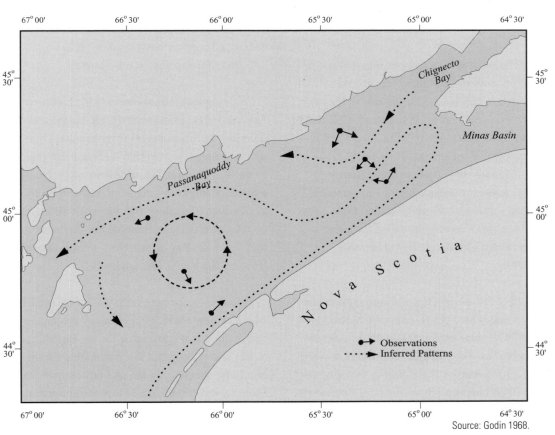

Source: Godin 1968.

The size distribution of bottom sediments ranges from coarse sand to fine mud in intertidal areas of the upper bay (Amos 1984) to areas where cobbles and boulders predominate such as Digby, Brier Island and Cape Split (Figure 2). Larger-size bottom sediments contain relatively little consolidated settled material. For example, settled material accounts for less than 5% of the total weight of sediment in the Annapolis Basin. This is reflected in very low trace metal concentrations in the sediment (Loring *et al.* 1996).

In the upper reaches of the Bay, the concentration of fine-grained material, such as silts and clays in sediment increases during summer months and is released to the water during winter months (Allen 1982). The increase in settled material is due to biological processes (Daborn *et al.* 1993). Concentrations of settled material in surface sediments can increase from less than 5% of the total weight of sediment during the winter and early summer months to more than 40% during the September bloom (T. Milligan, Fisheries and Oceans Canada, Dartmouth, NS, unpublished data).

In sheltered areas where bottom erosion is low due to low current velocities, fine sediment accumulation can be quite rapid due to the high concentration of suspended sediment. Courtney Bay and the finger piers in Saint John Harbour are examples of such areas. Parts of the Passamaquoddy Bay, such as in the Letang Harbour region, also have high concentrations of settled material, but accumulation rates are lower because of lower suspended sediment concentrations (Milligan 1994; Hargrave *et al.* 1995).

Perturbations of the Bay of Fundy system can result in major changes in sediment dynamics, especially in areas dominated by high sediment concentrations. For example, construction of the causeway at Windsor, Nova Scotia, in the upper bay resulted in restriction of the tidal flow which lead to extremely rapid (1 m.yr^{-1}) deposition of both fine and coarse sediment. Increased sedimentation in this area may continue for a long time. There is now an extensive mudflat on the seaward side of the causeway. This mudflat is rapidly on its way to becoming a marsh.

BIOLOGICAL RESOURCES

Two distinct finfish communities are found in the Bay (Dadswell *et al.* 1984; Dadswell 1997). An estuarine group, consisting of small-bodied species remain in the Bay year-round. This group includes a large number of anadromous species including smelt, tomcod, stickleback, silversides and flatfishes. Many of these species occur in immense numbers but have relatively low total biomass because of their small size. They dominate inner low-salinity regions such as the Cumberland Basin and Cobequid Bay (Figure 2). The second group consists mostly of large-bodied species which are summer migrants to the Bay. This group includes anadromous species such as American shad, Atlantic sturgeon, river herring (alewife, etc.), Atlantic salmon, striped bass; and oceanic fish such as cod, haddock, pollock, flatfishes, herring and sharks (dogfish, porbeagle, etc.). These fish make up the bulk of commercial fishes and support landings as high as 100 000 t for species such as herring.

During spring, summer and fall, anadromous fish tend to be found in the inner regions. In the outer bay oceanic species predominate. There is a mix of pelagic and benthic feeders in all regions of the Bay. Seasonal distributions are primarily as a result of migratory behavior, rather than food availability. Migratory fish generally follow the counter-clockwise currents of the Bay such that they enter the Bay on the Nova Scotia side and leave on the New Brunswick side.

Fishery management units do not isolate the Bay as a single unit; hence fishery production information is not available for the Bay of Fundy alone. Recent information is available for the outer bay, specifically the Quoddy (or the Fundy Isles) region (Figure 4). This area supports a productive and commercially valuable fishery.

Total landings in 1993 were 42 000 t with herring comprising 80% of the total (Chang *et al.* 1995). The value of total landings for all species in 1993 was $18 million. The most important commercial species are scallop, lobster and herring. Other valuable species include groundfish, especially cod and pollock, sea urchin (a relatively new fishery) and soft-shell clam.

Over the past two decades, intense and profitable growth in the Quoddy region aquaculture industry has made an increasingly important contribution to the value of commercial fisheries. In this region, cultured salmon production has increased to more than 90% of the salmon production of all eastern Canada (Chang *et al.* 1995). By 1993, salmon aquaculture production had grown to more than 10 000 t, with a value of almost $100 million.

Assessment of Chemical Contamination

SOURCES

Contaminant inputs into the Bay of Fundy are many and varied (i.e. industrial, terrestrial, airborne, municipal, domestic, agricultural). Major sources of contaminants include industrial (especially pulp and paper) and municipal effluents; non-point terrestrial inputs from river basins; ocean disposal of harbour dredge spoils; chronic oil discharges from the Saint John Refinery and related industrial operations and shipping; aquaculture wastes; and LRTAP or ocean currents. Butyltins and methyltins are frequently found in harbour sediments throughout the Atlantic provinces, especially in areas of heavy boating and shipping traffic (Maguire *et al.* 1986). The sources of contamination are well known (Eaton *et al.* 1994), but with the exception of dredge spoils, the extent and degree of their individual and cumulative impact on Bay ecosystems are largely unknown.

A 1991 inventory produced for the Gulf of Maine (NOAA 1994) indicated that most of the point sources of contamination to the Gulf of Maine are found in the U.S. The Bay of Fundy had 492 industrial facilities, 126 wastewater treatment plants (WWTP), and eight power plants. In 1991, 11% of the industrial facilities, 10% of the WWTP and 13% of the thermal power generating stations in the Bay of Fundy were found to be "major dischargers" in the Gulf of Maine. Only 10% of WWTPs, 1% of industrial facilities, and 13% of thermal power generating stations in the Bay of Fundy were considered to be significant "minor

Figure 4
Quoddy region of the Bay of Fundy.

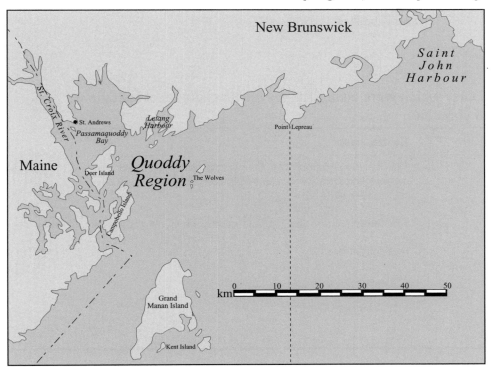

dischargers." In New Brunswick, major WTTP dischargers were those that served more than 10 000 people per day. Significant minor dischargers served more than 2 500 people per day. Major dischargers for industrial facilities discharged more than 660 kg.d^{-1} of BOD (biological oxygen demand) or TSS (total suspended solids). Minor point sources discharged more than 160 kg.d^{-1}. Major dischargers in Nova Scotia were identified by the Nova Scotia Department of the Environment. Minor dischargers released more than 7 300 000 L.d^{-1} of process flow water.

WWTP were responsible for the greatest portions of total nitrogen, total phosphorus, arsenic, cadmium, chromium, copper, iron, lead, mercury, oil and grease, and fecal coliform bacteria discharged into the Bay of Fundy/Gulf of Maine. Industrial facilities had higher discharges of process flow, biological oxygen demand, total suspended solids, and zinc.

Major industrial dischargers in the Bay of Fundy were concentrated in the Saint John River watershed and watersheds around the Minas Basin. The Avon River, which empties into the Minas Basin, contained 9% (143) of the industrial facilities in the Bay of Fundy/Gulf of Maine,

although most of these were minor dischargers (Figure 2). The Saint John River appears to be the greatest source of contaminants to the Bay of Fundy both from industrial facilities and from WWTPs. The Saint John River watershed, which is shared by the U.S. and Canada, had the greatest number of wastewater treatment plants at 69; 12% of these were major dischargers and 14% were significant minor dischargers. Other sites in the Bay of Fundy which were among the top 10 discharges of specific contaminants are listed in Table 1.

DISTRIBUTION, TRENDS AND EFFECTS
METALS AND ORGANOMETALLIC COMPOUNDS

Information on metals and related compounds in suspended particulate matter is limited. Showell and Gaskin (1992) examined cadmium and lead in suspended particulate matter (SPM) in the Quoddy Region of the outer bay (Figure 4). Mean concentrations of lead (19 to 28 μg.g^{-1}) and cadmium (1.2 to 1.5 μg.g^{-1}) in different size fractions of SPM in the early 1980s were similar to those reported for the lower St. Lawrence Estuary and Saguenay Fjord in the mid-1970s (lead 19 to 27 μg.g^{-1}; cadmium, 0.9 μg.g^{-1}; Cossa and Poulet 1976).

Table 1   Facilities in the Bay of Fundy which were among the top ten dischargers of specific contaminants in the Gulf of Maine.

Facility	Contaminant
Moncton Sewage Commision	Biological oxygen demand, total nitrogen, total phosphorus, arsenic, cadmium, chromium, copper, iron, lead, mercury, zinc, oil and grease, fecal coliform bacteria
Yarmouth Sewage Treatment Plant	Total phosphorus, cadmium, chromium, copper, iron, oil and grease, fecal coliform bacteria
Wolfville Sewage Treatment Plant	Fecal coliform bacteria
Irving Pulp and Paper Ltd., Saint John	Total suspended solids
St. Anne-Nackawic Pulp Co., Saint John River	Total suspended solids
Irving Ltd., Saint John	Zinc

Source: NOAA 1994.

A general survey of metal contaminants in sediments throughout the Bay has not been conducted since Loring's (1979) study in which copper, zinc, lead, cadmium, mercury, molybdenum, nickel, manganese, cobalt, chromium, vanadium, beryllium, arsenic, barium and selenium were measured in surficial sediments. Average concentrations were at or near background. Concentration of metals varied with sediment texture, such that lower concentrations were associated with sandy sediments in the central and eastern parts of the Bay. Higher concentrations were associated with fine-grained sediments in the Quoddy region where deposition of fine-grained material occurs. The finer sediments in the Quoddy region are a sink and accumulation area for metals, with concentrations decreasing seaward towards the south.

Elevated background concentrations of metals in other regions of the Bay were most frequently associated with weathering of bedrock, including areas along the Nova Scotian coast and those near Grand Manan Island (Loring 1982). Relatively high mercury concentrations were found in sediments adjacent to the coast between Passamaquoddy Bay and Point Lepreau (Figure 4). Higher metal concentrations occurred in sediments near the dredge disposal site in Saint John Harbour. More recently, Loring *et al.* (1996) found that 1988 samples of metal concentrations in sediments of the Annapolis Basin near Digby, were at background concentrations except for a few cases of higher concentrations of chromium which came from the erosion of nearby metal-bearing rocks.

Blue Mussel (*Mytilus edulis*)

Blue mussels are bivalve molluscs of the family Mytilidae. Living from high in the intertidal zone to depths of a few fathoms, blue mussels can be commonly found encrusting the submerged parts of buoys, piles, moorings and the hulls of boats. Their ability to grow profusely and rapidly has made them ideal candidates for aquaculture off the east coast of Canada where they are grown suspended from long lines (socks) hanging below the surface. Mussels capture their food, mainly phytoplankton, by pumping water through their gills. A commercial sized mussel (50 to 55 mm long) will pump up to 4 L of water an hour, depending on water temperature. Spawning in late spring or early summer, blue mussels grow rapidly, developing into free-swimming larva. These larvae feed on small plankton and possess the characteristic D-shaped shell only 48 hours after fertilization. Eventually, metamorphosing into a juvenile mussel with gills and a foot in by which to crawl, they slowly settle to the sea floor where they use secreted bysuss threads to anchor themselves to substrate. Preyed upon by crabs, starfish, and waterfowl, mussels have play an important role in the diets of coastal dwellers.

Bivalves such as blue mussels are useful bioindicators for monitoring environmental quality in coastal and estuarine ecosystems. They are widely distributed, live more than one year, are prominent members of intertidal communities, and are essentially non-motile through-out their life. They accumulate biologically available contaminants to concentrations which are orders of magnitude greater than those normally found in water or suspended particulate matter. Because of their ability to accumulate higher contaminant concentrations, chemical analyses of mussels tend to be easier to do and less expensive than samples of other media, such as water. Because they are cultured for human consumption, there is always a ready supply for use in the monitoring program. One of the longest established programs using mussels as sentinel monitors has been the Mussel Watch Program. First started in the United States, this monitoring program has been adopted by Canada and several other countries to provide long-term information on the concentrations of trace contaminants in coastal marine waters, and to identify areas where these compounds may be accumulated by living organisms in higher than expected concentrations. Such studies have shown promise in detecting and evaluating the presence and effects of a number of contaminants in coastal estuaries and nearshore marine environments.

In Saint John Harbour in the early 1980s, concentrations of copper, zinc, lead, cadmium, molybdenum, nickel, manganese and mercury measured in surficial sediments were found to be low and similar to the low mean values reported by Loring (1979) for uncontaminated sites in other parts of the Bay of Fundy (Ray and MacKnight 1984). Concentrations of metals were significantly lower at the ocean dumpsite in the outer harbour than in Courtney Bay, from which the most of the dredged material originates. Ray and MacKnight (1984) suggested that a combination of strong river current and tides and an open exposed harbour creates a dynamic regime which removes much of the fine-grained material, consequently preventing deposition of metals in the harbour.

Gulfwatch is a 10-year (1991-2000) monitoring program to measure both long-term and short-term exposure to selected organic and inorganic contaminants in the Gulf of Maine/Bay of Fundy ecosystem using the edible blue mussel. The sampling protocol is published as a technical report (Gulf of Maine Council on the Marine Environment (GOMC) 1992a). The first five years of Gulfwatch have recently been summarized (GOMC 1997) and individual annual reports have been published as technical reports (GOMC 1992b, 1994, 1996a, 1996b, 1996c). The following summary of the Gulfwatch program was prepared by G.C. Harding (personal communication).

The monitoring program found that year-to-year differences in contaminant concentrations were minor in comparison to the large differences between locations. The time series is currently too short to be able to detect long-term changes in contaminant concentrations. In the future, however, Gulfwatch data can be used to identify potential areas for establishing aquaculture facilities, indicate areas suitable for shellfish habitat restoration, and provide a framework for the monitoring of new issues such as sources of atmospheric mercury deposition or the spatial extent of endocrine disruptors.

Concentrations of copper, nickel, iron, aluminum and cadmium in blue mussel tissues were relatively uniform among regions within the Gulf of Maine. Higher concentrations of silver, lead, chromium and possibly zinc and mercury in mussel tissues were found the southern reaches of the Gulf of Maine. This is what one would expect based on the relatively higher population density and industrialization in Massachusetts and New Hampshire compared to that of Maine, New Brunswick and Nova Scotia. Three sites in the Bay of Fundy had elevated concentrations of metals relative to other sites in the northern Gulf of Maine. Mussels from Manawagonish Island in the Saint John Harbour had elevated concentrations of iron, aluminum, chromium and nickel. Mussels from Broad Cove and Argyle (near Yarmouth), Nova Scotia had above average concentrations of several metals (iron, cadmium, nickel and chromium; and iron, mercury, chromium and lead, respectively). Both of these locations are situated near small fishing communities with no obvious contaminant source. It is possible that the concentrations measured reflect naturally elevated metal concentration in the sediments due to the weathering of rock (see above).

Table 2 Mean concentrations of metals in muscle tissue of medium-sized scallops.

Sample Site	Concentration ($\mu g.g^{-1}$ wet weight)			
	Copper	Zinc	Cadmium	Lead
Bay of Fundy	2.9	66	3.1	0.4
Gulf of St. Lawrence				
Chaleur Bay	2.1	85	3.4	1.9
Northumberland Strait	2.7	70	3.2	0.3
Open Ocean				
Browns-Baccaro Bank and Georges Bank	3.2	97	6.1	0.6

Source: Ray and Jerome 1987.

Copper, zinc and lead concentrations in scallop muscle tissue collected from the Digby and Passamaquoddy Bay were comparable to those from non-contaminated areas (Ray and Jerome 1987; Table 2). Cadmium concentrations at the offshore sites of Georges Bank and Browns-Baccaro Banks were higher than those from Belledune in Chaleur Bay which has a known source of anthropogenic input of cadmium from smelting activities. The finding of high concentrations of cadmium in scallops from the offshore sites (Table 2), far removed from any known source of anthropogenic input, suggests that these elevated concentrations of cadmium are likely natural in origin.

Concentrations of cadmium, mercury, lead and 18 other metals in seabird tissues collected during the 1988 breeding season in Atlantic Canada, including two locations in the Bay of Fundy, were similar to those from other seabird species from different locations around the world (Elliott *et al.* 1992a). The highest mean concentrations of mercury (21 to 28 μg.g^{-1} dry weight) were found in the livers of Double-crested Cormorants from Saint John Harbour and highest mean lead concentrations (32 to 63 μg.g^{-1} dry weight) were found in bone tissue of Herring Gulls from three colonies in the Bay of Fundy. Leach's Storm Petrels had the highest cadmium concentrations in kidney tissue of all species studied. The highest mean concentration of cadmium was found in petrels from the Gulf of St. Lawrence (65 to 180 μg.g^{-1} dry weight), but petrels from the Bay of Fundy also had high concentrations (130 to 150 μg.g^{-1} dry weight) compared to other species (0.3 to 92 μg.g^{-1} dry weight). Histological examination of liver and kidney tissues did not reveal any evidence of tissue damage associated with elevated concentrations of metals. Elliott *et al.* (1992a) concluded that marine birds are able to carry appreciable body burdens of cadmium and mercury without any apparent adverse effects.

Mercury concentrations in blubber, muscle, liver and brain tissues of 12 harbour seals collected in 1971 were similar to concentrations found in porpoises in the early 1970s (Gaskin *et al.* 1973). However, total mercury concentrations in liver tissues of seals from the Bay of Fundy were considerably greater than those in seals from southern Maine.

Total mercury concentrations increased with age in muscle, brain, kidney and liver tissues of harbour porpoises taken during 1969 to 1977 in the Bay of Fundy and adjacent waters (Gaskin *et al.* 1972, 1979). However, mercury concentrations in various tissues, except the liver, did not show a strong relationship with either body length or weight of porpoises. Mercury concentrations in livers of both sexes decreased from 1970 to 1971, remained low for 3 years and increased from 1974 to 1977. Mercury in muscle was virtually all in the methylated form, while in the liver only 17% was in the methylated form. Other tissues had intermediate proportions of methylated mercury. These results are similar to those reported by Gaskin *et al.* (1973) for harbour seals and two species of whales from the Lesser Antilles.

Mean concentrations of mercury in Fundy porpoises sampled in 1989 (Johnston 1995) were significantly lower then those previously recorded for porpoises from the area (Gaskin *et al.* 1979), but were similar to those reported for porpoises in other locations (Teigen *et al.* 1993; Joiris *et al.* 1991). Other than mercury, no long-term studies of metal concentrations in porpoises exist.

Copper and zinc concentrations in liver, kidney and muscle tissues of Bay of Fundy porpoises sampled in 1989 (Johnston 1995) were similar to values previously published for similar species from other locations worldwide (Falconer *et al.* 1983) as well as for other cetaceans in Canadian waters (Wagemann *et al.* 1990). Mean cadmium concentrations in Fundy porpoises were somewhat higher than concentrations reported for porpoises in British waters (Law *et al.* 1992).

ORGANOMETALLIC COMPOUNDS

Tributyltin (TBT) is a biocide used in marine paints to prevent fouling of boats and ships and can cause detrimental effects on marine organisms. One effect, called imposex, is the masculinization (the development of a penis and *vas deferens*) in female neogastropod snails. Imposex leads to reproductive failure and consequently population decrease. In 1995 in eastern Canada, the occurrence of imposex in intertidal dogwhelks was examined at 34 harbour sites near areas of boating activity (N. Prouse personal communication). At 7 sites that were not near boating activity or other sources of TBT, no female dogwhelks had imposex. Dogwhelks with imposex were found at 13 sites and the frequency of affected females ranged from 20 to 100%. Two important harbours, Halifax and Sydney in Nova Scotia, had sites where all females were affected. The only site surveyed in the Bay of Fundy was Saint John Harbour where imposex frequency was 66%.

In other areas, such as the west coast of Canada and the United Kingdom, neogastropods have disappeared as a result of TBT contamination. Dogwhelks were absent from 14 harbour sites in eastern Canada with seemingly ideal habitat, possibly as a result of TBT.

ORGANIC CHEMICALS

Little if any published information exists on the concentration of organic contaminants in the waters, suspended particulate matter, or sediments of the Bay of Fundy.

ORGANOCHLORINE COMPOUNDS

No organochlorine compounds (DDT or derivatives) or PCBs were either detected or found in significant amounts (less than 5 $\mu g.g^{-1}$ wet weight) in soft-shelled clams sampled from two intertidal sites located in Annapolis Basin (Prouse *et al.* 1988), or in amphipod tissues (less than 1 $ng.g^{-1}$ wet weight) collected from the Minas Basin (Napolitano and Ackman 1989).

Average concentrations of PCBs in blue mussels from the Gulfwatch Monitoring Program exhibit a clear north to south increasing pattern in the Gulf of Maine. This pattern of distribution can

be attributed to local historical use around large population centres with long-term retention within the sediments and benthic community. DDTs were by far the predominant pesticide with a spatial distribution which closely paralleled that of PCBs. This also strongly suggests historic, local sources, rather than a more diffuse atmospheric fallout. It is surprising that higher concentrations were not found in mussel tissues further to the north, particularly off Maine and New Brunswick because these two jurisdictions used DDT extensively in forest spraying for spruce budworm in the 1950s and 1960s.

One of the longest running marine monitoring programs is that of the Canadian Wildlife Service of Environment Canada, which has been analyzing seabird eggs for organochlorine compounds at three sites (Manawagonish Island, Kent Island and Machias Seal Island in New Brunswick; Figure 2) in the Bay of Fundy since 1972 (N. Burgess, personal communication; Elliott *et al.* 1992b; Pearce *et al.* 1989). Eggs of the Double-crested Cormorant, Atlantic Puffin and Leach's Storm Petrel were sampled every four years. Contaminant concentrations in each species reflect contamination in different marine waters. Organochlorine compound concentrations in storm petrels reflect contamination of surface organisms in continental shelf regions; those found in the Atlantic Puffin reflect contamination of fish in deeper waters of the continental shelf; and those found in the cormorant reflect contamination of fish in inshore areas. However, since seabirds tend to range over considerable areas in their seasonal movements, it is sometimes difficult to pinpoint the exact geographic source of the contamination.

In cormorant and puffin eggs, PCBs make up more than two-thirds of the total organochlorine compound concentration; the DDT group makes up approximately 20%, and the other organochlorine compounds make up the rest (N. Burgess personal communication; Noble and Burns 1990).

Figure 5
Organochlorine compound concentrations in seabird eggs in the Bay of Fundy, 1972 to 1992.

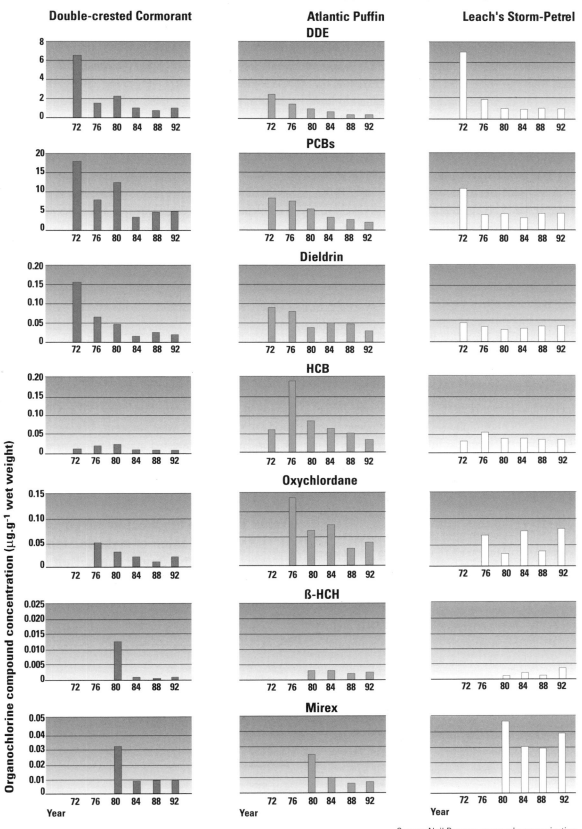

Source: Neil Burgess, personal communication.

Concentrations of DDE and PCBs in cormorant eggs have declined markedly since 1972, but plateaued at low levels from the mid 1970s to 1992 (N. Burgess personal communication; Elliott *et al.* 1992b; Pearce *et al.* 1989; Figure 5). These declines are similar to the trend reported in gannet eggs from the Gulf of St. Lawrence (Elliott *et al.* 1988). The decreases have been attributed to the cessation of PCB use in 1977 and to the ending of extensive use of DDT in New Brunswick forests to control spruce budworm (Pearce *et al.* 1989). In 1992, concentrations of DDE and PCBs in Bay of Fundy seabirds were higher than those off the coast of Newfoundland but comparable to those in the Gulf of St. Lawrence. Currently, the highest concentrations of all organochlorine compounds, except PCBs and DDE, are found in Leach's Storm Petrels (Figure 5). These concentrations may reflect trends in LRTAP of these compounds because this bird feeds on organisms found on the ocean's surface. Based on trends in marine birds, DDE, PCBs, and dieldrin residue concentrations in the near shore ecosystem of the North American continental shelf have generally declined in response to regulatory actions. In the open sea, dichlorodiphenyldichloroethylene (DDE) and PCBs appear to have decreased, although concentrations of other organochlorine compounds did not show any declining trend (Elliott *et al.* 1992b).

In 1979, concentrations of PCDD congeners in cormorants in the Bay of Fundy were 3 to 7 times lower than concentrations in cormorants from British Columbia in 1988 to 89 (Elliott *et al.* 1992b). This finding is consistent with the generally lower concentrations of PCDDs and PCDFs found in Atlantic ecosystems compared to Pacific ecosystems (See Pacific Marine and Freshwater Ecosystems Chapter). In contrast to DDT-group-related egg-shell thinning observed in Northern Gannet populations in the Gulf and cormorants in the Bay of Fundy in the 1970s, current polychlorodibenzo-*p*-dioxin (PCDD) concentrations pose no apparent threat to seabird populations.

The only information on organochlorine compound concentrations in seals comes from a study conducted in 1971 in which DDT-group pesticides, dieldrin and PCBs were measured in muscle, liver, blubber and brain tissues of harbour seals from the Bay of Fundy (Gaskin *et al.* 1973). Concentrations of total DDT were less than half those in porpoises sampled in the same year (Gaskin *et al.* 1971). In contrast to porpoises sampled at that time, relatively little dieldrin was found in seal blubber.

PCBs were the most prominent among 90 individual organic contaminants in blubber tissue of Fundy porpoises sampled during 1989 to 1991 (Table 3; Westgate *et al.* 1997). CHBs were the second most prominent organochlorine compound contaminant found in all porpoises in the western North Atlantic. The use of CHBs has been very limited in Canada. The presence of CHBs is thought to originate from LRTAP and ocean current transport from the southwest U.S.

Chlordane-related compounds (CHL) were found in lower concentrations than PCBs, CHBs and DDTs (Table 3; Westgate *et al.* 1997). Chlordanes were detected in most porpoise tissues in 1975-1977 in the Bay of Fundy, although their concentrations were much lower than those of

Table 3   Mean organochlorine compound concentrations in the blubber of male and female harbour porpoises from the Bay of Fundy/Gulf of Maine, 1989-1991.

Organochlorine compound	Concentration (μg.g^{-1} wet weight)	
	Male[1]	Female[2]
Total PCBs	17	11
Total CHB	12	8.4
Total DDT	7.7	6
Total chlordane	6.1	4
Total HCH	0.37	0.31
Total CBz	0.33	0.26

1 (N=55)
2 (N=53)

Source: Westgate *et al.* 1997.

PCBs (Gaskin *et al.* 1983). Chlordanes have also been detected in Arctic narwhal, Arctic and St. Lawrence belugas, Atlantic white-beaked dolphins and Pacific Dall's porpoise, indicating that this persistent organochlorine compound is widespread throughout different ecosystems (Muir *et al.* 1988, 1990, 1992).

Concentrations of organochlorine compounds in male and female harbour porpoises were strongly influenced by age (Gaskin *et al.* 1976, 1983; Westgate *et al.* 1997). Most organochlorine compound concentrations increased significantly

Atlantic Puffin
(*Fratercula arctica*)

The Atlantic Puffin is a member of the Alcidae or Auk family of birds. These birds are only found in the northern hemisphere and are the northern equivalent of penguins found in the southern hemisphere. In Canada, most Atlantic Puffins breed in grassy burrows and among rocks in colonies on islands primarily off the coasts of Newfoundland and Labrador. Smaller colonies and much smaller numbers breed on islands in the Gulf of St. Lawrence and off the coasts of Nova Scotia, New Brunswick and Maine. Since the 1880s, populations everywhere declined considerably as a result of excessive egg collecting, shooting, destruction of nesting habitat and by introduction of mammalian predators to small islands. At present, more than half the North American population breeds at Witless Bay, Newfoundland. This population has decreased substantially by 25 to 35% in the last few decades. The most likely factor in the decline seems to be changes in the availability of summer or winter food concurrent with the development of a capelin fishery in the region.

In summer, adult Atlantic Puffins feed primarily on fish. Their diet varies both geographically and seasonally. Atlantic Puffins tend to feed on small fish such as capelin, sand lance and herring which they pursue underwater to a depth of 30 m. They also may eat invertebrates.

In summer, the Atlantic Puffin is widely distributed over the continental shelf though it is only abundant near its breeding colonies. In winter, the birds are generally found far offshore in the western north Atlantic south of Newfoundland. Movements of various populations are perhaps more complicated than those of any other Atlantic alcid and they are not fully understood. The Atlantic Puffin was selected as a monitoring species, because it is a representative offshore subsurface feeder, it has an extensive breeding range, and because of the accessibility and abundance of its colonies.

Based on information provided in Noble and Elliott (1986).

with age in male porpoises while females had significant decreases. Female porpoises lose approximately 15% of their contaminant burden with each birth (Westgate *et al.* 1997).

Total PCB and DDT concentrations have significantly declined in Bay of Fundy porpoises between the 1970s (Gaskin *et al.* 1983) and 1989 to 1991 (Westgate *et al.* 1997). The magnitude of the decline was 4.3 and 17 times for total PCB and DDT, respectively; but due to changes in analytical methodology these should be regarded as relative indices rather than absolute values. This trend parallels that in seabirds and likely reflects similar changes in both the Bay of Fundy Marine Ecosystem and the northwest Atlantic.

In the 1970s, there were no regional differences in total PCBs or DDT within the Bay of Fundy or between the Bay of Fundy and Newfoundland or Rhode Island (Gaskin *et al.* 1983). From 1989 to 1991, porpoises from the Bay of Fundy had, in order of decreasing concentration PCBs>CHB>DDT>CHL, whereas porpoises from the Gulf of St. Lawrence and Newfoundland had CHB>PCBs>DDT>CHL (Westgate *et al.* 1997).

The significance of contaminant concentrations in porpoises is extremely difficult to determine. However, in hundreds of detailed necropsies of porpoises, none of the pathologies associated with contaminants in other marine mammals (i.e. St. Lawrence beluga) have been found in porpoises from the Bay of Fundy (Gaskin 1992).

PCBs had the highest concentration, followed by DDT-group pesticides in the blubber tissue of endangered right whales in the Bay of Fundy and on Browns-Baccaro Banks (Woodley *et al.* 1991). Relatively low concentrations found in adult females suggest that transmammary transfer of organochlorine compounds occurs during lactation. Concentrations of dieldrin, heptachlor epoxide, chlordanes, and hexachlorobenzene (HCB) were generally not detected or found in trace or very low amounts. Relative to other cetaceans, concentrations of organochlorine compounds in right whales were low (e.g., Gaskin 1992; Gaskin *et al.* 1983; Muir *et al.* 1988). Woodley *et al.* (1991) caution that concentrations may have been underestimated in their samples. This underestimate may have occurred as the blubber samples had a relatively low fat content and samples were obtained during the summer months. In the summer months, right whales are laying down fat reserves, with the result that existing concentrations of contaminants may have been diluted.

OTHER AROMATIC COMPOUNDS

Polycyclic aromatic hydrocarbons (PAHs) in blue mussels monitored in the Gulfwatch program exhibited a pattern similar to that observed for silver, lead and chromium. The highest concentrations and the most frequent detection of PAHs were found in the southern Gulf of Maine, particularly near large population centres.

Conclusions

The Bay of Fundy is an integral part of the larger Gulf of Maine ecosystem. It is a highly dynamic hydrologic system with strong tidal currents and some of the highest tides in the world. The system is a highly dispersive environment for both suspended sediment and contaminants. Sediment dynamics are relatively complex with extremely high suspended sediment loads in the upper bay and much lower suspended sediment loads in the lower bay.

Fisheries resources in the Bay, particularly the outer bay, are of substantial value. The outer bay has a thriving aquaculture industry. The upper bay is an important area for development and growth of young fish.

Metal contamination is generally of little concern in the Bay. Concentrations in sediments and blue mussels in most areas reflect what would be expected from local geological sources. Anthropogenic sources of metals do not appear to be sufficient to lead to local accumulation in the Bay's highly dispersive environment.

The downward trend of DDT-group and PCB concentrations in seabirds and porpoises indicates that regulatory controls have been effective in reducing input of these persistent compounds into the local environment. Some data on seabirds and porpoises indicate that input of persistent organic contaminants has shifted from riverine to atmospheric sources. Except for higher level predators such as seabirds and marine mammals, relatively little is known regarding contaminant distribution, fate and effects in biota in the Bay of Fundy.

Although source management has had a positive effect on concentrations of contaminants, there are nevertheless some important current and emerging issues. Only a few of the most important issues are described briefly below. For a comprehensive review of current and emerging issues in the Bay of Fundy, see Percy (1997) and Percy and Wells (1997).

Wastes discharged from municipalities and industries have been shown to have detrimental effects on shellfish and other aquatic life. Closures of shellfish harvesting areas are principally the result of bacterial contamination from human and animal wastes and agricultural run-off (Menon 1988). Since 1940, there has been a steady increase in the number of shellfish closures throughout the Bay of Fundy and Atlantic Canada (Machell and Menon 1992). Reversing this trend will require improved sewage treatment and improved effluent quality from industries and municipalities throughout the region (Menon 1988).

Few waste waters entering the Bay of Fundy have been characterized from a chemical contamination perspective. Contaminant discharge estimates for most industrial and municipal effluents in New Brunswick and Nova Scotia are incomplete and are frequently based on typical contaminant concentrations for each industry or municipality and not on monitoring data (NOAA 1994). It is particularly important to characterize contaminant discharge in multi-use industrialized harbours where local fisheries are significant. As local sources are controlled, the long-range transport and deposition of contaminants will assume greater relative importance.

The potential impact of the rapidly-growing salmon aquaculture industry is one of the major habitat concerns in the Bay of Fundy (Chang *et al.* 1995). There are more than 60 aquaculture sites in the Quoddy region. Aquaculture can have adverse impacts on marine habitat due to the discharge of wastes (e.g., food, feces) and chemicals (such as disinfectants, pesticides and therapeutants) that can affect bottom sediments, water quality and native wild species. Aquaculture operations can also obstruct fish migration routes, fishing activity, and navigation routes. Recent studies indicate that the aquaculture industry in some areas of the region may be approaching holding capacity, and that addition of new aquaculture sites may degrade the surrounding water quality and benthic environment (Strain *et al.* 1995).

Changes in sediment dynamics and hydrological processes, whether as a result of human activities or long-term global changes in sea level and climate, have the potential to make profound changes in the functioning of the Bay. The consequences of such changes on the fate and biological effects of chemical contaminants are largely unknown. Because most contaminants in the Bay are tied up in suspended sediments and bottom sediments, changes in sediment dynamics are expected to also affect the fate of contaminants.

Recent observations of changes in sediment dynamics and properties at the head of the Bay indicate large-scale erosion in some areas and deposition in others, implying movement of large amounts of sediment material in the upper Bay. Human alteration of the habitat (i.e. diking, building of causeways, upstream dams, river barrages and other river obstructions) has had measurable impact on local areas. The contribution of these developments to wide-scale sediment changes is only now being observed. Natural variability, such as long-term cycles in tidal amplitudes, water temperatures and sea levels, may also contribute to observed changes in sediment dynamics. Currently, there are plans for large-scale aggregate mining within the Bay of Fundy to remove gravel. It is not clear what impact this might have on the sedimentary dynamics of the Bay.

Changes in fish species assemblages and food web dynamics may affect the pathway of contaminants and their ultimate fate and effects in the ecosystem. For example, in recent years there has been a shift from the harvesting of cod and other groundfish to dogfish and skate in the Bay of Fundy. Fishers now exploit new resources such as seaweed and baitworms. Because relatively little is known about the biology and role in the food web of these species, their role in altering the fate of chemical contaminants is unknown.

References

ALLEN, J.R.L. 1982. Sedimentary Structures: Their Character and Physical Basis. Vol. 1. Elsevier, Amsterdam, 593 p.

AMOS, C.L. 1984. An overview of sedimentological research in the Bay of Fundy, p. 31-43. *In* D.C. Gordon Jr., and M.J. Dadswell [eds.] Update on the Marine Environment Consequences of Tidal Power Development in the Upper Reaches of the Bay of Fundy. Can. Tech. Rep. Fish. Aquat. Sci. 1256.

AMOS, C.L., and G.H.E. Joice. 1977. The sediment budget of the Minas Basin. Bedford Institute of Oceanography, Data Series BI-D-77-3: 411 p.

AMOS, C.L., and B.F.N. Long. 1980. The sedimentary character of the Minas Basin, Bay of Fundy, paper 80-10, p. 153-180. *In* S.B. McCann [ed.] The Coastline of Canada, Geological Survey of Canada.

CHANG, B.D., R.L. Stephenson, D.J. Wildish, and W. Watson-Wright. 1995. Protecting regionally significant habitats in the Gulf of Maine: a Canadian perspective. *In* Improving Interactions Between Coastal Science and Policy, Proceedings of the Gulf of Maine Symposium, Kennebunkport, Maine, November 1-3, 1994. National Academy Press, Washington, D.C.

CHARLIER, R.H. 1982. Tidal Energy. Van Nostrand Reinhold Co., New York, N.Y. 351 p.

COSSA, D., and S.A. Poulet. 1976. Survey of trace metal contents of suspended matter in the St. Lawrence estuary and Saguenay Fjord. Journal of Fisheries Research Board of Canada 35: 338-345.

DABORN, G.R. 1986. Effect of tidal mixing on the plankton and benthos of estuarine regions of the Bay of Fundy. *In* J. Bowman, M. Yentsch, and W.T. Peterson [eds.] Tidal Mixing and plankton Dynamics. Springer-Verlag, Berlin. 390-413 p.

DABORN, G.R. 1997. Fundy Marine Ecosystem Science Project: science overview. *In* J.A. Percy, P.G. Wells, and A.J. Evans [eds.] Bay of Fundy Issues: a scientific overview. Workshop Proceedings, Wolfville, N.S., January 29 to February 1, 1996. Environment Canada, Atlantic Region, Occasional Report No. 8, Environment Canada, Sackville, N. B. pp. 1-8.

DABORN, G.R., C.L. Amos, M. Brylinsky, H. Christian, G. Drapeau, R.W. Faas, J. Grant, B. Long, D.M. Paterson, G.M.E. Perillo, and M.C. Piccolo. 1993. An ecological cascade effect: migratory shorebirds affect stability of intertidal sediments. Limnology and Oceanography. 38:225-231.

DADSWELL, M.J. 1997. Fish. *In* J.A. Percy, P.G. Wells, and A.J. Evans [eds.] Bay of Fundy Issues: a scientific overview. Workshop Proceedings, Wolfville, N.S., January 29 to February 1, 1996. Environment Canada, Atlantic Region, Occasional Report No. 8, Environment Canada, Sackville, N. B. pp. 74-76.

DADSWELL, M.J., R. Bradford, A.H. Leim, D.J. Scarratt, G.D. Melvin, and R.G. Appy. 1984. A review of research of fishes and fisheries in the Bay of Fundy between 1976 and 1983, with particular reference to its upper reaches. Canadian Technical Report Fisheries and Aquatic Sciences 1256: 163-294.

EATON, P.B., A.G. Gray, P.W. Johnson, and E. Hundert. 1994. State of the Environment in the Atlantic Region Environment Canada, Atlantic Region. Minister of Supply and Services, Ottawa, Ont.

ELLIOTT, J.E., R.J. Norstrom, and J.A. Keith. 1988. Organochlorines and eggshell thinning in Northern Gannets (*Sula bassanus*) from eastern Canada, 1968-1984. Environmental Pollution 52: 81-102.

ELLIOTT, J.E., A.M. Schuehammer, F.A. Leighton, and P.A. Pearce. 1992a. Heavy metal and metallothionein concentrations in Atlantic Canadian seabird. Archives of Environmental Contaminants and Toxicology 22: 63 - 73.

ELLIOTT, J.E., D.G. Noble, R.J. Norstrom, P.E. Whitehead, M. Simon, P.A. Pearce, and D.B. Peakall. 1992b. Patterns and Trends of Organic Contaminants in Canadian Seabird Eggs, 1968-90, p. 181-194. *In* C.H. Walker and D.R. Livingston [eds.] Persistent Pollutants in Marine Ecosystems. Vol. 1. Pergamon Press, Oxford, New York.

FALCONER, C.R., I.M. Davies, and G. Topping. 1983. Trace metals in the common porpoise, *Phocoena phocoena*. Mar. Environ. Res. 8: 119.

GARRETT, C.J.R. 1972. Tidal resonance in the Bay of Fundy and Gulf of Maine. Nature 238: 441-443.

GARRETT, C.J.R., J.R. Keeley, and D.A. Greenberg. 1978. Tidal mixing versus thermal stratification in the Bay of Fundy and Gulf of Maine. Atmos.-Ocean 16(4): 403-423.

GASKIN, D.E. 1992. Status of the harbour porpoise (*Phocoena phocoena*), in Canada. Can. Field Nat. 106: 36-54.

GASKIN, D.E., M. Holdrinet, and R. Frank. 1971. Organochlorine pesticide residues in harbour porpoises from the Bay of Fundy region. Nature 223: 499-500.

GASKIN, D.E., K. Ishida, and R. Frank. 1972. Mercury in Harbour Porpoises (*Phocoena phocoena*) from the Bay of Fundy Region. J. Fish. Res. Board. Can. 29(11): 1644-1646.

GASKIN, D.E., R. Frank, M. Holdrinet, K. Ishida, C.J. Walton, and M. Smith. 1973. Mercury, DDT, and PCB in Harbour Seals (*Phoca vitulina*) from the Bay of Fundy and Gulf of Maine. J. Fish. Res. Board Can. 30: 471-475.

GASKIN, D.E., M. Holdrinet, and R. Frank. 1976. DDT residues in blubber of harbour porpoises *Phocoena phocoena* (L.) from eastern Canadian waters during the 5 year period 1969-1973. Mammals in the Seas. FAO Fisheries Series 5: 135-143.

GASKIN, D.E., K.I. Stonefield, and P. Suda. 1979. Changes in mercury levels in harbour porpoises from the Bay of Fundy, Canada, and adjacent waters during 1969-1977. Arch. Environ. Contam. Toxicol. 8(6): 733-762.

GASKIN, D.E., R. Frank, and M. Holdrinet. 1983. Polychlorinated biphenyls in harbour porpoises (*Phocoena phocoena* L.) from the Bay of Fundy, Canada, and adjacent waters, with some information on chlordane and hexachlorobenzene levels. Arch. Environ. Contam. Toxicol. 12(2): 211-219.

GODIN, G. 1968. The 1965 current survey of the Bay of Fundy - a new analysis of the data and an interpretation of the results. Manuscript Report Series No. 8, Marine Sciences Branch, Department of Energy, Mines and Resources, Ottawa.

GOMC (Gulf of Maine Council on the Marine Environment). 1992a. Gulfwatch Project: Standard procedures for field sampling, measurement and sample preparation. Gulfwatch Pilot Project Period 1991-1992. Gulf of Maine Council on the Marine Environment. Augusta, Maine.

____ 1992b. Evaluation of Gulfwatch: 1991 Pilot Project of Gulf of Maine Marine Environmental Monitoring Plan. Gulf of Maine Council on the Marine Environment. Augusta, Maine.

____ 1994. Evaluation of Gulfwatch: Second Year of the Gulf of Maine Environmental Monitoring Plan. Gulf of Maine Council on the Marine Environment. Augusta, Maine.

____ 1996a. Evaluation of Gulfwatch: Third Year of the Gulf of Maine Environmental Monitoring Plan. Gulf of Maine Council on the Marine Environment. Augusta, Maine.

____ 1996b. Evaluation of Gulfwatch: Fourth Year of the Gulf of Maine Environmental Monitoring Plan. Gulf of Maine Council on the Marine Environment. Augusta, Maine.

____ 1996c. Evaluation of Gulfwatch: Fifth Year of the Gulf of Maine Environmental Monitoring Plan. Gulf of Maine Council on the Marine Environment. Augusta, Maine.

____ 1997. The First Five Years of Gulfwatch, 1991-1995: A review of the program and results. Gulf of Maine Council on the Marine Environment. Monitoring Committee. Augusta, Maine.

GREENBERG, D.A. 1984. A Review of the Physical Oceanography of the Bay of Fundy, p. 9-30. *In* D.C. Gordon Jr., and M.J. Dadswell [eds.] Update on the Marine Environmental Consequences of Tidal Power Development in the Upper Reaches of the Bay of Fundy. Can. Tech. Rep. Fish. Aquat. Sci. 1256.

GREENBERG, D.A., B.D. Petrie, G.R. Daborn, and G.B. Fader. 1997. The physical environment of the Bay of Fundy. *In* J.A. Percy, J.A., P.G. Wells, and A.J. Evans [eds.] Bay of Fundy Issues: a scientific overview. Workshop Proceedings, Wolfville, N.S., January 29 to February 1, 1996. Environment Canada, Atlantic Region, Occasional Report No. 8, Environment Canada, Sackville, N. B. pp. 11-36.

HARGRAVE, B.T., G.A. Phillips, L.I. Doucette, M.J. White, T.G. Milligan, D.J. Wildish, and R.E. Cranston. 1995. Biogeochemical observations to assess benthic impacts of aquaculture in the Western Isles region of the Bay of Fundy, 1994. Can. Tech. Rep. Fish. Aquati. Sci. 2062: 159p.

JOHNSTON, D.W. 1995. Spatial and temporal differences in heavy metal concentrations in the tissues of harbour porpoises (*Phocoena phocoena* L.) from the western North Atlantic. M.Sc. thesis, University of Guelph, Guelph, Ont. 153 p.

JOIRIS, C.R., L. Holsbeek, J.M. Bourquegneau, and M. Bossicart. 1991. Mercury contamination of the harbour porpoise (*Phocoena phocoena*) and other cetaceans from the North Sea and the Kattegat. Water Air Soil Pollution 56: 283-293.

LAW, R.J., B.R. Jones, J.R. Baker, S. Kennedy, R. Milne, and R.J. Morris. 1992. Trace metals in the livers of marine mammals from the Welsh coast and the Irish Sea. Marine Pollution Bulletin 24: 296-304.

LORING, D.H. 1979. Baseline levels of transition and heavy metals in the bottom sediments of the Bay of Fundy, Nova Scotia, Canada. Proc. N.S. Inst. Sci. 29(4): 335-346.

LORING, D.H. 1982. Geochemical factors controlling the accumulation and dispersal of heavy metals in the Bay of Fundy sediments. Can. J. Earth Sci. 19(5): 930-944.

LORING, D.H., R.T.T. Rantala, and T.G. Milligan. 1996. Metallic contaminants in the sediments of coastal embayments of Nova Scotia. Can Tech. Rep. of Fish. Aquatic. Sci. 2111.

MACHELL, J.R., and A.S. Menon. 1992. Atlantic Shellfish Area Classification Inventory 1992. Environment Canada, Conservation and Protection, Atlantic Region, Dartmouth, N.S.

MAGUIRE, R.J., R.J. Tkacz, Y.K. Chou, G.A. Bengert, and P.T.S. Wong. 1986. Occurrence of organotin compounds in water and sediment in Canada. Chemosphere 15: 253-274.

MENON, A.S. 1988. Molluscan shellfish and water quality problems in Atlantic Canada. Toxicity Assessment 3: 679-686.

MILLIGAN, T.G. 1994. Suspended and bottom sediment grain size distributions in Letang Inlet, N.B., October 1990. Can. Tech. Rep. Hydrogr. Ocean Sci. iv: 55 pp.

MUIR, D.C.G., R. Wagemann, N.P. Grift, R.J. Norstrom, M.Simon, and J. Lien. 1988. Organochlorine chemical and heavy metal contaminants in white beaked dolphins (*Lagenorrhynchus albirostris*) and pilot whales (*Globicephala melaena*) from the coast of Newfoundland, Canada. Arch. of Environ. Contam. Toxicol. 17: 613-629.

MUIR, D.C.G., C.A. Ford, R.E.A. Stewart, T.G. Smith, R.F. Addison, M.E. Zinck, and P. Beland. 1990. Organochlorine contaminants in belugas, *Delphinapterus leucas*, from Canadian waters. *In*: T.G. Smith, D.J. St. Aubin, and J.R. Geraci [eds.] Advances in research on the beluga whale, *Delphinapterus leucas*. Canadian Bulletin of Fisheries and Aquatic Sciences 224: 165–190.

MUIR, D.C.G., C.A. Ford, N.P. Grift, R.E.A. Stewart, and T.F. Bidleman. 1992. Organochlorine contaminants in narwhal (*Monodon monoceros*) from the Canadian Arctic. Environmental Pollution 75: 307-316.

NAPOLITANO, G.E., and R.G. Ackman. 1989. Lipids and hydrocarbons in *Corophium volutator* from Minas Basin, Nova Scotia, Canada. Mar. Biol. 100(3): 333-338.

NOAA (National Oceanic and Atmospheric Administration). 1994. Gulf of Maine Point Source Inventory. The National Coastal Pollutant Discharge Inventory, Pollution Sources Characterization Branch, Strategic Environmental Assessments Division, Silver Spring, MD. 324 p.

NOBLE, D.G., and S.P. Burns. 1990. Contaminants in Canadian Seabirds. A State of the Environment Fact Sheet. Environment Canada, Conservation and Protection, Ottawa, Ont. EN1-12/90-1E. 12 p.

NOBLE, D.G., and J.E. Elliot. 1986. Environmental contaminants in Canadian seabirds. 1968-1985.: trends and effects. Can. Wildl. Serv. Tech. Rep. Ser. 13.

PEARCE, P.A., J.E. Elliott, D.B. Peakall, and R.J. Norstrom. 1989. Organochlorine contaminants in eggs of seabirds in the Northwest Atlantic, 1968-1984. Environ. Pollut. 56: 217-235.

PERCY, J.A. 1997. Marine resources of the Bay of Fundy. *In* J.A. Percy, P.G. Wells, and A.J. Evans [eds.] Bay of Fundy Issues: a scientific overview. Workshop Proceedings, Wolfville, N.S., January 29 to February 1, 1996. Environment Canada, Atlantic Region, Occasional Report No. 8, Environment Canada, Sackville, N. B. pp. 103-138.

PERCY, J.A., and P.G. Wells. 1997. Bay of Fundy ecosystem issues: a summary. *In* J.A. Percy, P.G. Wells, and A.J. Evans [eds.] Bay of Fundy Issues: a scientific overview. Workshop Proceedings, Wolfville, N.S., January 29 to February 1, 1996. Environment Canada, Atlantic Region, Occasional Report No. 8, Environment Canada, Sackville, N. B. pp. 139-150.

PROUSE, N.J., T.W. Rowell, P. Woo, J.F. Uthe, R.F. Addison, D.H. Loring, R.T.T. Rantala, M.E. Zinck, and D. Peer. 1988. Annapolis Basin soft-shell clam (*Mya arenaria*) mortality study: a summary of field and laboratory investigations. Can. Man. Rep. Fish. Aquat. Sci. 1987: 19 p.

RAY, S., and D. MacKnight. 1984. Trace metal distributions in Saint John Harbour sediments. Marine Pollution Bulletin 15: 12-18.

RAY, S., and V. Jerome. 1987. Copper, zinc, cadmium, and lead in scallops (*Placopecten magellanicus*) from the Maritimes. Can. Tech. Rep. Fish. Aquat. Sci. 1519: 29 p.

SHOWELL, M.A., and D.E. Gaskin. 1992. Partitioning of cadmium and lead within seston of coastal marine waters of the western Bay of Fundy Canada. Arch. Environ. Contam. Toxicol. 22(3): 325-333.

STRAIN, P.M., D.J. Wildish, and P.A. Yeats. 1995. The application of simple models of nutrient loading and oxygen demand to the management of a marine tidal inlet. Mar. Poll. Bull. 30: 253-261.

TEIGEN, S.W., J.U. Skaare, A. Bjorge, E. Degre, and G. Sand. 1993. Mercury and selenium in harbour porpoise (*Phocoena phocoena*) in Norwegian waters. Environmental Toxicology and Chemistry 12: 1251-1259.

WAGEMANN, R., R.E.A. Stewart, P. Béland, and C. Desjardins. 1990. Heavy metals and selenium in tissues of beluga whales (*Delphinapterus leucas*), from the Canadian Arctic and the St. Lawrence estuary. *In* T.G. Smith, D.J. St. Aubin, and J.R. Geraci [eds.] Advances in Research on the Beluga Whale (*Delphinapterus leucas*). Canadian Bulletin of Fisheries and Aquatic Sciences 224: 191-206.

WESTGATE, A.J., D.C.G. Muir, D.E. Gaskin, and M.C.S. Kingsley. 1997. Concentrations and accumulation patterns of organochlorine contaminants in the blubber of harbour porpoises, *Phocoena phocoena*, from the coast of Newfoundland, Gulf of St. Lawrence and Bay of Fundy/Gulf of Maine. Environmental Pollution 95: 105-119.

WOODLEY, T.H., M.W. Brown, S.D. Kraus, and D.E. Gaskin. 1991. Organochlorine levels in North Atlantic Right Whale (*Eubalaena glacialis*) blubber. Arch. Environ. Contam. Toxicol. 21: 141-145.

Personal Communications

BURGESS, N. Canadian Wildlife Service, Environment Canada, Sackville, N.B. September 1996.

HARDING, G.C. Fisheries and Oceans Canada, Bedford Institute of Oceanography, Dartmouth, N.S. March 1997.

PROUSE, N. Fisheries and Oceans Canada, Halifax, N.S. December 1996.

NORTHWEST Atlantic Marine Ecosystem

Highlights

- The Northwest Atlantic (NWA) marine ecosystem includes 10 000 km of coastline and a highly productive continental shelf. The area yields nearly one million tonnes of finfish and shellfish annually with a market value of about one billion dollars.

- There are generally low concentrations of contaminants present in NWA commercial fish species. This is due to the absence of significant sources of contamination.

- Inshore contaminant concentrations are moderate with significant sources of contaminants provided by chemical and biological input. Contaminant sources include industrial activity, untreated sewage discharges, municipal effluents, and road run-off.

- Overall, contaminant issues in the NWA are primarily related to effects on aquatic habitats, point source contaminant discharges and atmospheric sources.

- There are a number of emerging contaminant issues, including: biological effects of an immunotoxicological, physiological and reproductive nature, effects of metabolites and conjugates on fish health and the possible consequences of commercial oil production on the offshore environment.

Table of Contents

The Northwest Atlantic Marine Ecosystem

List of Figures

List of Tables

List of Sidebars

Introduction

The resources of the northwest Atlantic (NWA) are diverse and distributed over the entire east coast of Canada, an area containing the world's largest continental shelf. The ocean currents sweep the coast from north to south, with relatively clean waters from the open Atlantic and the Arctic Oceans. Consequently there is substantial dilution of contaminants that originate from land sources. The southern part of the NWA is downstream from inputs from the Gulf of St. Lawrence, while atmospheric transport brings contaminants to the NWA from eastern Canada and the U.S.

The NWA has a rich, long-standing history of small fishery-based communities. For the purpose of this report, the marine area extending from Hudson Strait in the north to Georges Bank in the south, but excluding the Gulf of St. Lawrence and Bay of Fundy, is defined as the Northwest Atlantic (NWA) marine ecosystem (Figure 1).

This chapter presents information regarding contaminants including organochlorine compounds, polycyclic aromatic hydrocarbons and metals in water, sediments, and tissues of marine finfish and other representative biota. In the past, there has been a serious lack of scientific information on the level and distribution of contaminants in the NWA.

MAJOR CONTAMINANT ISSUES

Contaminant concentrations in the NWA are generally lower than those reported for other marine areas of Canada. However, high concentrations of mercury commonly occur in large predatory fish such as tuna and

swordfish. Industrialized harbours have high concentrations of polycyclic aromatic hydrocarbons (PAHs), organochlorine compounds and metals in sediments and in some nearshore

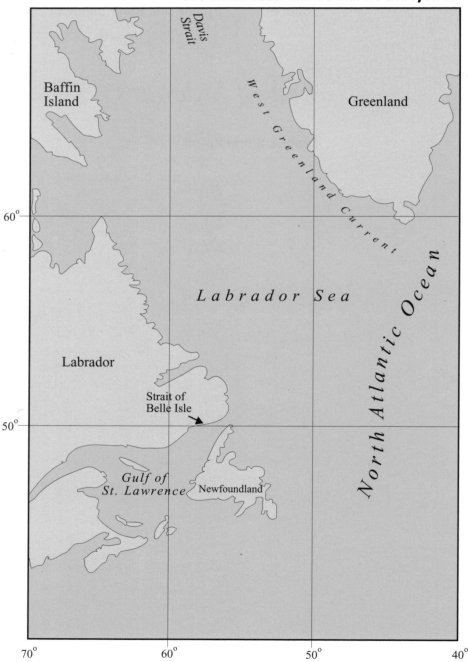

Figure 1
The Northwest Atlantic marine ecosystem.

Source: Adapted from Scarratt 1982.

biota. The concentrations of contaminants found offshore are generally low but the contaminants that are present are widespread. Large quantities of contaminants are shipped through the area, presenting a possible risk from spills. There are also large reserves of petroleum which will come into commercial production. Hibernia, Terra Nova, White Rose and Sable Island are all offshore developments which, under normal circumstances, pose little threat of contaminant effects on benthic communities except in the vicinity of the production platform or in the case of a spill. Effects on other aquatic organisms (particularly larval fish) may extend considerably beyond this distance, however, risks are not yet clearly defined.

The potential impacts of the nickel/cobalt mining operation at Voisey's Bay, Labrador, and of the associated smelter at Argentia, Placentia Bay have not yet been evaluated.

OCEANOGRAPHIC FEATURES OF IMPORTANCE TO CONTAMINANT DYNAMICS

The NWA area is characterized by a wide and shallow continental shelf, strong currents and a large riverine input. There are also large areas of seasonal ice cover with a strong storm-driven wave action.

The continental shelf stretches along an extensive coastline of about 10 000 km with width varying from 100 km off Labrador, to about 500 km on the Grand Banks. The depth varies from less than 50 m on the Georges and Sable Island Banks to 75 to 250 m over most of the area, with the exception of the Laurentian Channel where it is about 450 m. The shallow shelf provides good habitat for adult fish species, including: cod; halibut; haddock; redfish; flounder; skate; grenadier; hake; herring; eel; salmon; capelin; sandlance; and turbot as well as invertebrates; including shrimp; crab; lobster; scallops; squid; mussels; and clams.

The Labrador Current originates in the Arctic and flows south over the shelf off Labrador and Newfoundland. As it progresses over the shelf, the current splits into two branches. The westward, inshore branch sweeps the south coast of Newfoundland as it proceeds to the Gulf of St. Lawrence. The offshore branch follows the edge of the continental shelf around the Grand Banks and along the Scotian Shelf. The Labrador Current is very cold, with temperatures ranging from -1.6 to -1°C in winter and 10 to 15°C in summer. The cool waters and ice cover make this region much colder than the northeast Atlantic at similar latitudes. The volume of water carried by the Labrador Current is more than two orders of magnitude larger than the flow of the St. Lawrence River (Hildebrand 1984).

The Labrador Current carries nutrients and fresh water from terrestrial run-off of major Canadian rivers over the continental shelf. As the current passes over the shelf it mixes with riverine fresh water from the coastline as well as underlying and offshore waters, and gains heat from solar radiation. The mixing of nutrients from fresh water together with warming over a shallow shelf provides conditions for high plankton productivity and a good nursery for juvenile fish (Anon. 1984). The principal fisheries resources are located either at the areas of high primary productivity of plankton, or further downstream of these regions (Scarratt 1982).

In the south, the predominant currents on the shelf are waters emanating from the Gulf of St. Lawrence and the Nova Scotia Current flowing westwardly along the coast of Nova Scotia. The Nova Scotia Current increases in heat and nutrient content as it progresses southward due to mixing with offshore waters. The salinity of this water is less than that on the Grand Banks due to the freshwater from the St. Lawrence; while the inshore water mixes with warmer water from the Gulf Stream which originates in the tropics. Since this water is much warmer than that of the north, the marine environment in the southern part of the northwest Atlantic is different, in terms of species composition and growth rates, than that in the northern part.

BIOLOGICAL RESOURCES

In 1994 there were approximately 720 000 t of fish and shellfish caught and marketed in the NWA with a value of $1.1 billion. Groundfish accounted for about 48% of the catch tonnage, pelagic fish about 28%, and shellfish 24%. Shellfish amounted to 60% of the landed value of the fisheries, groundfish accounted for 33% (Anon. 1994).

The NWA groundfish stocks have shown dramatic declines in recent years. At the same time, shellfish stocks, particularly shrimp and crab, have increased. Cod normally represented about half of the groundfish catch. The remainder consisted of redfish and flatfish; including: three species of flounder; turbot in the Newfoundland Shelf and Grand Banks area; and haddock, pollack and flatfish in the Nova Scotia area (Anon. 1994).

Lobster has the highest landed value, accounting for more than one-quarter of the landed value of shellfish on the Atlantic coast. The next most valuable shellfish are crab, shrimp and scallops (Anon. 1994).

Marine mammals are a resource of considerable historic importance. Both whales and seals were once exploited extensively and whales are now highly valued by the ecotourism industry. During most of the past century, 250 000 seals a year (mostly harp seals) were harvested primarily for skins, fat, and to some extent, for the meat. Since 1982 less than half of the annual quota of 86 000 have been taken. Harp seals, the most abundant seals, are found throughout the area. They are seasonal migrants and most of their feeding is in the Atlantic and Arctic Oceans. Hooded seals are found in the same areas. Grey seals are found around Sable Island, the coast of Nova Scotia, and in the Gulf of St. Lawrence. Polar bears prey on seals on the sea ice and are common on the spring ice as far south as the east coast of Newfoundland. As the ice breaks up, bears swim north or in some cases come ashore in Labrador and return overland to the Arctic.

Marine birds are found in the areas of high pelagic fish abundance, including the Grand Banks, along the coast of Newfoundland and southern Labrador, off the southern coast of Nova Scotia, and off the coast of Cape Breton Island. The birds feed on fish or other marine organisms and bioaccumulate contaminants.

Assessment of Chemical Contamination

SOURCES

In the NWA, natural environmental processes result in the leaching of metals from rocks and soils while mining, smelting, other industrial activities and domestic sewage all add metals to the environment. These metals are transported to the sea either as run-off or through the atmosphere (Brandon and Yeats 1984).

DISTRIBUTION, TRENDS AND EFFECTS
METALS AND ORGANOMETALLIC COMPOUNDS

Several assessments have been made of the input of metals to the NWA from two major sources of fresh water, namely the Hudson Strait and the Gulf of St. Lawrence (Bewers and Yeats 1977; Brandon and Yeats 1984; Yeats 1993). Measurements of the metal content of some of the rivers of Labrador, Newfoundland, and Nova Scotia are available from the DOE ENVIRODAT environmental database. Measurements of trace elements in NWA seawater have been reported (Bruland 1983); however there are no published studies of metals in offshore Newfoundland sediment and water. Sediments from harbour dredging and other such harbour sampling in

Newfoundland show a range of 0.1 to 1 900 $\mu g.g^{-1}$ dry weight sediment for cadmium in Newfoundland harbour sediments. Elevated concentrations of cadmium are not known to exist offshore. Limited measurements are available for offshore Nova Scotia sediments and water (Buckley *et al.* 1974; Yeats 1987). Measurements have been made of metals in water and sediments of Nova Scotia harbours and estuaries (e.g., Buckley *et al.* 1974; Cranston *et al.* 1975; Windom *et al.* 1991), including a major study of Halifax Harbour (Nichols 1989). A chronological study on the metal contamination of sediment cores from Halifax Harbour indicated that the metals copper, zinc, lead and mercury increased from 1890 to 1980, with maximum concentrations among the highest reported from urban and industrialized coastal marine areas in the world (Gearing *et al.* 1991). Recently, metal concentrations in sediments were reported for a number of harbours in Nova Scotia (Loring *et al.* 1996) and high concentrations of metals were observed in several industrialized harbours (Figure 2). A large proportion of the cadmium in these sediments, and smaller proportions of the other metals, were found to be bioavailable.

A series of elements was examined in muscle, liver and gonad of various fish species inhabiting the inshore and offshore waters of the NWA (Hellou *et al.* 1992a, 1992b, and 1996a). In general, muscle was enriched in cesium and arsenic. Liver was enriched in silver, manganese, iron and molybdenum when compared to muscle and gonads. The above results in finfish tissue collected offshore provide a point of reference. They may be used to compare concentrations of elements in finfish inhabiting areas close to point sources of contamination with offshore fish. Hellou *et al.* (1996) found between 2 and 4 times higher concentrations of lead, aluminum, calcium and lithium in muscle of inshore yellowtail flounder than those found in the offshore species. For example, the concentration of aluminum was of 1.1 to 1.3 $\mu g.g^{-1}$ dry weight in muscle of inshore fish, as opposed to 0.1 to

0.8 $\mu g.g^{-1}$ dry weight in offshore fish. The concentration of lithium was 0.16 to 0.18 as opposed to 0.02 to 0.09 $\mu g.g^{-1}$ dry weight in these two groups.

High concentrations of methylmercury commonly occur in large predatory offshore, pelagic fish such as tuna and swordfish. Selenium counteracts some toxicological effects of the high mercury concentrations (Cappon 1994a; Cappon 1994b). Data for both swordfish and tuna from the NWA indicate that the mercury concentrations exceed the 0.5 $\mu g.g^{-1}$ wet weight limit recommended by the World Health Organization for human consumption (Freeman *et al.* 1978; Hellou *et al.* 1992a). However, Freeman *et al.* (1978) and Hellou *et al.* (1992a) show that there is enough selenium in swordfish and tuna to counteract the effects of the mercury. Mercury concentrations in the muscle of harp seals, which are also predatory, (although perhaps having a different diet than fish), do not exceed 0.28 $\mu g.g^{-1}$ (Botta *et al.* 1983). These data imply that although top predator fish are high in mercury, predatory mammals such as seals do not accumulate high concentrations of mercury in their flesh due either to dietary or metabolic differences or to the selenium in their diet which may control mercury accumulation.

Tributyltin (TBT) compounds are biocides used in marine anti-fouling paints. Although its use was regulated in 1989, TBT is still permitted for use on aluminum boats, naval vessels and vessels over 25 m in length. Prior to regulation, TBT was detected at < 0.1 $\mu g.L^{-1}$ in the water of Conception Bay, Newfoundland and in the harbour sediments of Argentia, Halifax and Port Hawkesbury at < 0.01 $mg.kg^{-1}$ dry weight as tin (Maguire *et al.* 1986). The same study found measurable TBT in sediments from Sydney Harbour and Conception Bay, at 0.01 and 0.03 $mg.kg^{-1}$ dry weight respectively. These concentrations were low relative to those measured in Vancouver Harbour (up to 11 $mg.kg^{-1}$ dry weight).

In spite of the 1989 controls regulating use, TBT concentrations are still high in large harbours. A study comparing concentrations in sediments collected from sites in Nova Scotia in 1988 and 1994 found that TBT had increased in large vessel harbours, e.g., Halifax, Liverpool and Port Mouton, but had decreased in recreational craft areas, e.g., Chester, Oak Island and Sambro (Ernst *et al*. 1995). Concentrations up to 4.9 mg.kg⁻¹ dry weight TBT were recorded from inner Halifax Harbour. Similar differences between recreational craft and commercial harbours are being found in the west coast (see Pacific Marine and Freshwater Ecosystems Chapter).

Figure 2
Metal concentrations in Nova Scotia harbour sediments.
Legend: Horizontal lines indicate background concentration for uncontaminated sediments.

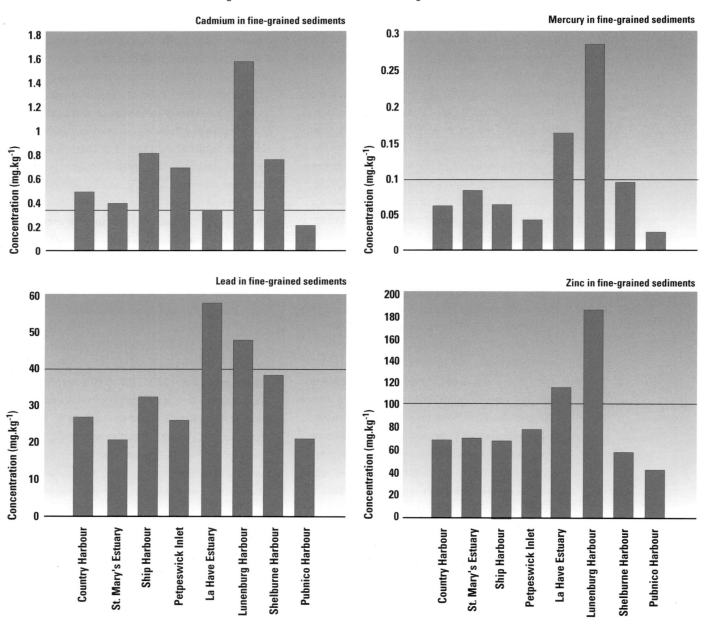

Source: Loring *et al*. 1996.

Imposex, that is the development of a penis and *vas deferens* of female neogastropods (marine snails) leading to reproductive failure, is used as a bioindicator of TBT contamination. A 1995 survey of eastern Canada found dogwhelks with imposex at 13 of 34 sites examined (Prouse and Ellis in preparation.). Included were sites in Halifax, Sydney, Chester and Lunenburg harbours in Nova Scotia and in Arnold's Cove and Come-by-Chance in Newfoundland. As dogwhelks prey on barnacles and mussels, the source of TBT could be dietary as well as via water or sediment ingestion.

ORGANIC CHEMICALS
ORGANOCHLORINE COMPOUNDS

The concentrations of organochlorine compounds have been investigated in muscle, liver and gonads of NWA cod and three flatfish species (turbot, American plaice and yellowtail flounder). Concentrations in liver followed the general trend: cod > turbot > American plaice > yellowtail flounder. The concentration and number of polychlorinated biphenyl (PCB) congeners also differed with the tissue studied, where liver > muscle > gonads. Females that were the same size as males had lower concentrations of organochlorine compounds in tissues (Hellou and Payne 1993; Hellou *et al.* 1993b; Hellou *et al.* 1995a; Hellou *et al.* 1995b). For example, *p,p'*-dichlorodiphenyldichloroethylene (*p,p'*-DDE) was present at a concentration of 8 to 110 ng.g^{-1} wet weight in liver of male American plaice as opposed to 3 to 9 ng.g^{-1} wet weight in female

liver. In muscle, concentrations were 2.5 to 18 ng.g^{-1} wet weight in males compared to 1.2 to 8.9 ng.g^{-1} wet weight in females. The total PCB and total dichlorodiphenyltrichloroethane (DDT) concentrations in liver and muscle were similar to those reported for various flounder from Maine (Hellou *et al.* 1995b). Hexachlorobenzene (HCB) was found at higher concentrations than in the Baltic Sea while concentrations of hexachlorocyclohexane (HCH), dichlorochlorophenylethane (DDD), DDT, and PCBs were lower (Hellou *et al.* 1995b). The PCB concentrations in turbot muscle from the Newfoundland area (24 ng.g^{-1} wet weight) were considerably lower than those from the Arctic (140 to 220 ng.g^{-1} wet weight, (See Arctic Marine Ecosystem Chapter).

In offshore Newfoundland cod, concentrations of organochlorine compounds are generally lower than or equal to values reported in the literature (Hellou *et al.* 1993b). PCBs were lowest in the NWA (0.15 ng.g^{-1} wet weight), while DDTs and HCHs were at the low end of the range reported for the North Sea in 1989 (Table 1).

Dieldrin concentrations in cod were lower than elsewhere, whereas HCB concentrations were within the range of values reported from the North Sea (deBoer 1988). The values for PCBs reported in 1980 off Nova Scotia were more than 10 times higher and those for DDTs were about 3 times higher than those from Newfoundland (Freeman *et al.* 1984). The cod investigated from Nova Scotia were inshore fish and the difference

Table 1: Organochlorine compound concentrations in cod liver.

Location and Year	Concentration (µg.g^{-1} wet weight)						
	α-HCH	γ-HCH	Chlordanes	HCB	Dieldrin	DDT	PCB
Newfoundland[a] (1991)	0.03	0.01	0.17	0.03	0.02	0.16	0.15
Nova Scotia[b] (1980)	0.06	NA	NA	0.02	0.06	0.53	1.7
Norway[c] (1982)	0.04	0.01	0.09	0.02	NA	1.1	0.5

Source: a. Data as complied in (Hellou *et al.* 1993b), b. (Freeman *et al.* 1984), c. (de Boer 1988).

may be due to the St. Lawrence and inshore sources. Of the polychlorodibenzo-*p*-dioxins (PCDDs) and polychlorodibenzo-furans (PCDFs) analyzed, ranging from tetra- to octachlorinated congeners, only tetrachlorodibenzo-furan (TCDF) was detectable in most tissues of the species analyzed (Hellou and Payne 1993; Hellou *et al.* 1993a). Comparison of the concentration observed in livers of cod from the NWA to cod from Norway and Finland showed lowest concentrations in the NWA fish.

The total PCB concentrations in flounder from non-urbanized Newfoundland bays (1 to 8 ng.g^{-1} wet weight; 140 to 1 000 ng.g^{-1} lipid) were similar to those reported for flounder from the Dutch Wadden Sea and lower than those reported for the New Bedford Harbor and the Hvaler Archipelago locations (Ray *et al.* 1997). The lack of significant differences in PCB concentrations between areas around Newfoundland may be indicative of atmospheric rather than local sources of input of these compounds (Ray *et al.* 1997). Aldrin, dieldrin, endrin, chlordane, oxychlordane, HCB and mirex were detected in all flounder samples, but were present at low concentrations (Ray *et al.* 1997). However, concentrations of PCBs, DDTs and chlordanes in muscle, liver and gonads of yellowtail flounder maintained in tanks receiving water from the vicinity of heavily urbanized St. John's Harbour, were higher than in offshore fish. The concentrations were up to 20 times higher in liver, up to 7 times higher in muscle and up to 4 times higher in ovaries of inshore compared to offshore fish (Hellou *et al.* 1995a). This difference was less pronounced in muscle tissue, where concentrations were less than 6 times higher in inshore compared to offshore samples. Concentrations of PCB congeners were also elevated in tissues of inshore flounders. For example, PCB congener 153 was present in a concentration of 0.7 to 3.8 ng.g^{-1} wet weight in liver of offshore male yellowtail flounder and 0.5 to 65 ng.g^{-1} wet weight in females and 14 to 28 ng.g^{-1} wet weight in inshore samples.

In the process of comparing dietary intake and water uptake of contaminants in offshore and inshore yellowtail flounder, inshore waters were shown to contain 100 times the concentration of offshore waters. Diet could have contributed from 2 to 10 times more contaminants to the inshore flounder (Hellou *et al.* 1997a and b).

Toxaphene was found to be widely distributed in Canadian east coast fish (Musial and Uthe 1983). The concentrations reported for herring fillets from the Halifax area were 0.4 µg.g^{-1} wet weight or 4.4 µg.g^{-1} lipid as compared to 1 µg.g^{-1} wet weight and 12 µg.g^{-1} lipid for herring from the Gulf of St. Lawrence (Musial and Uthe 1983). The concentrations in cod liver from the Gulf of St. Lawrence were 1.1 µg.g^{-1} wet weight and 2.4 µg.g^{-1} lipid (Musial and Uthe 1983). Since that time there have been no reported or known investigations of toxaphene concentrations in NWA fish. During the period from 1968 to 1984 the concentrations of toxaphene in storm petrel eggs increased from 0.9 to 1 mg.kg^{-1} to between 1.4 and 1.9 mg.kg^{-1}. These were the highest concentrations of a chlorinated pesticide found in storm petrel eggs from offshore Newfoundland (Elliott *et al.* 1992).

The concentration of organochlorine compounds in seabird eggs from Canada's eastern seaboard indicates that in general, the lowest concentrations are for birds nesting and fishing in Newfoundland waters (N. Burgess personal communication). The exception is that toxaphene concentrations are higher in storm petrel eggs from Newfoundland than from other areas (Elliott *et al.* 1992). Dolphins and pilot whales have considerably higher concentrations of organochlorine compounds than harp seals from the same area (Muir *et al.* 1988) (Figure 3).

Harp seals are migratory and feed throughout the area from the Gulf of St. Lawrence to the coast of Greenland, spending most of their lives in the NWA. The concentrations of PCBs and DDTs are dramatically lower in harp seals from the NWA than in comparable carnivores from the

Figure 3
Contaminant concentrations in NWA seals, dolphins and whales.

Units for all graphs are µg.g⁻¹

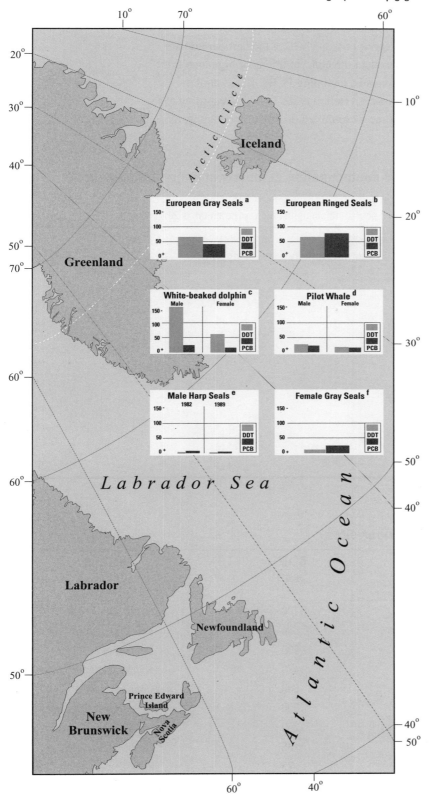

Sources: [a, b, e] Andersen *et al.* 1988, [c, d] Muir *et al.* 1988, [f] Addison *et al.* 1984.

European side of the north Atlantic (Anderson *et al.* 1988) (Figure 3). Lower PCB and DDT concentrations in NWA seals, compared to grey seals from offshore Nova Scotia and seals from Europe, mean that there are generally lower concentrations of organochlorine compounds off the coast of Newfoundland than off the coast of Nova Scotia. Seals from both areas are much lower in organochlorine compounds than those in European waters.

In 1982, the concentration of DDTs in NWA harp seals had declined to about 1/3 of the concentration found in 1974. The concentrations of PCBs, on the other hand, increased slightly during the same time (Addison *et al.* 1984). Burgess *et al.* (1995) found particularly high concentrations of DDE in eggs of Leach's Storm Petrels from Newfoundland waters in the 1972 sample with a dramatic drop in concentrations to about 1/3 by 1976, followed by stability in concentrations since 1980. As this species feeds on plankton, the DDE concentration probably reflects the rate of atmospheric deposition in the years following the widespread forest spraying of DDT, which ceased in 1968. DDE concentrations in fish-eating seabirds showed a slower decline (Pearce *et al.* 1989). The slower decline was probably due to the longer food chain of these birds. These data support the conclusion that DDT concentrations declined in biota from NWA due to the decline in use of the pesticide.

OTHER AROMATIC COMPOUNDS

Nova Scotia harbours were ranked for PAH contamination by assessing point sources, population, industrial and commercial activity, and international and domestic ship traffic, as well as a number of commercial fishing vessels (Prouse 1994; Table 2). Of the 11 harbours ranked, Sydney came first, followed by Halifax and Pictou. Concentrations of a series of PAH in sediments have been reported by Tay *et al.* (1992) for various locations in Halifax Harbour, where the highest concentrations were observed in Bedford Basin. The highest concentration of benzo[a]pyrene (B[a]P) was reported as being nearly 1 $\mu g.g^{-1}$ dry weight, compared to a mean value of 35 $\mu g.g^{-1}$ dry weight in Conception Bay, Newfoundland (Tay *et al.* 1992; Hellou *et al.* 1994d; Table 3). Concentrations of total di- and tri-aromatic compounds reported for sediments in Placentia Bay, on the south coast of Newfoundland were between 0.1 to 0.6 $\mu g.g^{-1}$ dry weight (Kiceniuk 1992). Hydrocarbons present in St. John's Harbour originate from petroleum and combustion sources, where 20 to 50% of larger-molecular-weight PAHs and aliphatic components are derived from crankcase oil sources (O'Malley *et al.* 1994, 1996, 1997; Bieger *et al.* 1996, 1997). In St. John's Harbour sediment, PAH concentrations vary within the harbour and are higher than the concentrations of organochlorine compounds, including PCBs, DDTs and chlordanes (Hellou *et al.* 1997b). According to the most recent sediment quality criteria study, most

Table 2: Scores for ranking the potential for PAH contamination of lobster in major harbours in the Maritime provinces.

The total score is the sum of (a) + (b) + (c) + (d) + (e). The maximum score, indicating the highest potential, is 20.

Harbour	(a) Point Sources	(b) Population	(c) Industrial Activity	(d) Shipping	(e) CFS	Total Score
Pubnico	1	1	1	1	3	7
Clark's Harbour	1	1	2	1	4	9
Shelburne	1	2	3	3	4	13
Lockeport	1	1	1	1	3	7
Liverpool	2	2	4	3	2	13
Lunenburg	1	2	4	2	2	11
Halifax	4	4	4	4	2	18
Port Hawkesbury	3	2	4	4	1	14
Sydney	4	4	4	4	3	19
Pictou	4	4	4	2	1	15

Source: Prouse 1994.

Table 3: Pyrene and B[a]P concentrations at various sites.

Location	Pyrene (ng.g⁻¹)	B[a]P (ng.g⁻¹)	Reference
Baltimore Harbour	58 000	1 100	Foster and Wright 1988
Sydney Harbour	33 000	28 000	Kieley *et al.* 1988
Halifax Harbour	5 500	2 600	Gearing *et al.* 1991
Severn Estuary (UK)	1 400	470	Thompson and Eglinton 1978
Tokyo Bay	600	500	Ohta *et al.* 1983
Lake Michigan	600	300	Helfrich and Armstrong 1986
St. John's Harbour	500	250	Hellou *et al.* 1997a
Conception Bay	30	35	Hellou *et al.* 1994a
Pacific Shelf	53	25	Prahl and Carpenter 1983

PAH concentrations in one analysed sample are in the "effect range low" of toxicity (Hellou *et al.* 1997a and b). This means that their concentration is associated with a low incidence of toxicity (Long *et al.* 1995).

Leach's Storm Petrel (*Oceanodroma leucorhoa*)

Petrels are small ocean birds that spend most of their lives on the ocean coming to land only to nest. Leach's Storm Petrels breed on uninhabited coastal islands on both Atlantic and Pacific coasts. On the Atlantic coast, they breed in burrows on coastal islands from southern Labrador to Maine. The largest population is found in Newfoundland. There, some colonies have disappeared as the result of introduction of ground predators. In the more southerly part of its range, the Leach's Storm Petrel population is decreasing because of the predation of large gulls. Population trends of this species are unknown because of a lack of reliable censuses. However, limited knowledge of their movements suggests that petrels disperse widely offshore in the winter in the western north Atlantic with subsequent mixing of populations. On the Pacific coast, storm petrels breed on coastal islands from the Queen Charlotte Islands to Vancouver Island. Virtually nothing is known about its their movements outside the breeding season.

Leach's Storm Petrels feed on zooplankton and on larval fish found on the surface or near surface of the ocean. The prey generally rises to the surface at night. The relatively high organochlorine compound concentrations in this species reflects its diet of invertebrates and larval fish. These are found in the oily layer of the ocean where airborne organochlorine compounds are deposited.

Based on information provided in Noble and Elliott (1986).

A study on the chronology of contamination of the northwest arm of Halifax Harbour indicated that in a dated core, hydrocarbon concentrations have increased 100 fold since 1900. Aliphatic contaminants characteristic of sewage and urban run-off had a steady, exponential growth over time, while combustion-derived PAHs have slightly declined from maximum values observed *circa.* 1950 (Gearing *et al.* 1991).

PAHs have been measured in commercial seafood (Dunn and Fee 1979). In feral bivalves and crustaceans, much higher concentrations of hydrocarbons were observed in the visceral mass or hepatopancreas than in muscle tissue (Hellou and Upshall 1993, 1994; Hellou *et al.* 1993a, 1994a, 1997c). Concentrations were nearly 5, 60, 70 and 250 times higher in the visceral mass of propeller clams, whelks, clams and scallops than in their respective muscle tissues. Biologically unsaturated hydrocarbons naturally derived from phytoplankton or bacteria were over 100 times more predominant than anthropogenically derived PAHs (from petroleum or combustion). Higher concentrations of biologically-derived unsaturated hydrocarbons were found than of anthropogenically-derived PAHs (Hellou *et al.* 1993a, 1994a; Hellou and Warren 1997).

Contrary to the case of invertebrates, low concentrations of various PAHs were detected in muscle of marine mammals and finfish (Hellou *et al.* 1990, 1991, 1994c; Hellou and Warren 1997; Hellou 1996). Higher concentrations of alkylated PAHs and sulphur-containing aromatics were observed in liver than in gonad or muscle of cod, American plaice and yellowtail flounder (Hellou *et al.* 1994c; Hellou and Warren 1997). More PAHs were present in cod, plaice and flounder collected on the Grand Banks than in other locations in the NWA (Hellou *et al.* 1994b, 1994c; Hellou and Warren 1997). Important observations include higher concentrations of PAHs in smaller (younger) fish of a species than in larger (older) fish, which is unlike organochlorine compounds for which concentrations normally increase with fish size. PAHs are also detectable in eggs and vary in concentration with season and location. In general, it could be stated that concentrations of PAHs were low and must be considered as background concentrations.

The concentrations of PAHs in sediments collected in a coal tar contaminated estuary correlated with the activity of various hepatic mixed function oxydase (MFO) enzymes in winter flounder (Vignier *et al.* 1994; Addison *et al.* 1994). Ethoxyresorufin *O*-deethylase (EROD) activity was shown to be 7 times higher at the most contaminated area, compared with the least contaminated location. The level of variability in the MFO response is illustrated in Figure 4. The activity of the MFO enzymes is shown to increase with proximity to the tar ponds. However, as illustrated, it is not equally induced in all fish. EROD activity in males and females correlated with the concentrations of PAHs in sediments.

Laboratory experiments involving the exposure of winter flounder to sediments contaminated with Hibernia crude oil allowed the determination of the concentration of oil in sediments (or water) at which bioaccumulation would occur (Hellou *et al.* 1994d, 1995c; Hellou and Upshall 1994, 1995; Payne *et al.* 1995; Hellou 1996). The bioelimination of PAH derivatives through the gall bladder bile was observed at a lower level of exposure than bioaccumulation (Hellou *et al.* 1994d, 1995c; Hellou and Upshall 1995). This indicates that oxidative enzymes are active at low levels of exposure producing more polar derivatives of PAH contaminants. The reactivity of these oxidized products can lead to biochemical disturbances if further reactions occur with macromolecules. Reactivity can also lead to the formation of conjugates which can be more easily eliminated from the organism than the less polar starting material. Correlations between the physical-chemical properties of PAH and their fate supports the bioaccumulation of the more water soluble, smaller PAHs. It also tends to confirm the higher reactivity of the larger molecular weight PAHs, as observed in other studies. Larger PAHs are more abundant in combustion sources of hydrocarbons, while smaller PAHs are more predominant in petroleum. The combination of assessment and experimental studies would indicate that a petroleum spill would be of more concern to the younger fish. These fish would be more susceptible to the effects of oil contamination than older fish since there would be more accumulation of hydrocarbons (Hellou *et al.* 1994d, 1995c; Hellou 1996; Hellou and Warren 1997).

The tar ponds at Sydney, which resulted from a coking operation, are probably the largest single source of non-alkylated aromatics, oxygen and nitrogen heterocyclics in the region as they contain between 2 000 000 and 4 000 000 kg of aromatic compounds (Vandermeulen 1989).

Figure 4
Individually plotted EROD activities in male and female winter flounder for sampling sites in Sydney Estuary and St. George's Bay (August 1989).

Legend: (j=juvenile, m=mature)

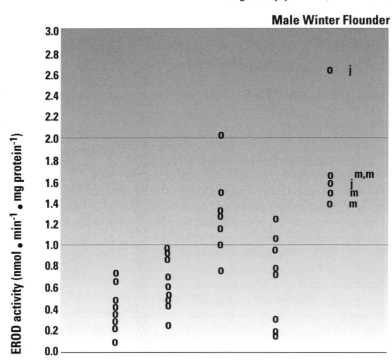

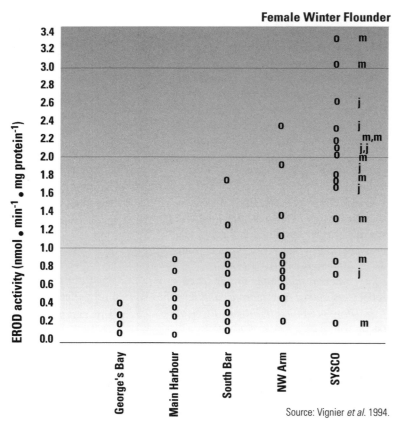

Source: Vignier *et al.* 1994.

Concentrations in the sediments of Sydney Harbour are highest near the ponds (about 2 000 mg.g^{-1} dry sediment) and decrease with distance away from the ponds. A similar gradient was reported for B[a]P concentrations found in the digestive glands of different lobsters from the same areas, with values as high as 930 ng.g^{-1} wet weight in the lobster from the south arm of the harbour compared to 1 to 1.6 ng.g^{-1} for lobster from the open coast near Port Morien (Uthe and Musial 1986).

The lobster fishery in the south arm of Sydney Harbour has been closed since 1982 due to the high concentrations (710 to 1 300 ng.g^{-1} wet weight) of the B[a]P in the digestive gland, and 27 to 37 ng.g^{-1} wet weight in the tail muscle of aromatic compounds found in lobsters from the area. The high concentrations remained at least until recent times (Vandermeulen 1989).

Conclusions

The assessment of the state of the NWA environment focuses mainly on determining the concentration of contaminants in various environmental matrices (biota, sediments and water). Contaminant concentrations are higher nearshore, near point sources of contamination, such as untreated sewage discharges, harbours, industrial inputs, dump sites and water run-off. Although chemical contaminants are widespread, it was found that their concentrations were relatively low in the NWA offshore sediments, water and biota.

The concentration of DDT compounds has declined in NWA biota since the mid 1970s. However toxaphene concentrations have increased. Dated sediment cores from Halifax Harbour have shown increases in contaminant concentrations over the past 100 years. Although some have been banned since the 1970s, organochlorine compounds are persistent and bioavailable in the NWA. The occurrence of organochlorine compounds in NWA biota is widespread but the concentrations are generally low. Higher concentrations are found in fattier

tissues, as well as older organisms within a species and at higher trophic levels in a food chain.

TBT occurs in sufficient concentrations in NWA harbours to cause imposex in dogwhelks. The effects of TBT on other biota, particularly commercially important species such as lobster and shellfish, have not been thoroughly investigated on the east coast. For example, imposex has been observed in the edible whelk in Europe (Ten Hallers-Tjabbes *et al.* 1994). This same species is caught commercially in the Gulf of St. Lawrence and the St. Lawrence estuary. The feasibility of a whelk fishery has been examined in Newfoundland (Flight 1987) and Nova Scotia (E. Kenchington personal communication). Studies on how TBT could affect this and other fisheries are needed as is a better understanding of TBT bioavailability and the extent and dynamics of its adverse effects on biota.

Biota inhabiting areas of the Grand Banks of the NWA slated for petroleum development have higher concentrations of PAHs in their tissues than other offshore biota. The first extensive study on the bioaccumulation of PAHs in finfish indicates that younger or smaller finfish would be more susceptible to contamination in the case of a spill. Large quantities of crude oil are currently shipped through the NWA. There are also a number of ongoing petroleum development projects. Due to the magnitude of the possible oil spills and the harsh environmental conditions prevailing in the NWA, there is a constant risk of detrimental effects on the various organisms.

Metals have been found at low concentrations in sediments and biota of offshore areas with higher concentrations in the harbours and the near inshore. High concentrations of mercury in offshore predatory fish are an exception to this general trend. There have been no measurable biological effects in the NWA except in the case of TBT, however, there have been closures of some fisheries. These include the fishery in the area around Long Harbour, closed due to a local elemental phosphorus contamination. The tuna and swordfish fishery were also closed for a period of time due to naturally occurring high concentrations of mercury in the fish. The potential effect of the Voisey's Bay mining activities and the associated smelter in Argentia need to be evaluated.

Case Study:

SYDNEY HARBOUR

Sydney Harbour is an example of an industrial harbour that continues to have high concentrations of several contaminants, both in sediments and in some aquatic biota, notably invertebrates. Although it is expected that controlling the sources of contaminants on land will reduce their input to the harbour, contaminants already in the sediments will likely persist for many years. Thus, the risk to fisheries from chemical contamination will exist for some time, necessitating careful management of the resource.

Sydney Harbour, on the north coast of Cape Breton Island is a Y-shaped embayment about 15 km long and 3 km at the mouth. South Arm has a shallow sill, 13 m deep, and increases in depth to about 18 m at the centre, thus forming a basin. Northwest Arm shallows gradually to its head. The water of the outer harbour is generally well mixed, primarily by tidal currents. Within south arm there is a two-layer estuarine circulation due to the input of fresh water from Sydney River. The combination of estuarine conditions with the sill results in the accumulation of organic matter, coming from the river and sewage from Sydney. Decomposition of organic matter produces anoxia in sediments, particularly during the summer, which is detrimental to marine life.

The cities around the harbour have a combined population of about 41 000 (1991 census). Major industries and commercial activities around the harbour area include shipping, a ferry terminal, ship repair, fish processing, coal processing, and a major steel mill (Vandermeulen 1989; Hildebrand 1984). Untreated sewage enters the harbour from the cities of Sydney on South Arm, Sydney Mines and North Sydney. The steel mill dumps effluent containing ammonia, phenols, cyanide, organic carbon, and other pollutants into South Arm (Hildebrand 1984). Aromatic hydrocarbons, nitrogen and oxygen heterocyclic compounds are also present in the harbour. These tar ponds are the residue from coke ovens that operated in the area until the mid-1980s. There are an estimated 1 million m^3 of contaminated sediment containing a total of between 2 and 4 million kg of aromatic material (Vandermeulen 1989) at concentrations from 600 to 30 000 $\mu g.g^{-1}$ dry weight (Vandermeulen 1989).

Runoff from Muggah Creek passes over the deposits and flows into South Arm. Some water samples from the harbour had the highest concentrations of PAHs found in surface waters in Atlantic Canada, falling in the range of 0.1 to 1.0 $\mu g.L^{-1}$. One sample exceeded the maximum acceptable concentration for drinking water B[a]P of 0.01 $\mu g.L^{-1}$ (O'Neill and Kieley 1992).

In the sediments the highest concentrations of aromatics, heterocyclic aromatics and trace metals were found at the source, the mouth of Muggah Creek, and declined seaward through South Arm and the outer harbour (Vandermeulen 1989). The concentrations of aromatic hydrocarbons were over 100 $\mu g.g^{-1}$ dry weight at the source and 30 to 100 $\mu g.g^{-1}$ in the middle of South Arm (Figure 5). Northwest Arm had lower concentrations of aromatics (3 to 10 $\mu g.g^{-1}$). Heterocyclic aromatic hydrocarbons were present at levels of 10 to 5 200 $ng.g^{-1}$ dry weight in the sediment of South Arm but rarely were detected outside South Arm (Kieley *et al.* 1988). Phthalate esters, chlorophenols and organochlorine compounds such as dieldrin, heptachlor epoxide, trans chlordane, DDT and mirex were also found in the sediments of Muggah Creek (Vandermeulen 1989).

Traditionally lobster were fished throughout the harbour and up both arms. Since the early 1980s South Arm has been closed to lobster fishing due to high concentrations of PAHs in the lobster tissues (Uthe 1979; Uthe and Musial 1986; Vandermeulen 1989; King *et al.* 1993a and b). Concentrations of PAHs ranged from 8 700 to 88 000 ng.g^{-1} wet weight of digestive gland of lobster from South Arm and were much higher than outside the harbour (Sirota *et al.* 1984; O'Neill and Kieley 1992).

Figure 5
PAH concentration contours in bottom sediments from Sydney Harbour

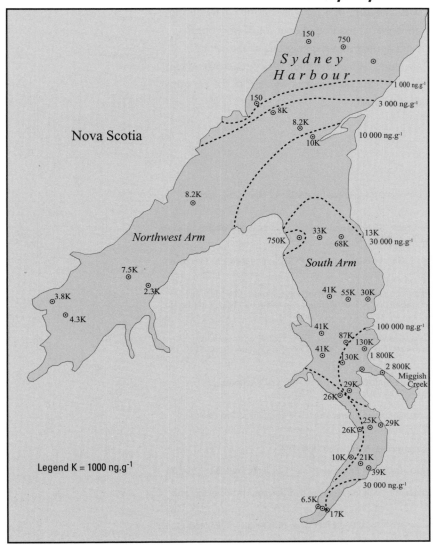

Legend K = 1000 ng.g^{-1}

Source: redrawn from Matheson *et al.* 1983.

Relatively high concentrations of PAHs and heterocyclics have persisted into the 1990s (King *et al.* 1993a and b). The highest concentrations of aromatics (approximately 39 000 ng.g^{-1} wet weight) were likewise found in mussels from South Bar in South Arm (Sirota *et al.* 1984). The hepta- and octachlorodibenzo-*p*-dioxins (HCDD and OCDD) were found at 4 to 13 pg.g^{-1} wet weight in the digestive gland of lobsters from South Arm but were not found in those at a clean site or at the mouth of the harbour. There were also high (300 to 590 pg.g^{-1}) concentrations for TCDFs and 35 to 53 pg.g^{-1} of PCDFs in the lobsters from South Arm compared to only background concentrations for lobsters from the harbour mouth (Clement et al. 1987). Data on more recent PCDD and PCDF concentrations in lobster from Sydney Harbour have been published (King *et al.* 1996). Chloro-biphenyls were analyzed in the digestive gland of lobsters from Sydney Harbour where concentrations expressed in terms of toxic equivalent factors exceed the 20 pg.g^{-1} (wet weight) Canadian guideline for 2,3,7,8,-tetrachlorodibenzo-dioxin (2,3,7,8-TCDD) (King *et al.* 1996). Flounder from the more contaminated parts of the harbour had the highest MFO concentrations (Vignier *et al.* 1994; Addison *et al.* 1994).

Case study:

CONTROLLING PHOSPHORUS CONTAMINATION IN LONG HARBOUR

A plant opened at Long Harbour, Newfoundland, in December 1968 to produce phosphorus from imported ore. Effluent from the plant was discharged to the sea without treatment. Regulations then in force did not require effluent treatment from a plant until it was demonstrated that the effluent could kill fish (Chamut *et al.* 1972). The adjoining area in Placentia Bay supported a seasonal herring, cod and lobster fishery which employed about 280 fishers. In February 1969 many dead and dying herring were found downstream of the plant. The herring typically exhibited extensive hemorrhaging, particularly around the gills. As a result of the redness of the herring, this occurrence was referred to as the "red herring" problem. Large numbers of other fish, mostly cod, were found dead on the bottom in Long Harbour towards the end of April. As a result of concerns over the fitness of fish for consumption, part of Placentia Bay near Long Harbour was closed to fishing in May. Due to the adverse publicity, the phosphorus plant was also closed for a short period. During operation the plant discharged 1 250 pounds of elemental phosphorus per day as well as considerable amounts of sulphur dioxide and fluoride, together with lesser amounts of cyanide and ammonia (Idler 1972). Subsequent scientific investigations found elemental phosphorus to be highly toxic to some fish, although less toxic to lobsters. This study experimentally exposed fish to known amounts of contaminants in water but also initiated the development of a sensitive analytical method for the analysis of the phosphorus in

water, sediment and tissue (Addison and Ackman 1970). It was found that upon exposure to phosphorus in water at concentrations as low as 2.5 $\mu g.L^{-1}$, the blood of fish tended to hemolyze thus allowing the hemoglobin (the red pigment in blood) to leak into tissues (Zitko *et al.* 1970). The other contaminants detected were of much lower toxicity. Phosphorus was thus identified as the causal agent. Large quantities of phosphorus were found in the sediments of Long Harbour (up to 2 000 $\mu g.g^{-1}$) (Addison *et al.* 1972). The operators of the plant agreed to remove the contaminated sediment from the harbour by suction dredge, contain the sediment and treat the water to remove the phosphorus prior to releasing the water in the bay. In the spring of 1973 there was a rupture of a pipeline on the wharf during phosphorus loading. About 5 t of phosphorus were spilled into the caissons of the wharf structure. When the plant was decommissioned in 1989, the phosphorus was left in the caissons of the wharf.

References

ADDISON, R.F., and R.G. Ackman. 1970. Direct determination of elemental phosphorus by gas-liquid chromatography. J. Chromatogr. 47:421-426.

ADDISON, R.F., M.E. Zinck, R.J. Ackman, P.S. Chamut, and A. Jamieson. 1972. Results of the elemental phosphorus program, Long Harbour, Newfoundland, Fisheries Research Board of Canada Tech. Rep. 303.

ADDISON, R.F., P.F. Brodie, M.E. Zinck, and D.E. Sergeant. 1984. DDT declines more than PCBs in eastern Canadian seals during the 1970s. Environ. Sci. Technol. 18: 935-937.

ADDISON, R.F., D.E. Willis, and M.E. Zinck. 1994. Liver microsomol mono-oxygenase induction in winter flounder (*Pleuronectes americanus*) from a gradient of sediment PAH concentrations at Sydney Harbour, Nova Scotia. Mar. Environ. Res. 37: 283-296.

ANDERSON, O., C.E. Linder, M. Olsson, L. Reutergardh, U.-B. Uvemo, and U. Wideqvist. 1988. Spatial differences and temporal trends of organochlorine compounds in biota from the northwest hemisphere. Arch. Environ. Contam. Toxicol. 17:755-765.

ANON. 1984. Health of the Northwest Atlantic. R.C.H. Wilson, and R.F. Addison [eds.] Department of Environment; Department of Fisheries and Oceans; Department of Energy, Mines and Resources, Ottawa. 174 p.

ANON. 1994. Overview of Canadian marine fisheries resources. Communications Directorate, Department of Fisheries and Oceans, Ottawa.

BEWERS, J.M., and P.A. Yeats. 1977. Oceanic residence times of trace metals. Nature 268: 595-598.

BIEGER, T., J. Hellou, and J. Abrajano. 1996. Petroleum biomarkers as tracers of lubricating oils. Mar. Poll. Bull. 32: 270-274.

BIEGER, T., T.A. Abrajano, and J. Hellou. 1997. Isotopic characterisation of marine biogenic hydrocarbons in sediments and biota of Newfoundland coastal waters. Org. Geochem. In Press.

BOTTA, J. R., E. Arsenault, and H.A. Ryan. 1983. Total Mercury Content of Meat and Liver from Inshore Newfoundland-caught Harp Seal (*Phoca groenlandica*). Bull. Environ. Contam. Toxicol. 30:28-32.

BRANDON, E. W., and P. A. Yeats. 1984. Contaminant Transport Through the Marine Environment. *In* Health of the Northwest Atlantic. R. C. H. Wilson, and R. F. Addison, [eds.] Department of the Environment; Department of Fisheries and Oceans; Department of Energy, Mines and Resources, Ottawa, Ont.

BRULAND, K. W. 1983. Trace elements in sea-water. Chem. Oceanogr. 8:157-169.

BUCKLEY, D. E., E.H. Owens, C.T. Schafer, G. Vilks, R.E. Cranston, M.A. Rashid, F.J.E. Wagner, D.A. Walker, and B.R. Pelletier. 1974. Canso Strait and Chedabucto Bay: a multidisciplinary study of the impact of man on the marine environment. Geol. Survey Can. Pap. 74-30:1133-1160.

CAPPON, C. J. 1994a. Mercury and Organomercurials. *In* Analysis of Contaminants in Edible Aquatic Resources. J. W. Kiceniuk, and S. Ray [eds.] VCH, New York. 175-204.

CAPPON, C. J. 1994b. Selenium and organic selenides. *In* Analysis of Contaminants in Edible Aquatic Resources. J. W. Kiceniuk, and S. Ray [eds.] VCH, New York. 205-224.

CHAMUT, P.S., G. Pond, and V.R. Taylor. 1972. Red herring pollution crisis in Placentia Bay, Newfoundland. A general description and chronology. *In* Effects of elemental Phosphorus on marine Life. P.N. Ganjaard [ed.] Atlantic Regional Office, Research and Development, Fisheries Research Board of Canada, Halifax, NS. 29-43.

CLEMENT, R.E., H.M. Tosine, V. Taguchi, C.J. Musial, and J.F. Uthe. 1987. Investigation of American lobster, (*Homarus americanus*) for the presence of chlorinated dibenzo-p-dioxins and dibenzofurans. Bull. Environ. Contam. Toxicol. 39:1069-1075.

CRANSTON, R.E., R.A. Fitzgerald, and G.V. Winters. 1975. Geochemical data: LaHave River and Estuary. BIO Data Report BI-D-75-4.

DE BOER, J. 1988. Trends in chlorobiphenyl contents in livers of Atlantic cod (*Gadus morhua*) from the North Sea, 1979-1987. Chemosphere 17:1811-1819.

DUNN, B. P., and Fee, J. 1979. Polycyclic aromatic hydrocarbon carcinogens in commercial seafoods. J. Fish. Res. Board Can. 36:1469-1476.

ELLIOT, J.E., D.G. Noble, R.J. Norstrom, P.E. Whitehead, M. Simon, P.A. Pearce, and D.B. Peakall. 1992. Patterns and Trends of Organic Contaminants in Canadian Seabird Eggs, 1968-90. *In* Persistant Pollutants in Marine Ecosystems. C.H. Walker, and D.R. Livingstone [eds.] Pergamon Press, Oxford.

ERNST, W.R., G. Julien, P. Hennigar, and J. Hanson. 1995. Changes to butyltin residue concentrations in sediments from Atlantic Canada between 1988 and 1994. 22nd Annual Aquatic Toxicity Workshop, Oct. 1-4, 1995. St. Andrews. N.B.

FLIGHT, J. 1987. Whelk Fishery in the Newfoundland Region. Can. Tech. Rep. Fish. Aquat. Sci. 1630: vi + 43p. Department of Fisheries and Oceans, Ottawa, Ont.

FOSTER, G.D., and D.A. Wright. 1988. Unsubstituted polynuclear aromatic hydrocarbons in sediments, clams, and clam worms from Chesapeake Bay. Mar. Pollut. Bull. 19: 459-465.

FREEMAN, H.C., G. Shum, and J.F. Uthe. 1978. The selenium content in swordfish (*Xiphias gladius*) in relation to total mercury content. J. Environ. Sci. Health A13:235-240.

FREEMAN, H.C., J.F. Uthe, and P.J. Silk. 1984. Polychlorinated biphenyls, organochlorine pesticides and chlorobenzenes content of livers from Atlantic cod (*Gadus morhua*) caught off Halifax, Nova Scotia. Environ. Monit. Assess. 4:389-394.

GEARING, J.N., D. E. Buckley, and J.E. Smith. 1991. Hydrocarbon and metal contents in a sediment core from Halifax Harbour: A chronology of contamination. Can. J. Fish. Aquati. Sci. 48: 2344-2354.

HELFRICH, J., and D.E. Armstrong. 1986. Polycyclic aromatic hydrocarbons in sediments of the southern basin of Lake Michigan. J. Great Lakes Res. 12: 192-199.

HELLOU, J. 1996. Polycyclic aromatic hydrocarbons in marine mammals, finfish and mammals. *In* N. Beyer, and G. Heinz [eds.] Interpreting concentrations of environmental contaminants in wildlife tissues. SETAC Special Publication, Pergamon Press. 229-250.

HELLOU, J., and J.F. Payne. 1993. Polychlorinated dibenzo-p-dioxins and dibenzofurans in cod (*Gadus morhua*) from the Northwest Atlantic. Mar. Environ. Res. 36: 117-128.

HELLOU, J., and C. Upshall. 1993. Effect of biogenic impurities on polycyclic aromatic compounds analysis by fluorometry. Int. J. Environ. Anal. Chem. 53: 249-259.

HELLOU, J., and C. Upshall. 1994. Fate of hydrocarbons in *Pseudopleuronectes americanus* exposed to Hibernia crude oil. Can. Tech. Rep. Fish. Aquat. Sci. 2050: 5-12.

HELLOU J., and C. Upshall. 1995. Monocyclic aromatic hydrocarbons in bile of flounder exposed to a petroleum oil, Int. J. Environ. Anal. Chem. 60: 101 -111.

HELLOU, J., and W. Warren. 1997. Polycyclic aromatic compounds and saturated hydrocarbons in tissues of flatfish: insight on environmental exposure. Mar. Environ. Res. 43: 11-25.

HELLOU, J., G. Stenson, I-H. Ni, and J.F. Payne. 1990. Polycyclic aromatic hydrocarbons in muscle tissue of marine mammals from the Northwest Atlantic. Mar. Poll. Bull. 21: 469-473.

HELLOU, J., C. Upshall, I.-H. Ni, J.F. Payne, and Y.S. Huang. 1991. Polycyclic aromatic hydrocarbons in harp seals (*Phoca groenlandica*) from the Northwest Atlantic. Arch. Environ. Contam. Toxicol. 21: 135-140.

HELLOU, J., L. Fancey, and J.F. Payne. 1992a. Concentrations of 24 elements in bluefin tuna (*Thunnus thynnus*) from the Northwest Atlantic. Chemosphere. 24: 211-218.

HELLOU, J., W.J. Warren, J.F. Payne, S. Belkhode, and P. Lobel. 1992b. Heavy metals and other elements in three tissues of cod (*Gadus morhua*) from the Northwest Atlantic. Mar. Poll. Bull. 24: 452-458.

HELLOU, J., C. Upshall, J.F. Payne, S. Naidu, and M.A. Paranjape. 1993a. Total unsaturated compounds and polycyclic aromatic hydrocarbons in molluscs collected from waters around Newfoundland. Arch. Environ. Contam. Toxicol. 24:249-247.

HELLOU, J., W.G. Warren, and J.F. Payne. 1993b. Organochlorines including polychlorinated biphenyls in muscle, liver and ovaries of cod (*Gadus morhua)*. Arch. Environ. Contam. Toxicol. 25: 497-505.

HELLOU, J., C. Upshall, D. Taylor, P. OKeefe, V. O'Malley, and A.T. Abrajano. 1994a. Unsaturated hydrocarbons in muscle and hemolymph of two crab species, *Chionecetes opilio* and *Hyas coarctatus*. Mar, Poll, Bull. 28: 492-488.

HELLOU, J., C. Upshall, J.F. Payne, and P.V. Hodson. 1994b. Polycyclic aromatic compounds in cod (*Gadus morhua*) from the Northwest Atlantic and the St-Lawrence Estuary. Sci. Tot. Environ. 145: 71-79.

HELLOU, J., J.F. Payne, and C. Hamilton. 1994c. Polycyclic aromatic hydrocarbons in Northwest Atlantic cod (*Gadus morhua*). Environ. Poll. 84 - 197-202.

HELLOU, J., J.F. Payne, C. Upshall, L.L. Fancey, and C. Hamilton. 1994d. Bioaccumulation of polycyclic and monocyclic aromatic hydrocarbons, from sediments: a dose-response study with flounder (*Pseudopleuronectes americanus*). Arch. Environ. Contam. Toxicol. 27: 477-485.

HELLOU, J., G. Mercer, and L.L. Fancey. 1995a. Organochlorines in inshore versus offshore yellowtail flounder (*Pseudopleuronectes ferruginea*) the effect of a small urban population on the environment. Int. J. Environ. Anal. Chem. 61: 275-284.

HELLOU, J., W. G. Warren, and G. Mercer. 1995b. Organochlorine contaminants in *pleuronectides*: comparison between three tissues of three species inhabiting the Northwest Atlantic. Arch. Environ. Contam Toxicol. 29: 302-309.

HELLOU, J., D. Mackay, and B. Fowler. 1995c. Bioconcentration of polycyclic aromatic hydrocarbons from sediments to muscle of finfish. Environ. Sci. Technol. 29: 2555-2560.

HELLOU, J., V. Zitko, J. Friel, and T. Alkanaifi. 1996. Distribution of elements in tissues of yellowtail flounder *Pleuronectes ferruginea*. Sci. Tot. Environ. 181: 137-146.

HELLOU, J., D. MacKay, and J. Banoub. 1997a. Dietary and aqueous uptake of organochlorines in finfish: a case study. Submitted to Archives of Environmental Contamination and Toxicology.

HELLOU, J., D. MacKay, and J. Banoub. 1997b. Artificial weathering of sewer sludge from a harbour and uptake of contaminants by finfish. Submitted.

HELLOU, J., W. Warren, C. Andrews, and G. Mercer. 1997c. Long-term fate of crankcase oil in rainbow trout: time and dose-response. Environ. Toxicol. Chem. In Press.

HILDEBRAND, L.P. 1984. Oceanographic Setting. *In* Health of the Northwest Atlantic. R.C.H. Wilson, and R.F.Addison [eds.] Department of the Environment/Department of Fisheries and Oceans/Department of Energy, Mines, and Resources. Bull. Environ. Contam. Toxicol. 50: 907-914.

IDLER, D.R. 1972. Coexistance of fishery on a major industry in Placentia Bay, *In* Effects of Elemental Phosphorus on Marine Life. P.N. Ganjaard [ed.], Atlantic Regional Office, Research and Development, Fisheries Research Board of Canada, Halifax, NS. 126.

KICENIUK, J.W. 1992. Aromatic hydrocarbon concentrations in sediments of Placentia Bay, Newfoundland. Can. Tech. Rep. Fish. Aquat. Sci. 1888: 1-9.

KIELEY, K.M., P.A. Hennigar, R.A.F. Matheson, and W.R. Ernst. 1988. Polynuclear aromatic hydrocarbons and heterocyclic aromatic compounds in Sydney Harbour, Nova Scotia. A 1986 survey. Environment Canada, Environmental Protection Report EPS-5-AR 88-7.

KING, T. L., J.F. Uthe, and C.J. Musial. 1993a. Polycyclic aromatic hydrocarbons in the digestive glands of the American lobster, *Homarus americanus*, captured in the proximity of a coal-coking plant. Bull. Environ. Contam. Toxicol. 50:907-914.

KING, T.L., J.F. Uthe, and C.J. Musial. 1993b. Rapid screening of fish tissue for polychlorinated dibenzo-p-dioxins and dibenzofurans. Analyst 118: 1269-1275.

KING, T.L., B.K. Haines, and J.F. Uthe. 1996. Non-, mono-, and di-o-chlorobiphenyl concentrations and their toxic equivalents to 2, 3, 7, 8,-tetrachlorodibenzo[p]dioxin in Aroclors and digestive glands from American lobsters (*Homarus americanus*) captured in Atlantic Canada. Bull. Environ. Contam. Toxicol. 57:465-472.

LONG, E.R., D.D. McDonald, S.M. Smith, and S. D. Calder. 1995. Incidence of adverse biological effects within ranges of chemical concentrations in marine and estuarian sediments. Environmental Management. 19: 81-97.

LORING, D.H., R.T.T. Rantala, and T. Milligan. 1996. Metallic contaminants in the sediments of coastal embayments of Nova Scotia. Can. Tech. Rep. Fish. Aquat. Sci. 2111.

MAGUIRE, R.J., R.J. Tkacz, Y.K. Chao, G.A. Bengert, and P.S.T. Wong. 1986. Occurrence of organotin compounds in water and sediment in Canada. Chemosphere. 15:253-274.

MATHESON, R.A.F., G.L. Trider, W.R. Ernst, K.G. Hamilton, and P.A. Hennigar. 1983. Investigation of polynuclear aromatic hydrocarbon contamination of Sydney Harbour, Nova Scotia. Surveillance Rept EPS-5-AR-83-6. Environment Canada.

MUIR, D.C.G., R. Wageman, N.P. Grift, R.J. Norstrom, M. Simon, and J. Lien. 1988. Organochlorines chemical and heavy metal contaminants in white-beaked dolphins (*Lagenorhynchus albrostris*) and pilot whales (*Globicephala melaeans*) from the coast of Newfoundland Canada. Arch. Environ. Contam. Toxicol. 17:613-629.

MUSIAL, C.J., and J.F. Uthe. 1983. Widespread occurrences of the pesticide toxaphene in Canadian East coast marine fish. Intern. J. Environ. Anal. Chem. 14: 117-126.

NICHOLS, H.B. 1989. Investigations of the marine environmental quality in Halifax Harbour. Can. Tech. Rep. Fish. Aquat. Sci., 1693.

NOBLE, D.G., and J.E. Elliot. 1986. Environmental contaminants in Canada seabirds. 1968-1985: trends and effects. Canadian Wildlife Service Technical Report Series. No. 13:1-275

OHTA, K., N. Handa, and T. Matsomoto. 1983. Trends and factors governing polycyclic aromatic hydrocarbon levels in Tokyo Bay sediments. Geochem. Cosmochim. Acta 47: 1651-1654.

O'MALLEY, V.P., T.A. Abrajano, and J. Hellou. 1994. Determination of the $^{13}C/^{12}C$ ratio of individual PAH from environmental samples: Can PAH sources be apportioned? Org. Geochem. 21: 809-822.

O'MALLEY, V.P., T.A. Abrajano, and J. Hellou. 1996. Stable carbon isotopic apportionment of individual polycyclic aromatic hydrocarbons in St. John's Harbour, Newfoundland. Environ. Sci. Technol. 30: 634-639.

O'MALLEY, V.P., T.A. Abrajano, A. Stark, J. Hellou, and L. Winsor. 1997a. Tracing polycyclic aromatic hydrocarbons in the environment using $^{13}C/^{12}C$ ratios. Int. J. PAH. In Press.

O' NEILL, H.J., and K.M. Kieley. 1992. Polynuclear aromatic hydrocarbons: Atlantic regional data summary and review. Conservation and Protection Atlantic Region Technical report series No. 92-01. 64p.

PAYNE, J.F., L.L. Fancey, J. Hellou, M.J. King, and G.L. Fletcher. 1995. Aliphatic hydrocarbons in sediments: a chronic toxicity study with flounder (*Pseudopleuronectes americanus*) exposed to oil well drill cuttings. Can J. Fish. Aquat. Sci. 52: 2724-2735.

PEARCE, P. A., J.E. Elliott, D.B. Peakall, and R.J. Norstrom. 1989. Organochlorine contaminants in eggs of seabirds in the Northwest Atlantic, 1968-1984. Environ. Pollut. 56:217-235.

PRAHL, F.G., and R. Carpenter. 1983. Polycyclic aromatic hydrocarbon (PAH) phase associations in Washington coastal sediments. Geochem. Cosmochim. Acta 47: 1013-1023.

PROUSE, N. 1994. Ranking harbours in the Maritime Provinces of Canada for potential to contaminate American lobster (*Homarus americanus*) with polycyclic aromatic hydrocarbons. Can. Tech. Report. Fish. Aquat. Sci. No. 1960.

PROUSE, N.J., and D. J. Ellis. A baseline survey of dogwhelk (*Nucella lapillus*) imposex in eastern Canada (1995) and interpretation in terms of tributyltin (TBT) contamination. In Press.

RAY, S., M. Bailey, G. Paterson, T. Metcalfe, and C. Metcalfe. 1997. Comparative levels of organochlorine compounds in flounder from the Northeast coast of Newfoundland and an offshore site. 1997. Chemosphere. In Press.

SCARRATT, D. J. 1982. Canadian Atlantic offshore fishery atlas. Can. Spec. Publ. Fish. Aquat. Sci 47. Ottawa, Ont.

SIROTA, G.R., JF. Uthe, D.G. Robinson, and C.J. Musial. 1984. Can. MS Rep. Fish. Aquat. Sci. 1958.

TAY, K.-L.,K.G. Doe, S.J. Wade, J.D.A. Vaughan, R.E. Berrigan, and M.J. Moore. 1992. Sediment bioassessment in Halifax Harbour. Environmental Toxicology and Chemistry 11:1567-1581.

TEN HALLERS-TJABBES, C.C., J.F. Kemp, and J.P. Boon. 1994. Imposex in whelks (*Buccinum undatum*) from the open North Sea: relation to shipping traffic intensities. Mar. Pollut. Bull. 28-311-313.

THOMPSON, S., and G. Eglinton. 1978. Composition and sources of pollutant hydrocarbons in the Severn Estuary. Mar. Pollut. Bull. 9: 133-136.

UTHE, J. F. 1979. The Environmental Occurrence and Health Aspects of Polycyclic Aromatic Hydrocarbons. Tech. Rep. Fish. Mar. Serv. (Can.). 914:33.

UTHE, J. F., and Musial, C. J. 1986. Polycyclic aromatic hydrocarbon contamination of American lobster, *Homarus americanus*, in the proximity of a coal-coking plant. Bull. Environ. Contam. Toxicol. 37:730-738.

VANDERMEULEN, J. H. 1989. PAH and heavy metal pollution of the Sydney estuary: Summary and review of studies to 1987. Can. Tech. Rep. Hydrogr. Ocean Sci. No.108.

VIGNIER, V., J.H. Vandermulen, J. Singh, and D. Mossman. 1994. Internal mixed function oxidase (MFO) activity in winter flounder (*Pleuronectes americanus*) from a coal tar contaminanted estuary. Can. J. Fish. Aquat. Sci. 51: 1368-1375.

WINDOM, H.J., J. Byrd, R. Smith, M. Hungspeugs, S. Dharvmvanij, W. Thumkul, and P. Yeats. 1991. Trace metal nutrient relationships in estuaries. Mar. Chem. 32: 177-194.

YEATS, P.A. 1987. Trace metals in Eastern Canadian coastal waters. Can. Tech. Rep. Hydrogr. Ocean Sci. No. 96.

YEATS, P.A. 1993. Input of metals to the North Atlantic from two large Canadian estuaries. Mar. Chem. 43: 201-209.

ZITKO, V., D.E. Aikens, S.N. Tibbo, K.W.T. Hesch, and J.M. Anderson. 1970. Toxicity of yellow phosphorus to herring (*Clupea harengus*), Atlantic salmon (*Salmo salar*), lobster (*Homarus americanus*), and beach flea (*Gammarus oceanicus*). J. Fish. Res. Board Can. 27:21-29.

Additional reading

ANDERSON, D. P., O. W. Dixon, and W. B. van Muiswinkel. 1990. Reduction in the numbers of antibody-producing cells in rainbow trout, *Oncorhynchus mykiss*, exposed to sublethal doses of phenol before bath immunization. In Aquatic Toxicology and Risk Assessment: Thirteenth Volume, ASTM STP 1096. W. G. Landis, and W. H. van der Schalie, [eds.] American Society for Testing and Materials, Philadelphia. 331-337.

BUCKLEY, D.E., J.N. Smith, and G.V. Winters. 1995. Accumulation of contaminant metals in marine sediments of Halifax Harbour, Nova Scotia: environmental factors and historical trends. Applied Geochem. 10: 175-195.

BURGESS, N. M., N. Garrity, and B.M. Braune. 1995. Trends in organochlorine contaminants in seabird eggs from Atlantic Canada, 1968-1992. (In Press).

GOGAN, N. J., and J.W. Kiceniuk. 1997. Changes in chemical speciation of dissolved copper, zinc, cadmium and lead upon mixing of freshwater effluents with seawater. Tech. Rep. Fish. Aquat. Sci. (In Press).

Personal Communication

BURGESS, N. Canadian Wildlife Service, Environment Canada, Sackville, N.B., n.d.

KENCHINGTON, E. Department of Fisheries and Oceans, Halifax, N.S., n.d.

The GREAT*Lakes* Ecosystem

Highlights

■ The Great Lakes form the world's largest freshwater ecosystem and contain 80% of the North American supply of fresh water. More than 33 million people live in the Great Lakes Basin, including approximately nine million Canadians.

■ In 1992, Canadians harvested over 17 000 t of fish from the Great Lakes worth $36 million. The recreational fishery was worth nearly $1.0 billion.

■ Over 360 anthropogenic chemicals have been identified in the Great Lakes ecosystem, including 32 metals, 68 pesticides and over 260 other chemicals. About one-third of them can have lethal or sublethal effects on plant and animal life. Lake Ontario is generally the most contaminated of all the Great Lakes, while Lake Superior, the largest lake, is the least contaminated.

■ Since the initiation of large-scale monitoring programs in the 1970s, there has been a general decline in contaminant concentrations in Great Lakes biota such as fish and the eggs of fish-eating birds. In some areas, particularly in the lower Great Lakes, the decline in chemical inputs and subsequent lessening of contamination in habitats has not always resulted in a continuing decrease in contaminants in fish. This is especially true of fish such as lake trout which are at the top of the aquatic predatory food chain.

■ Indicators of aquatic ecosystem health, including reproductive impairment, damage to immune systems, and increased activation of enzymes in fish populations as a result of exposure to chronic concentrations of contaminants, have been identified in 11 species of Great Lakes biota.

Table of Contents

The Great Lakes Ecosystem

List of Figures

List of Tables

List of Sidebars

Introduction

The Great Lakes (Figure 1) form the largest surface freshwater system in the world, with 20% of the planet's fresh water. Millions of people rely on the lakes and their tributaries for drinking water. The Great Lakes commercial fishery provides 9 000 work-years of employment (Government of Canada 1991a). Canadian commercial fishers of the Great Lakes landed 18 000 t in 1992, a catch valued at $36 million (DFO 1994).

Over the past quarter century, the sport fishery has become one of the most productive and valuable freshwater resources of its kind in the world. Direct spending by sport fishermen is estimated at $2.8 billion annually, nearly a billion of that in Canada. There are one million craft registered for recreational purposes (Allardice and Thorpe 1995).

MAJOR CONTAMINANT ISSUES

Although studied since the 1970s, contaminant issues in the Great Lakes are complicated by the variety of chemicals in the ecosystem, the complexity of food webs and differences among individual lakes with respect to inputs from the surrounding population. Over 360 anthropogenic

Figure 1
The Great Lakes Drainage Basin.

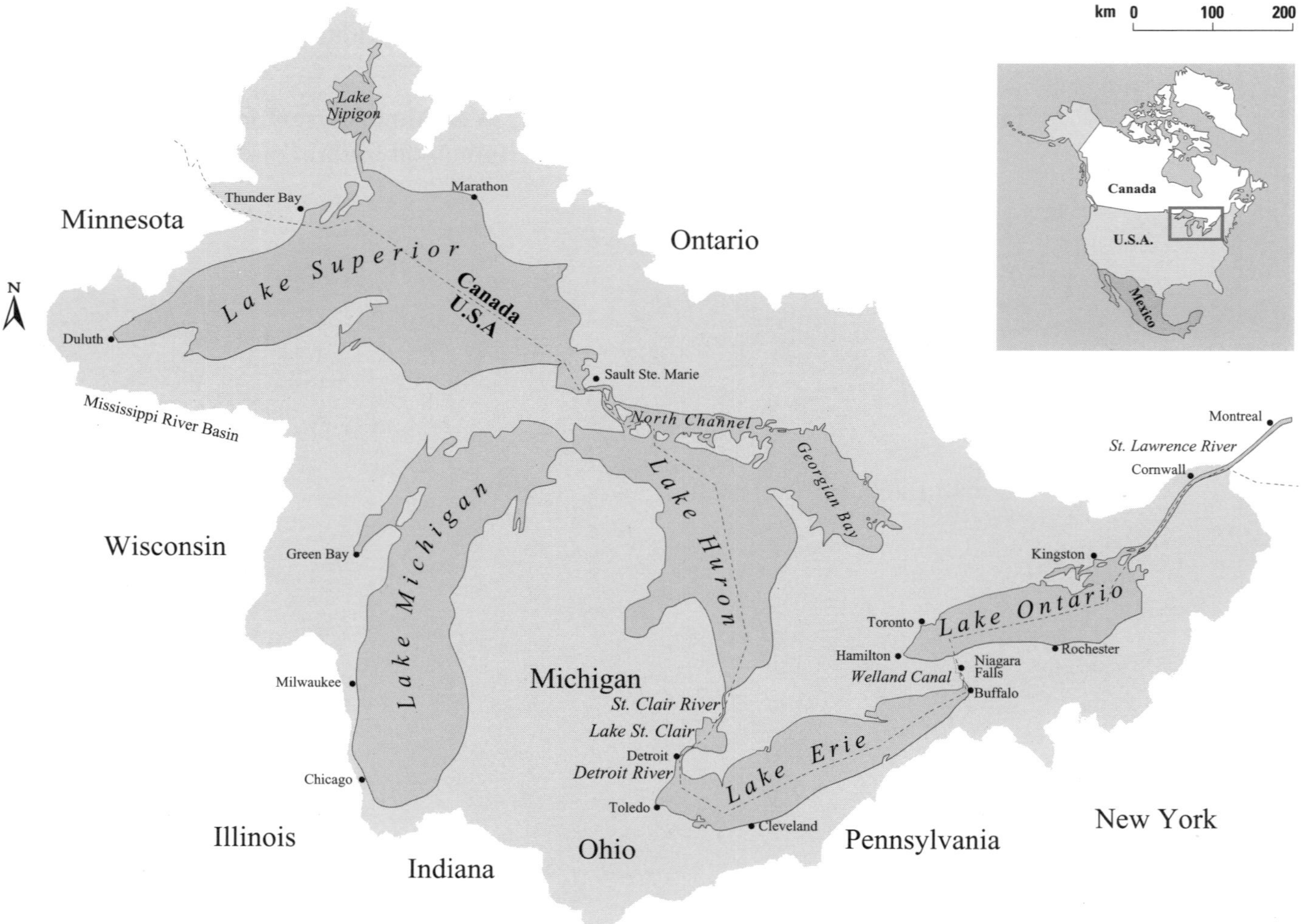

km 0   100   200

Lake Nipigon

Minnesota

Thunder Bay

Marathon

Ontario

Canada

U.S.A.

Lake Superior

Canada U.S.A.

Duluth

Sault Ste. Marie

Mississippi River Basin

North Channel

Georgian Bay

Montreal

St. Lawrence River

Cornwall

Wisconsin

Green Bay

Lake Michigan

Lake Huron

Kingston

Toronto

Lake Ontario

Rochester

Hamilton

Niagara Falls

Welland Canal

Buffalo

Milwaukee

Michigan

St. Clair River

Lake St. Clair

Detroit

Detroit River

Lake Erie

Chicago

Toledo

Cleveland

New York

Illinois

Indiana

Ohio

Pennsylvania

Source: Government of Canada 1991a.

Table 1 Selected criteria, action levels or guidelines for designated critical contaminants in the Great Lakes ecosystem.

Contaminant	Health Canada (1)	IJC (2)	IJC (3)
2,3,7,8-TCDD	20 pg.g^{-1}	–	–
DDT	5 µg.g^{-1}	0.003 µg.L^{-1}	1.0 µg.g^{-1}(a)
PCB	2 µg.g^{-1}	–	0.1 µg.g^{-1}(a)
Mercury	0.5 µg.g^{-1}	0.2 µg.L^{-1}	0.5 µg.g^{-1}
Lead	–	25 µg.L^{-1}	–

1. Consumption guidelines for edible portions of fish.
2. and 3. Objectives for protection of wildlife and aquatic life.
(a) Whole fish

Source: International Joint Commission 1987.

chemicals have been identified, including 32 metals, 68 pesticides and over 260 others, of which one-third are present in quantities dangerous to plant and animal life. Table 1 shows selected criteria, action levels and guidelines for critical contaminants in the Great Lakes for the protection of aquatic life and human consumers (International Joint Commission 1987). The Canada-Ontario Agreement Respecting the Great Lakes Basin Ecosystem (1994) lists a series of toxic chemical "critical pollutants" (Table 2) in the Great Lakes. These are either targeted for virtual elimination (Tier I), or have the potential for causing widespread impacts, or have already caused local adverse effects on the Great Lakes environment (Tier II).

Contamination, over-fishing, habitat destruction and the invasion or deliberate introduction of exotic species has abetted the decline of the Great Lakes commercial fishery, which now lands less than half the catch harvested at the turn of the century.

The presence of chemical contaminants resulted in restrictions on the consumption and harvest of fish. As a consequence, millions of dollars of investment in pollution abatement practices were made by several industrial sectors. This has led to decreasing concentrations of contaminants in habitats, especially from identifiable point sources. Mercury, dichlorodiphenyltrichloroethane (DDT), mirex, and polychlorinated biphenyls (PCBs) have generally declined once the source of each contaminant was curtailed or eliminated. However, as indicated in more detail later in this chapter, the decline in chemical inputs and subsequent lessening contamination of habitats has not always resulted in a corresponding decrease in contaminant concentrations in fish. This is especially true of fish which are at the top of the aquatic predatory food chain, such as lake trout. Because of this occurrence, contamination

Table 2 Great Lakes Critical Pollutants - Tier I and Tier II.

Tier I	Tier II
Aldrin/Dieldrin	Anthracene
B[a]P	1,4-Dichlorobenzene
Chlordane	3,3-Dichlorobenzene
DDT	Dinitropyrene
Hexachlorobenzene	Hexachlorocyclohexane
Alkyllead	4,4-Methylenebis (2-chloranaline)
Mercury	Pentachlorophenol
Mirex	Tributyltin
Octachlorostyrene	17 other PAHs
PCBs	
PCDDs	
PCDFs	
Toxaphene	

Source: Canada-Ontario Agreement Respecting the Great Lakes Basin Ecosystem 1994.

remains even after point sources have been eliminated. Indeed, there have been increases recently in the number of consumption advisories across the basin.

Further complications in dealing with chemical contaminants and their effects in the Great Lakes Basin result from atmospheric loadings to the Great Lakes. In some cases, compounds whose usage was previously banned within the Great Lakes have recently accumulated to concentrations high enough to trigger advisories on the consumption of fish by humans. For example, toxaphene, an insecticide banned for general use in Canada in 1974 and restricted in the U.S. in 1984, was found in Lake Superior fish in 1994 at concentrations which restrict sport fish consumption (Government of Ontario 1995). The most likely source of this compound is via long range atmospheric transport (LRTAP) and subsequent deposition within the Lake Superior basin.

Controlling the eventual fate of these still-present contaminants is made complex by the multiple processes of ecosystem stress and the inherent complexity of the Great Lakes ecosystem. The success of system-wide control of contaminants is not only important to the basin's ecosystem, but has a measurable effect on the health and function of the St. Lawrence River and Gulf of St. Lawrence ecosystems downstream.

As there is likely more documentation on the effects of chemical contaminants for the Great Lakes than for other Canadian aquatic systems, this chapter relies heavily on summary documents on Great Lakes contaminants and their effects prepared throughout the 1990s. Significant reports of this type include a three-volume report on Toxic Chemicals in the Great Lakes and Associated Effects (Government of Canada 1991b), The State of Canada's Environment (Government of Canada 1991a), State of the Great Lakes (Environment Canada/U.S. EPA 1995), and various reports to the International Joint Commission by the Water Quality Board

and the Science Advisory Board. This chapter focuses on chemicals and their effects in Canadian waters of the Great Lakes, which comprise portions of Lake Superior, Lake Huron, Lake Erie and Lake Ontario, recognizing nonetheless that Lake Michigan, entirely in U.S. territory, is an integral part of the entire ecosystem.

LIMNOLOGICAL FEATURES OF IMPORTANCE TO CONTAMINANT DYNAMICS

The Great Lakes hold 80% of North America's supply of surface fresh water (LTI, Limno-Tech

Lake Trout (*Salvelinus namaycush*)

Native to North America, the lake trout is one of the world's largest freshwater fish and is highly regarded by commercial, sport and subsistence fishermen. A member of the char group of the salmon family, it has a large head with well developed teeth on the jaws, tongue and roof of the mouth. It is distinguished by its deeply forked tail and dappled colour. Lake trout often grow to weigh 23 kg with the largest recorded at 46 kg and over 120 cm in length. The average weight in current catches, however, is less than 5 kg. Lake trout spawn in the fall on clean rocky shoals where the eggs hatch in 4 to 5 months. They mature at about 5 to 8 years of age and live for 15 to 25 years. Some lake trout have been known to live as long as 60 years. Lake trout will eat almost any food but the larger trout that prefer to eat fish generally grow faster and live longer. Preferring cold water of about 10° C they mainly inhabit large deep lakes although they are also found in shallow tundra lakes and deep rivers.

Lake trout are a useful indicator of concentrations of toxic chemicals in the environment. This is because they are long lived and fatty fish. Those characteristics make them excellent accumulators of fat soluble contaminants and many synthetic organic chemicals. Metals such as mercury and lead may be bioaccumulated by lake trout over their lifetime. Lake trout occupy the top trophic level of the aquatic foodchain and therefore reflect the contaminant accumulations in the rest of the foodchain. Because of their prevalence throughout Canada, they provide an excellent indicator of spatial variation in contaminant concentrations in aquatic ecosystems. Lake trout have been used in the Great Lakes since 1977 as the principal monitoring organisms to track the temporal and spatial trends of toxic chemicals throughout fish communities in the basin.

Inc. 1993) and cover an area of nearly 250 000 km^2, about the size of the United Kingdom. The lakes have 17 000 km of coastline and drain adjacent lands totaling more than half a million square km, via 750 000 km of tributaries and more than 80 000 small upland lakes. The system includes five of the world's 14 largest freshwater lakes (Environment Canada/U.S. EPA 1995). Portions of four of the five Great Lakes are in Canada. Only Lake Michigan is entirely within the borders of the U.S. Table 3 shows the maximum depth, mean depth, surface area, volume, shoreline length and retention time of each lake.

The flushing time of the lakes is very long, leading to the accumulation of contaminants in the system. Less than 1% of the volume of the Great Lakes exits the system each year out through the St. Lawrence River. While in the Great Lakes system, contaminants tend to bind to particles and sink to the bottom. There they become part of the lake sediments, or they are ingested by organisms and incorporated into the food web. Contaminants can be retained for many years, and can pass back and forth between water and other components of the ecosystem (Government of Canada 1991b).

Lake Superior is the largest body of fresh water in the world and is the deepest of the Great Lakes. It contains more than half the water in the Great Lakes system, a volume which takes about 190 years to flow into Lake Huron via the St. Mary's River (Environment Canada/U.S. EPA 1995).

Lake Huron is the world's third-largest body of fresh water and drains an area of 134 000 km^2, the largest drainage basin of all the Great Lakes. It is the second shallowest of the Great Lakes. Lake Huron's waters flow via the St. Clair River, Lake St. Clair, and the Detroit River into Lake Erie (Environment Canada/U.S. EPA 1995).

Lake Erie is by far the shallowest of the Great Lakes, containing the smallest volume of water. Lake Erie drains into Lake Ontario via the Niagara River and the Welland Canal (Environment Canada/U.S. EPA 1995).

Lake Ontario is the smallest of the Great Lakes in surface area, but it is nonetheless the world's fourteenth-largest freshwater lake by volume. It has the smallest drainage basin, 64 000 km^2. On average however, it is the second deepest of the lakes. Lake Ontario's water volume takes six years to flow into the Atlantic Ocean via the St. Lawrence River (Environment Canada/U.S. EPA 1995).

BIOLOGICAL RESOURCES

In 1992 commercial fisheries accounted for a total of about 18 000 t with a value of $36 million (DFO 1994). Eighty percent of the commercial catch of the Great Lakes fishery, representing 80% of the total earnings, was harvested from Lake Erie. Lake Huron, with 12% of the catch and 16% of the income, was the source of most of the remainder (DFO 1994). Smelt (41%), yellow pickerel or walleye (23%), white perch (14%), yellow perch (14%), and lake whitefish (12%)

Table 3   Physical characteristics of the Great Lakes.

Lake	Maximum depth (m)	Mean depth (m)	Surface area (km²)	Volume (km³)	Shoreline length (km)	Retention time (yr)
Superior	410	150	82 000	12 000	4 800	190
Michigan	280	85	58 000	5 000	2 700	99
Huron	230	59	60 000	3 500	5 100	22
Erie	64	19	26 000	480	1 400	3
Ontario	250	86	19 000	1 600	1 200	6

Source: Government of Canada 1991a.

accounted for the largest proportions of the catch. Pickerel was the most valuable species landed ($13 million), followed by yellow perch ($11 million), lake whitefish ($4.5 million), smelt ($2.7 million) and white perch ($2.2 million) (DFO 1994). The success of recreational fishing, which accounts for a total (Canada/U.S.) annual value of $2.8 billion, is based on stocking of desired species. It is a "put-and-take" fishery, dependent on annual inputs of lake trout, splake, rainbow trout, coho, chinook and Atlantic salmon (LTI, Limno-Tech Inc. 1993).

Assessment of Chemical Contamination

SOURCES

The Great Lakes are used for industry, agriculture, domestic water supply, sewage disposal, transportation and navigation, fishing and power production. Industries employing more than two million Canadian workers, and accounting for half of Canada's industrial capacity, are located near the shores of the Great Lakes. The Great Lakes Basin industrial activity accounts for 20% of U.S. industry if all the lakes, especially the heavily industrialized basin of Lake Michigan, are included.

Agricultural activities in the Great Lakes Basin encompass approximately 11 million ha in the U.S. As well, the vast majority of Ontario's approximately 5.4 million ha of farmland is located within the basin (Allardice and Thorpe 1995).

Land uses include urbanized areas and industrial facilities, agricultural and private recreational land. The surrounding population is distributed unevenly. Three-quarters of the inhabitants are concentrated in the Lake Erie and Lake Michigan basins, and another fifth in the

Lake Ontario basin, leaving the Lake Huron and Lake Superior basins sparsely populated. Of the nearly 8.5 million Canadians in the region, nearly two-thirds (5.5 million) live in the Lake Ontario basin. Just six urban areas contain 75% of the Canadian Great Lakes Basin population (Allardice and Thorpe 1995). The most significant trend in the region has been urban sprawl of lower density residential and other developments. These developments are more environmentally stressful than crowded urban forms (Environment Canada/U.S. EPA 1995).

The large water surfaces of the five Great Lakes and the other lakes in the basin make them particularly vulnerable to atmospheric deposition of contaminants (Government of Canada 1991b). Table 4a shows loadings of hexachlorobenzene (HCB) and total polychlorodibenzo-*p*-dioxins (PCDDs) and polychlorodibenzo-furans (PCDFs) to the Great Lakes from both air and water sources. Table 4b provides estimates of the inputs and losses for total PCBs in the Lake Superior basin from various sources. The upper Great Lakes — Superior, Michigan and Huron — receive a significant fraction of their lead, PCBs, and DDT directly from the atmosphere (IJC 1988). Compared to the lower lakes, they have larger surface areas and fewer local contaminant point sources. Therefore, the predominant atmospheric source of toxic chemical loadings to the upper Great Lakes is more difficult to reduce than the major point sources which are important to the lower Great Lakes. Due to local sources,

Table 4a **Amounts of HCB, PCDDs and PCDFs contributed to the Great Lakes from air and water sources.**

	HCB loadings (kg.yr⁻¹)		PCDD and PCDF loadings (g.TEQ.yr⁻¹)	
	Air	**Water**	**Air**	**Water**
Superior	11	0.1	5.6	1.4
Huron	16	0.6	8.6	1.4
Erie	15	<72	7.3	11
Ontario	23	35	6.4	<3.9

Source: Cohen *et al.* 1995.

total PCB inputs to the lower Great Lakes are 4 times higher than to the upper Great Lakes (IJC 1988).

Municipal and industrial direct discharges contribute sizable quantities of chemical contaminants to the Great Lakes and their tributaries. Industrial dischargers to the Great Lakes surface water system include pulp and paper mills; major chemical plant complexes on the Ontario side of the St. Clair River; and on the American side of the Niagara River, petroleum refineries, and iron and steel mills. Other sources of contaminants are automotive plants, wood preserving facilities, metal processing and finishing plants, municipal sewer systems, leaking municipal and industrial dumps, and urban and agricultural run-off (IJC 1991).

Throughout the Great Lakes Basin, 42 "Areas of Concern" (AOC) have been identified. In these areas, chemical contamination has impaired the functioning of the ecosystem. Of the 42 sites, 17 are totally within Canadian jurisdiction. Some of these sites, which are often harbours or embayments, act as point sources of contaminants to the main lake. A study by Fox *et al.* (1996) identified Hamilton Harbour, an AOC, as a net source of polycyclic aromatic hydrocarbons

(PAHs) and PCBs to Lake Ontario. However, the total loading was not significant in comparison to the atmospheric input which is estimated to be as much as 100 times greater. Often, point source loadings in the Great Lakes are small in comparison to atmospheric inputs which sometimes originate from sources outside the basin.

For some chemical contaminants, such as lead, mercury and PCBs, there are no large and easily identifiable individual sources discharging directly to the Great Lakes, and the impact from emissions from individual sources to the atmosphere is not known. Most recognizable sources have already been subjected to controls. Remaining sources tend to be widely dispersed and, in the case of air emissions, often outside the Great Lakes Basin.

Lead, mercury and PCBs, are subject to regional and long distance transport and may then be deposited from the atmosphere. In addition, the so-called "grasshopper effect," (successive emission, transport, deposition and re-emission processes from sources within and beyond the Great Lakes Basin), may contribute significantly to the loading of contaminants to the Great Lakes (IJC 1993). For a more detailed description of the "grasshopper effect" see the discussion of the Cold Condensation Theory in the Arctic Marine Ecosystem Chapter.

Remote and local sources contribute to the atmospheric mercury fallout into the Great Lakes. Some of the atmospheric deposition of lead in the Great Lakes comes from sources outside the basin. For example, preliminary calculations performed by the U.S. Environment Protection Agency (EPA) indicate that sources located 500 to 1 000 km south of Lake Superior contributed more than 30% of the lead deposited into the lake in 1985 (IJC 1993).

There has been a marked reduction in lead concentrations in the atmosphere in the last

Table 4b   Estimates of total PCB loadings and losses for Lake Superior 1992.

		Amount (kg.yr^{-1})
Input	Wet and dry deposition	85
	Tributary input	54
	Direct discharges	40
	Vapour deposition	320
Output	Sediment burial	-110
	River outflow	-43
	Volatilization	-2 000
Total mass balance[1]		-1 700

[1] Total amount is not the sum of individual amounts due to rounding.

Source: Hoff *et al.* 1996.

decade. This is expected to result in lower lead concentrations in Great Lakes water, surficial sediments and biota. This decrease is due primarily to reduced emissions caused by the introduction of unleaded gasoline (IJC 1993).

Point and diffuse sources of PCB contamination of the Great Lakes have been identified. Atmospheric deposition is also a major source of PCBs (L'Italien 1993). Analysis of suspended solids in streams entering the Great Lakes from 1974 to 1976 found the highest concentrations of PCBs in the Humber and Niagara rivers, both of which flow into Lake Ontario (Frank *et al.* 1981).

Spottail shiners taken from inshore Ontario waters at the western end of Lake Erie had higher concentrations of PCBs than those from the eastern end of the lake, suggesting that the Detroit River is a significant source of PCBs (See Figure 2). Spottail shiners collected near the Love Canal toxic waste dump (102nd Street and Cayuga Creek. sites) in Niagara Falls, N.Y., had high PCB concentrations, suggesting the Love Canal as another significant source of PCBs. Figure 2 also shows that PCB concentrations in these forage fish collected from Etobicoke Creek and the Humber and Don Rivers in Metro Toronto were higher than fish collected from nearby sites in Lake Ontario, indicating watershed inputs of PCBs to Lake Ontario (Suns *et al.* 1993).

The primary source of the most toxic forms of PCDDs and PCDFs (tetrachloro-) in the upper Great Lakes (Superior and Huron) was pulp and paper mill effluents from operations which employed chlorine bleaching. The primary source in the lower Great Lakes is from old chlorophenol production sites and chemical waste dumps (Whittle *et al.* 1992).

DISTRIBUTION, TRENDS AND EFFECTS

Concentrations of chemical contaminants in the Great Lakes are generally dramatically lower now than they were when measurements were first taken in the mid-1970s. However, declines in

PCBs, DDT and some other organochlorine compounds in biota have ceased, and even reversed, in the past decade. The reason for this is uncertain, and factors other than contaminant loadings may be responsible. One possible reason is that changes in the food web may have altered contaminant pathways to the top predator fish species and made these fish more vulnerable to excessive contaminant accumulation (See Case Study: Increases in Contaminant Burdens Related to the Invasion of Lake Erie by Zebra Mussels).

Overall, concentrations of several chemical contaminants in water and aquatic biota exceed values recommended for the protection of human health, resulting in fish consumption advisories for each of the Great Lakes. With local exceptions, Lakes Ontario, Michigan, and Erie and the connecting water bodies continue to be more severely contaminated than Lakes Superior and Huron. Lake Ontario, situated at the end of the system, is the most contaminated as it receives the cumulative loading from all water bodies upstream. In the past the highly contaminated Niagara River has accounted for more than half of Lake Ontario's contaminant loading (Government of Canada 1991).

The Ontario Sport Fish Contaminant Monitoring Program, in operation since the early 1970s, provides consumption advice for anglers for more than 1 600 locations throughout the province including Lakes Ontario, Erie, Huron, and Superior. Dorsal muscle samples are analyzed for a range of metals, pesticides and industrial organic chemicals. Table 5 shows the percentage of sites with consumption restrictions in any size class of Great Lakes top predator fish in 1993 and 1995. The long term trend observed by the monitoring program is a decline of contaminant concentrations in fish once the source of contamination is eliminated or curtailed.

METALS AND ORGANOMETALLIC COMPOUNDS

Analysis of Canadian Great Lakes sediments in the early 1970s found the lowest mean

Figure 2
Total PCB concentrations in young-of-the-year spottail shiners taken from the Great Lakes and connecting channels during 1990.

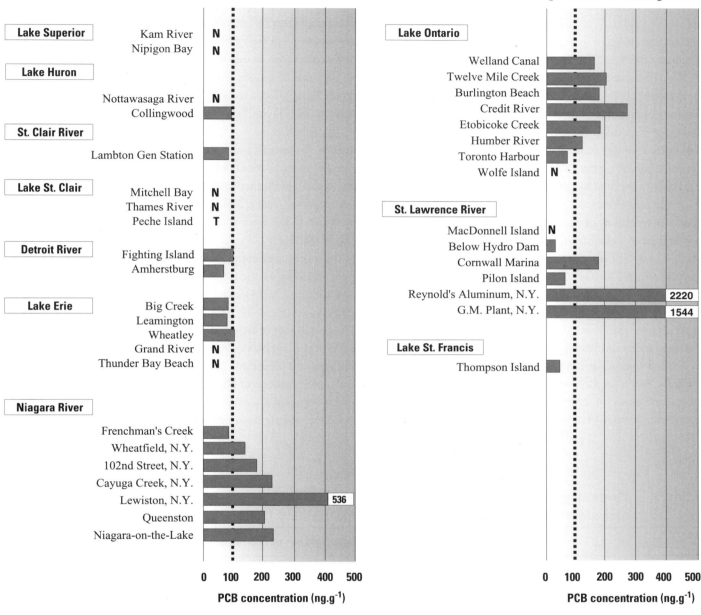

PCB concentration (ng.g^{-1})

PCB concentration (ng.g^{-1})

The International Joint Commission Life Guideline for PCB concentration is 100 ng.g^{-1}.
N = Not detected
T = Trace

Source: Suns *et al.* 1993.

concentrations of mercury in Lake Superior (83 ng.g^{-1}), and the highest in Lake St. Clair (630 ng.g^{-1}), Lake Erie (610 ng.g^{-1}), and Lake Ontario (650 ng.g^{-1}) (Thomas 1974).

Concentrations of mercury in fish have generally declined, although data for some Lake Huron fish have also shown minimal increases in the early 1990s. Analysis of lake trout show similar concentrations of mercury (about 0.1 µg.g^{-1} wet weight) in all four Canadian Great Lakes for samples taken in 1994 (Figure 3). Rainbow smelt, a prey fish species, has been routinely monitored by the Department of Fisheries and Oceans since 1977. Results showed that concentrations of mercury in smelt samples are highest in Lake Superior, and have declined significantly in Lakes Superior, Huron, Erie and Ontario (Figure 4) (De Vault *et al.* 1995). Suns *et al.* (1985) reported that a survey of more than 80 Great Lakes sites, using spottail shiners as the indicator organism, identified the highest concentrations of mercury in fish from a collection site on the Niagara River prior to its entry into Lake Ontario. Overall mercury concentrations at most sites sampled declined significantly from 1976 to 1982.

Most lead entering the lakes settles to the bottom (IJC 1993). An Environment Canada analysis of metals in Great Lakes sediments collected from 1991 to 1993 found median lead concentrations higher in Lake Ontario (55 µg.g^{-1} dry weight) and Georgian Bay (47 µg.g^{-1} dry weight) than in the other Canadian Great Lakes (Superior 23 µg.g^{-1}; Huron 22 µg.g^{-1} dry weight; Erie 25 µg.g^{-1} dry weight) (Reynoldson *et al.* 1997).

Because lead is not significantly bioaccumulated, it is not routinely monitored in fish tissue. Special studies indicate very low concentrations in most Great Lakes fish (Hodson *et al.* 1984a). Much of the concern about lead contamination in the Great Lakes is related to alkyllead, an organometallic compound. Sources of alkyllead include sites on the Detroit and St. Clair Rivers (Chau *et al.* 1985) and the St. Lawrence River (Wong *et al.* 1988). Alkyllead compounds were detected in fish, sediment, macrophyte and water samples collected near these sites. Subsequent reductions in environmental concentrations of alkyllead compounds were reported after the closure of production facilities arising from the 1985

Table 5   Percentage of sites with consumption restrictions in any size class of Great Lakes top predator fish in 1993 and 1995.

		Lake trout	Rainbow trout	Coho salmon	Chinook salmon	Walleye
			Percentage (number of sample sites)			
Ontario	1993	100 (8)	55 (11)	67 (6)	90 (10)	83 (12)
	1995	10 (6)	86 (7)	100 (3)	100 (8)	83 (6)
Erie	1993	0 (4)	0 (4)	0 (7)	-	43 (14)
	1995	100 (2)	0 (2)	75 (4)	-	70 (10)
Huron	1993	44 (9)	0 (13)	0 (1)	0 (11)	79 (19)
	1995	57 (7)	75 (4)	-	83 (6)	80 (5)
Superior	1993	54 (35)	0 (6)	-	0 (6)	100 (8)
	1995	91 (11)	50 (2)	-	83 (6)	100 (4)

Source: Government of Ontario 1995.

banning of lead additives in gasoline (Wong *et al.* 1989). Previous studies have linked exposure of Great Lakes fish to specific contaminants such as lead to biochemical changes such as inhibition of activity of the enzyme δ-amino levulinic acid dehydratase (ALAD) (Hodson *et al.* 1984b).

A three-year (1991 to 1993) Environment Canada, basin-wide survey found median concentrations of cadmium in sediments relatively similar throughout the Great Lakes: in Georgian Bay 1.0 µg.g⁻¹ dry weight and Lake Ontario 0.9 µg.g⁻¹ dry weight; in Superior 0.4 µg.g⁻¹ dry weight Huron 0.2 µg.g⁻¹ dry weight; and Erie 0.5 µg.g⁻¹ dry weight. (Reynoldson *et al.* 1997).

Other metals such as arsenic, selenium and zinc have also been detected in Great Lakes fish. These metals have generally not been a significant problem with respect to biological effects in aquatic biota or restrictions on human consumption.

In 1989, Canada joined with other countries to prohibit the use of organotins (such as tributyltin) in anti-fouling paints for vessels less than 25 m in length with the exception of those with aluminum hulls. Prior to this date the use of tributyltin/formulated paints was extensive in the recreational boating sector. In a 1985 survey of 265 sites across Canada, Maguire *et al.* (1986)

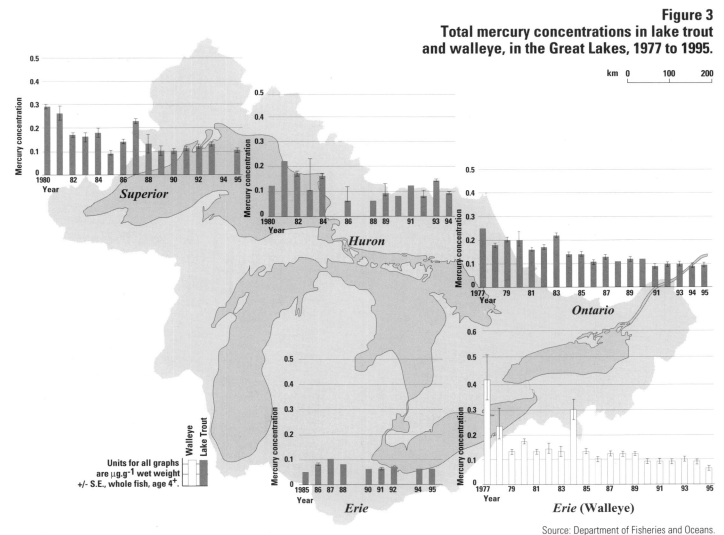

Figure 3
Total mercury concentrations in lake trout and walleye, in the Great Lakes, 1977 to 1995.

Source: Department of Fisheries and Oceans.

found high concentrations of tributyltin in subsurface water samples in the many recreational boating harbours throughout the Great Lakes. Highest concentrations were found in Port Hope Harbour (2.3 µg.L^{-1}) and Whitby Harbour (1.7 µg.L^{-1}) on Lake Ontario. Additional surveys in 1989 by Wong *et al.* (1994b) detected elevated concentrations of tributyltin in sediments (390 ng.g^{-1} dry weight), freshwater clams (66 ng.g^{-1} soft tissue wet weight), and northern pike (240 ng.g^{-1} wet weight) collected near a recreational boating harbour on Lake Huron. A repeated survey of this site in 1992 showed that

organotin compounds in the sediments peaked in May to June at a mean of 790 ng.g^{-1} dry weight and generally decreased with time through October to 580 ng.g^{-1} dry weight.

The spring increase in organotins measured in the sediments corresponded to the spring launching of large numbers of pleasure craft with hulls recently treated with anti-fouling paints containing organotins. The peak in concentrations was probably a consequence of the organotin being leached from the hulls of these small vessels.

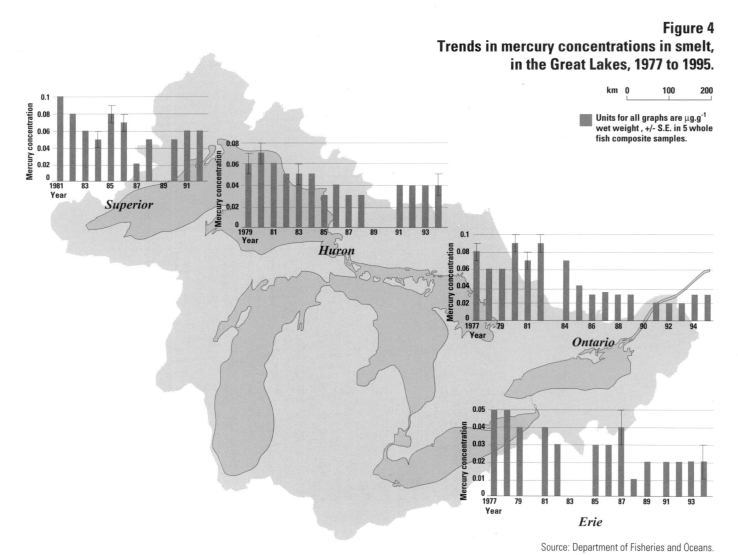

Figure 4
Trends in mercury concentrations in smelt, in the Great Lakes, 1977 to 1995.

Units for all graphs are µg.g^{-1} wet weight , +/- S.E. in 5 whole fish composite samples.

Source: Department of Fisheries and Oceans.

ORGANIC CHEMICALS
ORGANOCHLORINE COMPOUNDS

Current atmospheric concentrations of PCBs in the Great Lakes Basin range from 0.1 to 1.0 ng.g^{-1} PCBs are delivered by wet and dry deposition (rain, snow and dust particles) to the Great Lakes and their watersheds. Once in the lakes, PCBs enter the food chain, evaporate into the atmosphere or settle to the bottom with subsequent burial in sediments (IJC 1993).

Of the lakes, the waters of Lake Ontario and Lake Erie are those most contaminated by PCBs. Lake Superior is the least contaminated. In Lake Ontario, localized point sources were responsible for significantly elevated PCB concentrations in the Niagara River plume, along the shores of the Niagara Peninsula and in the eastern basin of Black River Bay and in the western and eastern basins of Lake Erie (Stevens and Neilson 1989).

Water samples from each of the Canadian Great Lakes, particularly Lake Ontario, have often exceeded the New York State Aquatic Resource Protection Criteria of 1.0 ng.g^{-1} (L'Italien 1993). These limits were set by the State of New York primarily for the protection of the health of wildlife consumers of aquatic life.

Figure 5
Trends in PCB concentrations in lake trout and walleye, in the Great Lakes, 1977 to 1995.

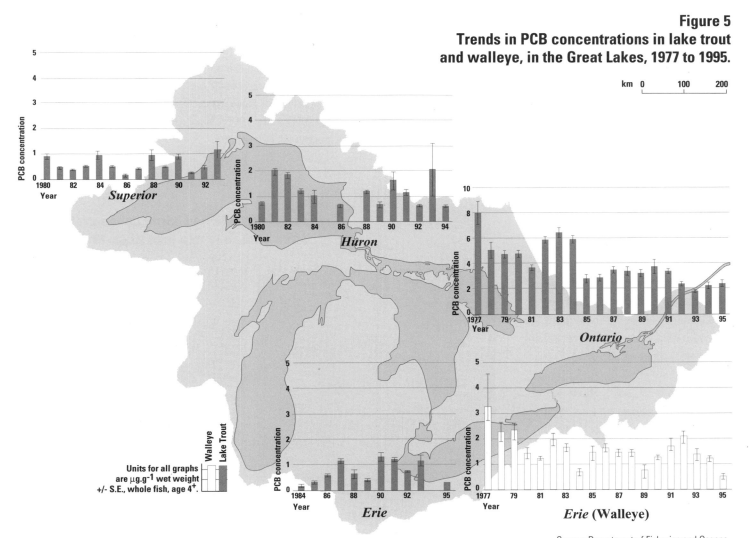

Source: Department of Fisheries and Oceans.

The technology required to measure contaminants at the low concentrations of specific PCBs (i.e. PCB congeners) found in open water in the Great Lakes has become widely available only recently. As a result, meaningful trend data are limited. However, data from Lake Superior show a near-consistent year-to-year decline of total PCBs from 1.7 ng.L^{-1} in 1978 to 0.18 ng.L^{-1} in 1992.

While unconfined use of PCBs was eliminated in 1972, they are still consistently found at higher concentrations in Lake Ontario fish than in fish from the other lakes. Concentrations declined in the 1970s, and by 1990 leveled off at concentrations which still exceed the IJC Great Lakes Water Quality Agreement (GLWQA) objective for fish of 0.1 μg.g^{-1}.

Since annual open lake monitoring data was initiated in 1977, the lowest concentrations of total PCBs in Lake Ontario lake trout whole fish occurred in samples collected in 1993 (1.8 μg.g^{-1}). However, annual mean concentrations have remained relatively stable or risen slightly since then (Figure 5). The leveling off of PCB concentrations in lake trout is puzzling in light of continued declines in PCB concentrations in the water in which the trout live. One general explanation for this phenomenon is that because top predators such as lake trout receive more than 90% of their PCBs from food, changes in the composition of the food web may affect concentrations in trout (De Vault *et al.* 1995).

As shown in Figure 6, Huestis *et al.* (1996) presented data on total PCB and 61 PCB congeners in Lake Ontario whole lake trout samples collected from 1977 to 1993. They reported that total PCB and PCB congener concentrations showed slight increases in 1982, 1988

and annually from 1991 through 1993 after periods of steady decline. The authors reported no significant temporal trends for dichlorodiphenyldichloroethylene (*p,p'*-DDE) or mirex after 1978. When lake trout eggs collected from the Great Lakes were analyzed, different PCB congener profiles were found in fish from different lakes which suggests that inputs of PCB contamination originate from different point sources (Mac *et al.* 1993).

PCB concentrations measured in edible portions of sport fish species from the Great Lakes have generally declined since monitoring began in 1972. Concentrations fluctuated in the 1980s and early 1990s but are currently a fraction of those measured in 1972. An example of this trend is found in the long-term monitoring program of Pacific salmon in Lake Ontario. PCB concentrations in coho salmon in the Credit River, a tributary of Lake Ontario, dropped dramatically, with a decline of about 80%, between 1972 and 1978, and have declined slowly or fluctuated since then (Government of Ontario 1995). Similar patterns are seen for pesticides and other

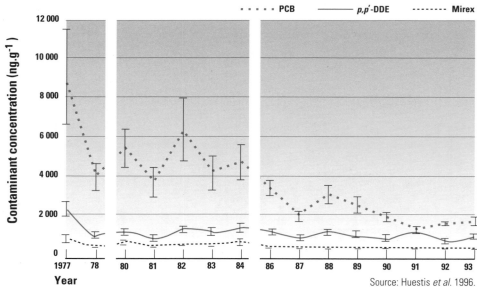

Figure 6
Trends in PCB, DDE and mirex concentrations in Lake Ontario whole lake trout, 1977 to 1993.

· · · · · PCB   —— *p,p'*-DDE   - - - - - - Mirex

Source: Huestis *et al.* 1996.

industrial organic chemicals in sport fish monitored by this program throughout the Great Lakes Basin.

Since rainbow smelt is a prey fish, using this type of fish species as a monitoring tool provides an indication of chemical accumulation at the base of the fish community food chain. Results, as indicated in Figure 7, show concentrations of PCBs are highest in Lake Ontario, and have declined significantly over time. Concentrations in smelt from Lakes Superior, Huron and Erie are significantly lower (De Vault *et al.* 1995).

Young-of-the-year spottail shiners are monitored by the Ontario Ministry of

Environment and Energy to detect localized contaminant exposure in the Great Lakes Basin. At most collection sites, PCBs declined significantly from the mid 1970s to 1990. This decline follows a series of regulatory actions. Even so, PCBs in spottail shiners exceeded the IJC, GLWQA Objective of 0.1 μg.g^{-1} at most of the monitoring sites on the Niagara River (86% of fish sampled) and in Lake Ontario (75% of fish sampled) (Suns *et al.* 1993).

PCB concentrations in Herring Gull eggs, the result of contaminant biomagnification in the Great Lakes aquatic ecosystem, have been monitored since 1974 by the Canadian Wildlife Service of Environment Canada. Concentrations

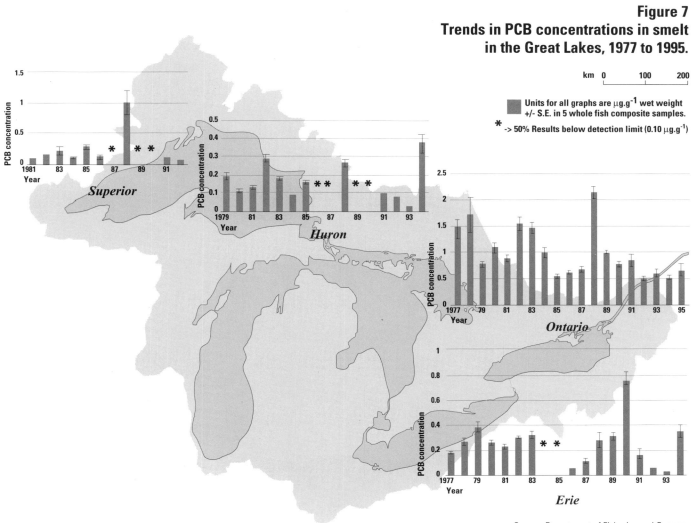

Figure 7
Trends in PCB concentrations in smelt in the Great Lakes, 1977 to 1995.

Units for all graphs are μg.g^{-1} wet weight +/- S.E. in 5 whole fish composite samples.

* -> 50% Results below detection limit (0.10 μg.g^{-1})

Source: Department of Fisheries and Oceans.

have declined significantly between 1974 and the mid 1980s at most colonies sampled. Since then, the rate of decrease has been much slower. PCB concentrations in eggs were generally higher in Lakes Erie and Ontario, although one colony in Lake Huron contained the highest concentrations (De Vault *et al.* 1995).

Lake Ontario is the Great Lake most contaminated by DDE (*p,p'*-DDE), DDT's principal breakdown product. DDE is also present in the waters of Lake Huron and Lake Erie. This is a result of previous use (Stevens and Neilson 1989). Both DDT and DDE were found at numerous sampling points in Lake Ontario, and at a few sites in Lake Erie. L' Italien (1993)

reported that DDT was detected in water samples taken near Toronto in 1988. DDE was found at several sites on Lake Ontario near Port Dalhousie, and Port Weller Harbour (in 1986); as well as near Kingston, at the mouths of the Niagara and Oswego Rivers, Port Weller and in Hamilton Harbour (in 1988). DDT or DDE was not detected in 1990 Lake Ontario samples. The 1986 samples from near Monroe and Toledo at the western end of Lake Erie also contained detectable concentrations of DDE (L'Italien 1993).

Trends in DDT concentrations in top predator fish are similar to those found for PCBs, with dramatic declines from the mid 1970s to the mid 1980s, followed by steady levels. In the case of

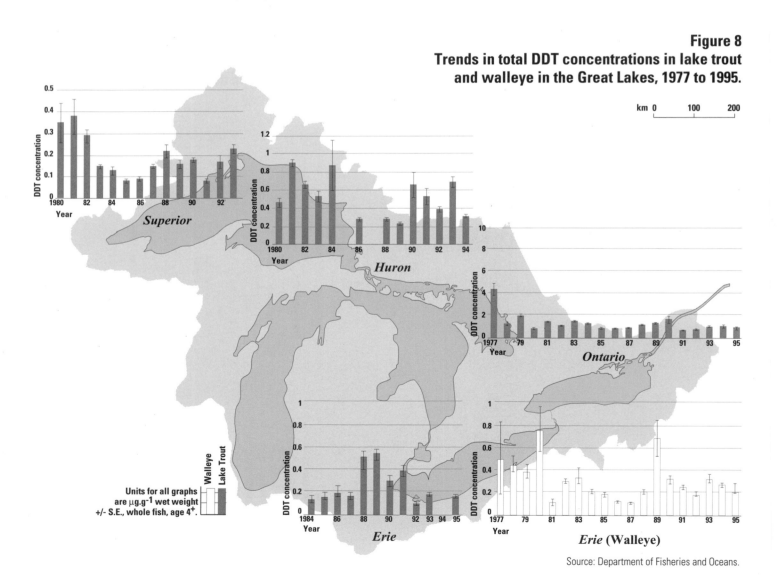

Figure 8
Trends in total DDT concentrations in lake trout and walleye in the Great Lakes, 1977 to 1995.

Source: Department of Fisheries and Oceans.

DDT, present concentrations are just below the IJC Objective of 1.0 $\mu g.g^{-1}$ in Lake Ontario. Between 1977 and 1990, DDT concentrations in fish in Lakes Superior, Huron, Ontario and Erie also declined. The rate of reduction leveled off in Lake Superior lake trout in the mid 1980s, and in Lake Erie fish in the early 1980s (See Figure 8; (De Vault *et al.* 1995).

Rainbow smelt concentrations of DDT are highest in Lake Ontario, and have declined significantly in Lakes Superior, Huron, Erie and Ontario, as shown in Figure 9 (De Vault *et al.* 1995). For young-of-the-year spottail shiner annual collections, DDT declined significantly for most of the sites from the mid 1970s to 1990 (De Vault *et al.* 1995).

Total DDT and *p,p'*-DDE concentrations in Herring Gull eggs declined significantly between 1974 and the mid 1980s at most colonies sampled. Since then, the rate of decrease has slowed (De Vault *et al.* 1995; Figure 10). Ewins *et al.* (1994) reported that a survey of DDE, PCBs, and mirex in Caspian Tern eggs from sites on Lakes Ontario, Huron, and Michigan indicated markedly lower concentrations in the 1991 collections than reported by other authors for previous surveys at the same sites in the 1970s and from 1980 to 1981.

Mirex was initially detected in biota and sediments from Lake Ontario and downstream in the St. Lawrence River (Kaiser 1974, 1978). Later

Figure 9
Trends in total DDT concentrations in smelt in the Great Lakes, 1977 to 1995.

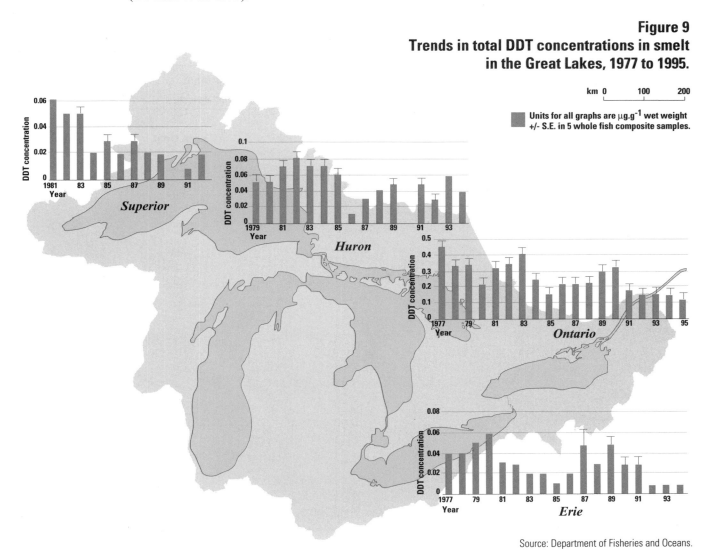

Source: Department of Fisheries and Oceans.

studies found low concentrations of mirex in fish and sediments of the upstream lakes — Erie, Huron and Superior (Sergeant *et al.* 1993; Swackhamer and Hites 1988). Concentrations measured in Lake Ontario have declined significantly since the compound was originally detected in the mid 1970s and have recently stabilized. However, concentrations still exceed the IJC GLWQA Objective of 0.1 $\mu g.g^{-1}$ for fish.

Hexachlorobenzene is present in fish throughout the Great Lakes, with the highest concentrations found in samples from Sarnia, Ontario.

Concentrations of the pesticide toxaphene in fish are highest in Lake Superior and have recently triggered fish consumption advisories. Concentrations of the pesticide dieldrin in fish have not decreased as quickly as other pesticides have, with decreases noted in samples from Lakes Ontario and Superior, but no clear trend in samples from Lake Huron. The explanation for this is still unknown.

An analysis of fish collected annually between 1977 and 1990 from each of the Great Lakes for the most toxic isomers of PCDDs and PCDFs found that concentrations in fish from Lake Ontario were higher than the concentrations

Table 6   PCDD and PCDF isomer concentrations in Great Lakes fish in 1990.

	Isomer concentration (pg.g⁻¹ whole fish wet weight)			
	Lake Ontario		Lake Superior	
Isomer	**Lake trout**	**Forage fish**	**Lake trout**	**Forage fish**
2,3,7,8-T_4CDD	44	9.6	2.8	<2.0
1,2,3,7,8-P_5CDD	<2.0	<2.0	<1.0	<2.0
1,2,3,4,7,8-H_6CDD	6.7	<3.0	<1.0	<3.0
1,2,3,6,7,8-H_6CDD	<3.0	<3.0	<2.0	<3.0
1,2,3,4,6,7,8-H_7CDD	<4.0	<4.0	<2.0	<4.0
OCDD	<7.0	<7.0	<4.0	<7.0
2,3,7,8-T_4CDF	72	30	21	2.5
1,2,3,7,8-P_5CDF	14	6.1	1.6	<1.0
2,3,4,7,8-P_5CDF	39	11	7.5	<1.0
1,2,3,4,7,8-H_6CDF	18	8.3	<1.0	<2.0
2,3,4,6,7,8-H_6CDF	2.6	<2.0	<1.0	<2.0
1,2,3,4,6,7,8-H_7CDF	<3.0	<3.0	<2.0	<3.0
OCDF	<6.0	<6.0	<3.0	<6.0

Source: Whittle *et al.* 1992.

detected in Lake Superior fish. Table 6 shows PCDD and PCDF isomer concentrations in top predator fish collected from both lakes in 1990. Top predator fish (lake trout) had significantly higher concentrations of PCDDs and PCDFs than the prey fish they feed on, suggesting biomagnification occurs as these contaminants move up the food chain (Whittle *et al.* 1992). Retrospective analysis of archived samples of whole lake trout from 1978 through 1993 detected no significant decline in the concentrations of 2,3,7,8,-tetrachlorodibenzo-dioxin (2,3,7,8-TCDD) as shown in Figure 11. Niagara River water samples tested during the

same period showed decreased concentrations of total organic contaminant loadings, including PCDDs (Kuntz 1993). One possible explanation for the lack of decline of PCDDs in fish is that remobilization of in-place contaminants may be occurring (Whittle *et al.* 1992). Another is that shifts in the composition of food webs may affect contaminant transfer up the food chain (De Vault *et al.* 1995).

In general though, declines of PCDDs in whole lake trout samples collected from 1978 to 1992 have been observed in many of the other Great Lakes (Table 7a and 7b). PCDF concentrations

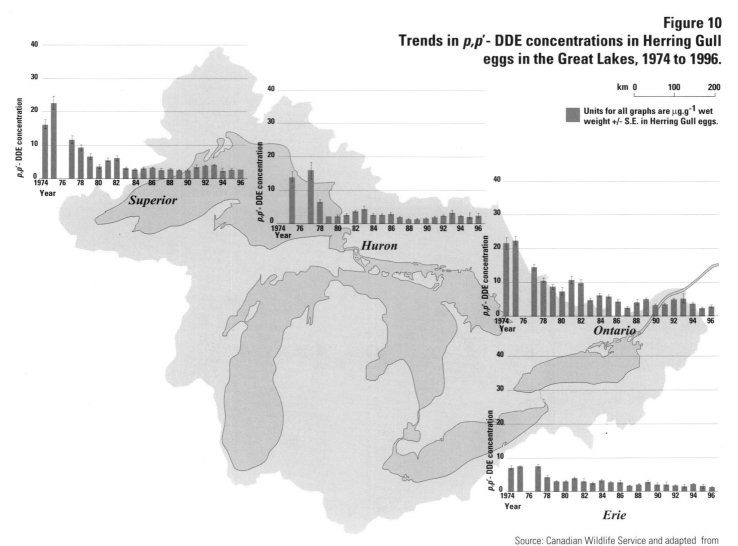

Figure 10

Trends in *p,p'*- DDE concentrations in Herring Gull eggs in the Great Lakes, 1974 to 1996.

km 0   100   200

Units for all graphs are μg.g^{-1} wet weight +/- S.E. in Herring Gull eggs.

Source: Canadian Wildlife Service and adapted from Bishop *et al.* 1992, and Petit *et al.* 1994.

follow the same general pattern, except for Lakes Superior and Ontario in 1990 and 1992.

The Canadian Wildlife Service of Environment Canada found detectable concentrations of nine isomers of PCDDs and PCDFs in eggs taken from 15 Herring Gull colonies from 1981 to 1991 (Hébert *et al.* 1994). The test locations included colonies in all five Great Lakes and the St. Lawrence River. Among the isomers, the most toxic form of PCDD — 2,3,7,8-TCDD — was found present in all samples tested (Figure 12).

PCDD concentrations in Herring Gull eggs were found to have generally declined between 1971 and 1984. There was little change noted thereafter. PCDFs which were first tested in 1984 declined in some colonies between 1984 and 1987. PCDF concentrations remained relatively unchanged thereafter, except that they increased steadily at Gull Island (Lake Michigan) between 1987 and 1990. The short term fluctuations observed tended to be found for all isomers and lakes, and were therefore attributed to climate fluctuations affecting food availability for gulls.

OTHER AROMATIC COMPOUNDS

PAHs, such as fluoranthene and the carcinogen benzo[a]pyrene (B[a]P), are most concentrated near the industrialized areas of the Great Lakes, and are particularly associated with coking facilities as reported by Baumann and Whittle (1988). The greatest concentration of PAHs in the waters of the Great

Lakes is in Hamilton Harbour on Lake Ontario. An analysis of large volume surface water samples showed concentrations of three PAHs (phenanthrene, B[a]P and fluoranthene) exceeded guidelines for the protection of aquatic life in both 1988 and 1990 at the majority of sampling stations in Hamilton Harbour. The Niagara and Oswego Rivers, Black River Bay and Toronto Harbour were other major sources of PAHs to Lake Ontario (L'Italien 1993). The 1988 sampling found lakewide concentrations of PAHs in water to be six times higher in Lake Ontario than in Lake Huron (L'Italien 1993). This finding is probably related to the prevalence of combustion sources near the higher concentrations. Sampling in 1990 found PAH concentrations offshore of Grimsby, Ontario, near Hamilton, to be very high. However, concentrations of phenanthrene, B[a]P and fluoranthene in four other Lake Ontario sites showed dramatic reductions between 1988 and 1990 (L'Italien 1993).

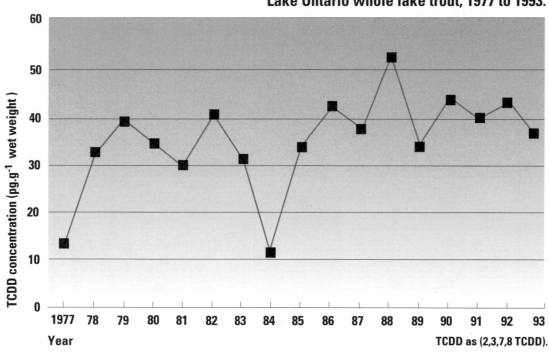

Figure 11
Trends in TCDD concentrations in Lake Ontario whole lake trout, 1977 to 1993.

TCDD as (2,3,7,8 TCDD).

Source: Department of Fisheries and Oceans unpublished data.

Because B[a]P is metabolized by fish and other vertebrates, it is not routinely monitored in fish tissue and studies indicate very low concentrations in most Great Lakes fish (De Vault *et al.* 1995). Studies have indicated that increased frequency of neoplasms or tumours in fish may be associated with exposure to elevated concentrations of PAHs (Maccubbin *et al.* 1985). In 1988, Baumann and Whittle reported that the stomach contents (mainly small clams and snails) of white suckers contained total PAH concentrations in excess of 700 ng.g^{-1} at sites with a high frequency of neoplasms in bottom feeding fish from the same Great Lakes sites.

ECOLOGICAL CONSEQUENCES

Surveys have described the use of biochemical indicators such as the activity of mixed-function oxidases (MFO), particularly the aryl hydrocarbon hydroxylase (AHH) enzyme in fish, as a response indicator of exposure to hydrocarbons (Luxon *et al.* 1987). That study described a 6 to 62-fold higher AHH enzyme activity in lake trout from the industrialized basin of Lake Ontario compared to enzyme activities measured in other less contaminated basins of Lake Huron, Lake Superior and an isolated pristine control lake within the Great Lakes Basin. Other studies have focused on the biochemical and genotoxic effects associated with the exposure to effluents from specific industries such as pulp mills in the Great Lakes Basin (Wong *et al.* 1994). Studies by Munkittrick *et al.* (1992, 1994) (See Case Study: Impact of Pulp and Paper Mill Discharges in Jackfish Bay, Lake Superior) documented a series of physiological and biochemical effects in fish

Table 7a   **2,3,7,8-TCDD concentrations in whole lake trout from Lakes Superior, Michigan, Huron and Ontario and in walleye from Lake Erie.**

2,3,7,8 TCDD concentration (pg.g^{-1})				
1978	**1984**	**1988**	**1990**	**1992**
Superior 2	1	DL	3	2
Michigan 7	5	3	NA	3
Huron 22	9	20	NA	9
Erie 3	2	DL	NA	2
Ontario 79	49	22	44	40

Table 7b   **2,3,7,8-TCDF concentrations in whole lake trout from Lakes Superior, Michigan, Huron and Ontario and in walleye from Lake Erie.**

2,3,7,8-TCDF concentration (pg.g^{-1})				
1978	**1984**	**1988**	**1990**	**1992**
Superior 33	15	7	21	24
Michigan 27	40	13	NA	16
Huron 32	23	11	NA	12
Erie 25	11	8	NA	16
Ontario 55	19	9	73	40

NA = Not analyzed
DL = Below limit of detection

Source: DeVault *et al.* 1995.

communities downstream of bleached kraft pulp mill discharges. Bottom-dwelling fish such as brown bullhead and white sucker have developed liver tumours. The specific causes are unknown but there is strong circumstantial evidence that tumour frequency increases at sites contaminated with known carcinogens (See Case Study: The Occurrence of Tumours in White Suckers from the Great Lakes). In addition, PAHs may have the potential to induce cancer in wildlife at low concentrations (L'Italien 1993).

Fish stocks near urban areas have declined during this century in variety and species mix and numbers due to extensive habitat loss, over harvesting, eutrophication and the acute and sub-lethal effects of chemical contamination (Egerton 1985). Recent studies on lake trout populations indicated that it was unlikely that concentrations of major Great Lakes contaminants such as PCBs, DDT, PCDDs and PCDFs were the principal cause of early life stage mortalities (Fitzsimons 1996).

In general, it has not been possible to separate the ecological effects of each chemical from the complex mixtures of chemicals in the Great Lakes. There are also biological and physical stresses which interact with the chemically based stresses to affect aquatic organisms. But chemical contaminants may be involved in reproductive failure, tumour development, other pathological anomalies, and a large number of physiological and biochemical changes whose biological significance is not yet fully understood.

Conclusions

The Great Lakes ecosystem is complex and diverse. It contains a waterway shared by two countries and large urban population centres with waste treatment facilities and major concentrations of industrial and manufacturing activities. The region generates considerable amounts of effluents containing a wide variety of chemical contaminants. The Great Lakes also serve as a transportation corridor for a high volume of international shipping.

More than 300 anthropogenic chemicals have been identified in the waters, sediment and fish of the Great Lakes. More than one-third of these are classified as toxic and can have both lethal and sublethal effects on plant and animal life. Lake Ontario and Lake Erie are the most heavily industrialized of the Great Lakes and are generally the most contaminated. The upper lakes, including Lake Huron and Lake Superior, are less contaminated. Major sources of contamination, besides industrial and municipal effluent discharges, include agricultural run-off and materials lost from historical chemical waste dumps. Atmospheric sources and transboundary

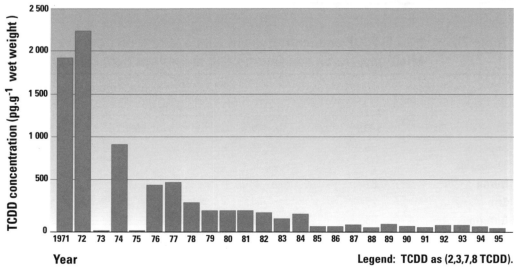

Figure 12
Trends in TCDD concentrations in Herring Gull eggs in eastern Lake Ontario, 1971 to 1995.

Legend: TCDD as (2,3,7,8 TCDD).

Source: C. Weseloh, Canadian Wildlife Service, Hull, Quebec.

movement of chemicals from U.S. based industrial complexes and chemical waste storage facilities also add contaminants to the Great Lakes ecosystem.

The pulp and paper industry has a concentration of mills on the shores of the upper Great Lakes. Formerly, a large volume of chemicals was discharged from these facilities. However, recent pollution abatement measures have reduced the amounts of chemicals released from the mills. There are also ongoing joint studies involving the industry as well as environmental agencies to minimize or eliminate the discharge of those chemicals considered toxic.

Point source loadings of chemical contaminants to the Great Lakes have declined over the past two decades. The marked improvement is a result of implementation of pollution abatement measures required by binational agreements such as the Canada-U.S. GLWQA and the Niagara River Toxic Chemical Management Plan. These declines have been seen for industrial chemicals such as PCBs, pesticides (including DDT and mirex) and metals such as mercury.

The rapid decline in contaminant concentrations in the lower Great Lakes system has recently slowed. However, present concentrations are still many times lower than those measured in the 1970s. The slowing decline or the reversal of the downward trend at some sites is reason for continued assessment of the situation. This general downward trend in contaminant burdens in fish is similar to long term patterns of reductions in consumption advisories for sport fish. However, despite this general downward trend, there has been a modest increase in the number of advisories issued for some sport fish species in all lakes since 1993.

For many contaminants of the Great Lakes, documented point source loadings have declined. Increases in concentrations in the ecosystem must therefore result from either internal redistribution or from sources located outside the Great

Lakes Basin. One explanation for the internal redistribution of contaminants is that the invasion and proliferation of zebra mussels has caused dramatic changes in the composition and functioning of some Great Lakes ecosystems (See Case Study: Increases in Contaminant Burdens Related to the Invasion of Lake Erie by Zebra Mussels). The large volume of international shipping traffic on the Great Lakes-St. Lawrence River system has provided a significant vector for the introduction of exotic species via the exchange of ballast water within the systems. As a result, contaminant dynamics within these systems have changed. Previously unavailable contaminant sources in sediments and elsewhere are now being incorporated into the food web. These changes in pathways have resulted in slight increases of contaminant concentrations in some fish communities.

The proliferation of accidentally introduced fish species such as the ruffe and goby have also disrupted the function of some Great Lakes ecosystems. It is unclear at this time what impact these accidental introductions will have on the movement of chemical contaminants within the ecosystem and their effects.

In 1995, consumption advisories were issued for some species of Lake Superior fish as a result of toxaphene contamination. Since toxaphene is not registered for use in Canada, and identified sources exist outside the Great Lakes Basin, this contaminant has been most likely introduced into the Lake Superior system by LRTAP.

Experiments identified reproductive effects on fish exposed to complex organic effluents. Further studies will need to be directed at identifying the specific causative agents. Additional efforts should focus on the complex issue of endocrine disruptors.

Continued research efforts will be required to identify specific chemicals and describe the significance of their effects on the form and function of the Great Lakes freshwater ecosystem.

Remediation efforts to restore ecosystems are underway at many sites throughout the Great Lakes Basin. Much of this remediation effort is directed at eliminating or minimizing the impact of chemical contaminants present at these sites.

Many of the contaminant issues documented within the Great Lakes ecosystem have the potential to impact on the downstream aquatic communities of the St. Lawrence River, the St. Lawrence Estuary, and ultimately the Gulf of St. Lawrence (See the St. Lawrence Marine Ecosystem Chapter).

The effects of climate warming, the continued introduction and proliferation of exotic species, and the cumulative impact of as yet unidentified chemicals are all part of the future of the Great Lakes ecosystem. These remaining undefined impacts on the ecosystem will influence the direction of both scientific investigations and management of chemical contaminants.

Case study:

INCREASES IN CONTAMINANT BURDENS RELATED TO THE INVASION OF LAKE ERIE BY ZEBRA MUSSELS

The invasion and rapid proliferation of zebra mussels has caused dramatic changes in the physical, chemical and biological functioning of the Lake Erie ecosystem. It is thought that zebra mussels were introduced into Lake St. Clair in 1986 via ballast water discharge from a European vessel. By the time the mussel was discovered in 1988, it was abundant throughout Lake St. Clair and northern Lake Erie. Zebra mussels adapted readily to their new environment and quickly spread to many of the Great Lakes. Of the Great Lakes, Lake Erie provided the most ideal habitat for these mussels. Abundant plankton (the food of zebra mussels) and rocky reefs (the preferred substrate of mussels) enabled these molluscs

to reach the highest recorded population densities in the world.

Most changes to the Lake Erie ecosystem attributed to zebra mussels are due to the filter-feeding activity of large mussel populations, which remove plankton and other particulate matter from the water column. Vast numbers of mussels deposit feces and pseudofeces, the products of digestion and rejected food particles, on the lake bottom. These carbon-rich deposits provide food for bottom dwelling (benthic) organisms, and the population of the benthic organisms has increased. Unfortunately in Lake Erie, the plankton digested and excreted by mussels also contain chemical contaminants (Bruner *et al.* 1994a). Some of these contaminants accumulate in the tissues of mussels and some get deposited in feces and pseudofeces only to be ingested by benthic organisms feeding on these deposits (Bruner *et al.* 1994b). As a food source for top predators, including fish, benthic organisms act as a biological conduit for contaminants.

As a result of these processes, Lake Erie has been transformed from a system where a significant portion of the chemical contaminants were carried in the water columns to a system dominated by benthic organisms and benthic pools of contaminants. Chemical contaminants, which previously were flushed rapidly through the lake, are scavenged from the water column and trapped in a flourishing benthic layer. This benthic layer then becomes a focal point for contaminant accumulation and transfer of contaminants into the food web. The fish community in Lake Erie is shifting towards lower valued benthic feeding fish while valued top predators are declining in numbers and becoming more contaminated. The increased contaminant burdens in fish have been caused by a change in the pathway of contaminants through the ecosystem, rather than by an increase in contaminant loadings.

Herring Gull
(Larus argentatus)

Through most of its Canadian range and in the northeastern United States, the Herring Gull is a very common bird. In Canada, it can be found in every province from the Atlantic to the Pacific Ocean as well as inland along lakes and rivers. Herring Gulls nest in colonies at a variety of sites. On offshore islands they frequently occupy flat ground. On the coastal mainland they tend to nest on cliffs, mainly in an effort to avoid predatory mammals. In some places where food from human activities is very abundant, they have begun to nest on roofs and window ledges of buildings. In most areas, a clutch of three eggs has been laid by mid May and incubation begins. Although eggs are well protected, egg and chick mortality is usually high. Herring Gulls seek their food as close to their breeding colony as possible but they may travel great distances if a food source is plentiful. Herring Gulls eat almost anything, including clams, small fish, floating dead animals, young and adults of other nesting birds, bread, french-fried potatoes, and so on. Individual Herring Gulls, however, tend to specialize on particular types of food and feeding techniques, with some birds regularly visiting garbage dumps while others feed entirely on fish and crabs found along the shore.

In the mid 1960s Herring Gulls were one of the first wildlife species found on the Great Lakes to be showing severe biological effects of toxic chemicals. During the 1960s and early 1970s Herring Gulls in Lakes Michigan and Ontario showed elevated concentrations of embryonic mortality and severely depressed productivity; reproductive failure was widespread. In the early 1970s, because they were year-round residents of the Great Lakes and because they accumulated toxic chemicals to relatively high concentrations, Herring Gulls were selected as the main indicator species for a long-term monitoring program of contaminant concentrations and their effects on wildlife. Through this program dramatic declines in contaminant concentrations (up to 90% for some compounds) were well documented in the upper levels of the Great Lakes food web in the mid to late 1970s. Analysis of Herring Gull eggs in 1980 led to the discovery of 2,3,7,8 TCDD in the Great Lakes ecosystem. By the 1980s, however, Herring Gull productivity had returned to normal with concentrations of many compounds continuing to decrease while others stabilized. The Herring Gull Egg Monitoring Program continues to serve as an indicator of responses by wildlife to contaminants in the environment.

Case study:

IMPACT OF PULP AND PAPER MILL DISCHARGES IN JACKFISH BAY, LAKE SUPERIOR

Studies conducted in Jackfish Bay, Lake Superior during the late 1980s found that fish exposed to primary treated effluent of the bleached kraft pulp mill at Terrace Bay exhibited a variety of reproductive changes, including an increased age to maturity, smaller gonads, lower fecundity with age and an absence of secondary sex characteristics in males. In addition, females failed to show an increase in egg size with age. These findings represented early indications that chemical contaminants could disrupt the hormonal systems of fish. Ongoing studies are trying to evaluate the ecological significance of these biochemical changes.

Many Canadian pulp and paper mills are implementing process changes to comply with new regulations which came into effect in 1993. These regulations placed stricter limits on effluent discharges, including limits on acute toxicity and biological oxygen demand. To meet these regulations, many of the older Canadian pulp and paper mills are installing secondary treatment systems to treat the effluent.

The pulp mill at Terrace Bay initiated secondary treatment of their effluent in October of 1989. However, no improvement in reproductive parameters in the first few years after secondary treatment was detected. This finding prompted a study of 10 additional Ontario mills which had a variety of pulping processes and waste treatment strategies. These studies found that downstream of several pulp mills induction of liver detoxification enzymes was stimulated and sex steroid hormone levels were decreased. These changes were also seen at some mills that did not use chlorine bleaching, and at mills which had secondary effluent treatment.

Field testing conducted in 1993 identified the disappearance of steroid hormone depressions in fish exposed to effluent collected from Terrace Bay, presumably linked to some changes in mill processes. These findings were confirmed in laboratory tests conducted in the fall of 1993 and in the spring of 1994. Several studies are continuing in an attempt to determine why the disappearance of steroid depressions took place.

Case study:

THE OCCURRENCE OF TUMOURS IN WHITE SUCKERS FROM THE GREAT LAKES

Prior to 1970, there were occasional reports of tumors in Great Lakes fish, but no systematic surveys had been conducted to determine if the presence of tumors in fish represented a risk to fish populations. There was very little background information and it was not understood how tumor prevalence was influenced by fish age, gender, diet and chemical exposure. Scientists suspected that some fish were sensitive to chemical carcinogens, and postulated that fish tumour surveys would be a useful tool for assessing the health of fish populations. Such surveys could also serve to identify sites contaminated with chemical carcinogens.

Early surveys reported a high prevalence of skin tumours in walleye, pike, muskellunge, brown bullhead and white suckers. Viruses were implicated as the agents responsible for skin tumours in walleye, pike and muskellunge, but there was no obvious cause for the tumours in the two bottom-dwelling species. Surveys of tumor incidence in white sucker populations in Lakes Ontario and Huron conducted in the late 1970s, showed an increase in lip and body papillomas in western Lake Ontario (30 to 50%). Papilloma frequency increased with age and was independent of gender. There were no adverse affects of the papilloma on fish growth, year class strength, reproduction or migratory behavior. White sucker populations were affected with liver disease and at least two types of liver tumours. Both tumours are thought to be caused by chemical carcinogens.

Surveys are being repeated for the second time in 10 years to determine if their frequency has changed since the first observations. Most of the individuals present in the original survey have been replaced by the next generation of white suckers and presumably reflect changes in the environmental quality that have occurred over the past 15 years. The completed study will identify changes in tumour frequency and sites contaminated with chemical carcinogens. Results from Hamilton Harbour indicated that the frequency of lip papillomas has not changed from the high rate reported in the late 1970s. The frequency of body papillomas increased from 3% in 1979 to 23% in 1994.

References

ALLARDICE, D.R., and S. Thorpe. 1995. A changing Great Lakes economy and environmental linkages. State of the Lakes Ecosystem Conference. Background Paper. Environment Canada/United States Environmental Protection Agency. EPA 905-R-95-017: 50 p.

BAUMANN, P.C., and D.M. Whittle. 1988. The status of selected organics in the Laurentian Great Lakes: an overview of DDT, PCBs, dioxins, furans and aromatic hydrocarbons. Aquatic Toxicology 11: 241-257.

BISHOP, C.A., D.V. Weseloh, N.M. Burgess, J. Struger, R.J. Norstrom, R. Turle, and K.A. Logan. 1992. An Atlas of contaminants in eggs of fish-eating colonial birds of the Great Lakes. (1970-1988) Vol I. Technical report Series No. 152, Canadian Wildlife Service, Ontario Region.

BRUNER K.A., S.W. Fisher, and P. F. Landrum. 1994a. The role of the zebra mussel, *Dressenia polymorpha*, in contaminant cycling I: The effect of body size and lipid content on the bioconcentration of PCBs and PAHs. J. Great Lakes Res. 20(4): 725-734.

BRUNER K.A., S.W. Fisher, and P. F. Landrum. 1994b. The role of the zebra mussel, *Dressenia polymorpha*, in contaminant cycling II: zebra mussel contaminant accumulation from algae and suspended particles, and transfer to the benthic invertebrate, *Gammarus fasciatus*. J. Great Lakes Res. 20(4): 735-750.

CANADA-Ontario Agreement Respecting the Great Lakes Basin Ecosystem. 1994. 11 p.

CHAU, Y.K., P.T.S. Wong, G.A. Bengert, J.L. Dunn, and B. Glen. 1985. Occurrence of alkyllead compounds in the Detroit and St. Clair Rivers. J. Great Lakes Res. 11 (3): 313-319.

COHEN, M., B. Commoner, H. Eisl, P. Bartlett, A. Dickar, C. Hill, J. Quigley, and J. Rosenthal. 1995. Quantitative estimation of the entry of dioxins, furans and hexachlorobenzene into the Great Lakes from airborne and waterborne sources. Center for the Biology of Natural Systems. Queens College, New York. Report Submitted to Joyce Foundation.

DE VAULT, D.S., P. Bertram, D.M. Whittle, and S. Rang. 1995. Toxic contaminants in the Great Lakes. State of the Lakes Ecosystem Conference. Background Paper. Environment Canada/United States Environmental Protection Agency. EPA 905-R-95-016: 59 p.

DEPARTMENT of Fisheries and Oceans. 1994. Annual Summary of Commercial Freshwater Fish Harvesting Activities 1992. Winnipeg, Manitoba.

EGERTON, F.N. 1985. Overfishing or pollution? Case history of a controversy on the Great Lakes. Great Lakes Fishery Commission. Technical Report No. 41: 28 p.

ENVIRONMENT Canada and the U.S. Environmental Protection Agency. 1995. State of the Great Lakes. ISBN 0-662-61887-4/Catalogue En40-11/35. Burlington, Chicago.

EWINS, P.J., D.V. Weseloh, R.J. Norstrom, K. Legierse, H.J. Auman, and James P. Ludwig. 1994. Caspian Terns on the Great Lakes: organochlorine contamination, reproduction, diet, and population changes, 1972-1991. Environment Canada, Canadian Wildlife Service, Occasional Paper Number 85. Ottawa, Ontario.

FITZSIMONS, J. F. 1996. A critical review of the effects of contaminants on early life stage (ELS) mortality of lake trout in the Great Lakes. J. Great Lakes Res. 21 (Supplement 1): 267-276.

FOX, M.E., R.M. Khan, and P.A. Theissen. 1996. Loadings of PCBs and PAHs from Hamilton Harbour to Lake Ontario. Water Qual. Res. J. Canada 3: 593-608.

FRANK, R., H. Thomas, H. Holdrinet, R.K. McMillan, H.E. Braun, and R. Dawson. 1981. Organochlorine residues in suspended solids collected from the mouths of Canadian streams flowing into the Great Lakes 1974-1977. J. Great Lakes Res. 7 (4): 363-381.

GOVERNMENT of Canada. 1991a. The State of Canada's Environment. Chapter 18.

GOVERNMENT of Canada. 1991b. Toxic Chemicals in the Great Lakes and Associated Effects. 3 volumes: 755 p.

GOVERNMENT of Ontario. 1995. Guide to Eating Ontario Sport Fish 1995-1996. Queen's Printer for Ontario. 173 p.

HEBERT, C.E., R.J. Norstrom, M. Simon, B.M. Braune, D.V. Weseloh, and C.R. Macdonald. 1994. Temporal trends and sources of PCDDs and PCDFs in the Great Lakes: herring gull egg monitoring, 1981-1991. Environmental Science and Technology 28 (7): 1268.

HODSON, P.V., D.M. Whittle, P.T.S. Wong, U. Borgmann, R.L. Thomas, Y.K. Chau, J.O. Nriagu, and D. J. Hallet. 1984a. Lead contamination of the Great Lakes and its potential effects on aquatic biota, p.336-364. *In* J.O. Nriagu, and M.S. Simons [eds.] Toxic contaminants in the Great Lakes. John Wiley and Sons, Inc.

HODSON, P.V., B.R. Blunt, and D.M. Whittle. 1984b. Suitability of biochemical methods for assessing the exposure of feral fish to lead, p. 389-405. *In* W.E. Bishop, R.D. Cardwell, and B.B. Heidolph [eds.] Aquatic Toxicology and Hazard Assessment: Sixth Symposium. ASTM STP802.

HOFF, R.M., W.M.J. Strachan, C.W. Sweet, C.H. Chan, M. Shackleton, T.F. Bidleman, K.A. Brice, D.A. Burnison, S. Cussion, D.F. Gatz, K. Harlin, and W.H. Scroeder. 1996. Atmospheric deposition of toxic chemicals to the Great Lakes: A review of data through 1994. Atmospheric Environment. 30: 3505-3527.

HUESTIS, S.Y, M.R. Servos, D.M. Whittle, and D.G. Dixon. 1996. Temporal and age related concentration trends for polychlorinated biphenyl congeners and organochlorine contaminants in Lake Ontario lake trout. J. Great Lakes Res. 22: 310-330.

INTERNATIONAL Joint Commission. 1987. 1987 Report on Great Lakes Water Quality. Windsor, Ontario.

INTERNATIONAL Joint Commission. 1988. Mass balancing of toxic chemicals in the Great Lakes: the role of the atmosphere. Windsor, Ontario.

INTERNATIONAL Joint Commission. 1991. Cleaning up the Great Lakes: a report from the Water Quality Board to the IJC on toxic substances in the Great Lakes Basin. Windsor, Ontario.

INTERNATIONAL Joint Commission. 1993. Report of the Virtual Elimination Task Force to the International Joint Commission. Windsor, Ontario. 2 volumes.

KAISER K.L.E. 1974. Mirex: an unrecognized contaminant of fishes from Lake Ontario. Science 185: 523-525.

KAISER, K.L.E. 1978. The rise and fall of Mirex. Environ. Sci. Technol. 12: 520-524.

KUNTZ, K.W. 1993. Trends in the contaminant levels in the Niagara River. A State of the Environment Fact Sheet. No. 93-2. Environment Canada, Burlington, Ontario. 12 p.

L'ITALIEN, S. 1993. Organic Contaminants in the Great Lakes 1986-1990. Report No: EQB/IWD-OR/93-02-1. Environment Canada, Environmental Quality Branch, Ontario Region, Burlington, Ontario.

LTI, Limno-Tech, Inc. 1993. Great Lakes Environmental Assessment. Ann Arbor, Michigan. 268 p.

LUXON, P.L., P.V. Hodson, and U. Borgmann. 1987. Hepatic Aryl Hydrocarbon Hydroxylase Activity of Lake Trout (*Salvelinus namaycush*) as an indicator of organic pollution. Env. Tox. and Chem. Vol 6: 649-657.

MAC, M.J., T.R. Schwartz, C.C. Edsall, and A.M. Frank. 1993. Polychlorinated biphenyls in Great Lakes lake trout and their eggs: Relations to survival and congener composition 1979-1988. J. Great Lakes Res. 19(4): 752-765.

MACCUBBIN, A.E., P. Black, L. Treciak, and J.J. Black. 1985. Evidence of polynuclear aromatic hydrocarbons in the diet of bottom feeding fish. Bull. Environ. Contam. Toxicol. 34: 876-862.

MAGUIRE, R.J., R.J. Tkacz, Y.K. Chau, G.A. Bengert, and P.T.S. Wong. 1986. Occurrence of organotin compounds in water and sediment in water and sediment in Canada. Chemosphere 15, 253-274.

MUNKITTRICK, K.R., G.J. Van Der Kraak, M.E. McMaster, and C. Portt. 1992. Reproductive dysfunction and MFO activity in three species of fish exposed to bleached kraft mill effluent at Jackfish Bay, Lake Superior. Water Pollut. Res. J. of Can. 27(3): 161-168.

MUNKITTRICK, K.R., G.J. Van Der Kraak, M.E. McMaster, C.B. Portt, M.R. van den Heuvel, and M.R. Servos. 1994. Survey of receiving water environmental impacts associated with discharges from pulp mills. 2. Gonad size, liver size, hepatic EROD activity and plasma sex steroid levels in white sucker. Environ. Toxicol. Chem. 13: 1089-1101.

PETIT, K.E., C.A. Bishop, D.V. Weseloh, and R.J. Norstrom. 1994. An atlas of contaminants in eggs of fish-eating colonial birds of the Great Lakes (1989-1992) Vol I. Technical Report Series No. 193, Canadian Wildlife Service, Ecosystems Health Branch, Ontario Region.

REYNOLDSON, T.B., K.E. Day, and T. Pascoe. 1997. A summary report on Biological Sediment Guidelines for the Laurentian Great Lakes. NWRI Report No. 97-134.

SERGEANT, D.B., M. Munawar, P.V. Hodson, D.T. Bennie, and S.Y. Huestis. 1993. Mirex in the North American Great Lakes: New detections and their confirmation. J. Great Lakes Res. 19(1): 145-157.

STEVENS, R.J.J., and M.A. Neilson. 1989. Inter- and intra-lake distributions of trace organic contaminants in surface waters of the Great Lakes. J. Great Lakes Res. 15 (3): 377-393.

SUNS, K., G.E. Crawford, D.D. Russell, and R.E. Clement. 1985. Temporal and Spatial Trends of Organochlorine and Mercury Residues in Great Lakes Spottails shiners (1975-1983). Ontario Ministry of the Environment, Water Resources Branch and Laboratory Services Branch Report. 43 p.

SUNS, Karlis R., G. G. Hitchin, and D. Toner. 1993. Spatial and temporal trends of organochlorine contaminants in spottail shiners from selected sites in the Great Lakes (1975-1990). J. Great Lakes Res. 19(4): 703-714.

SWACKHAMER, D.L., and R.A. Hites. 1988. Occurrence and bioaccumulation of organochlorine compounds in fishes from Siskiwit Lake Isle Royale, Lake Superior. Environ. Sci. Technol. 22: 543-548.

THOMAS, R.L. 1974. The distribution and transport of mercury in the sediments of the Laurentian Great Lakes System. Proceedings of the International Conference on Transport of Persistent Chemicals in Aquatic Ecosystems. Ottawa, Ontario. 16 p.

WHITTLE, D.M., D.B. Sergeant, S.Y. Huestis, and W.H. Hyatt, 1992. Foodchain accumulation of PCDD and PCDF isomers in the Great Lakes aquatic community. Chemosphere 25: 181.

WONG, P.T.S., Y.K. Chau, J. Yaromich, P. Hodson, and M. Whittle. 1988. Alkyllead Contaminations in the St. Lawrence River and St. Clair River (1981-1987). Canadian Technical report of Fisheries and Aquatic Sciences. No. 1602.

WONG, P.T.S., Y.K. Chau, J. Yaromich, P. Hodson, and M. Whittle. 1989. The analysis of alkyllead compounds in fish and environmental samples in Ontario, Canada (1981-1987). Applied Organometallic Chemistry 3: 59-70.

WONG, P.T.S., Y.K. Chau, N. Ali, and D.M. Whittle. 1994a. Biochemical and genotoxic effects in the vicinity of a pulp mill discharge. Environ. Toxicol. Water Qual. 9: 59-70.

WONG, P.T.S., Y.K. Chau, M. Brown, and D.M. Whittle. 1994b. Butyltin compounds in Severn Sound, Lake Huron, Canada. Applied Organometallic Chemistry 8: 385-391.

Additional reading

HODSON, P.V., K.R. Munkittrick, R. Stevens, and A. Colodey. 1996. A tier-testing strategy for managing programs of environmental effects monitoring. Water Pollut. Res. J. of Can. 31:215-224.

KIRILUK, R.M., M.R. Servos, D.M. Whittle, G. Cabana, and J.B. Rasmussen. 1995. Using ratios of stable nitrogen and carbon isotopes to characterize the biomagnification of DDE, mirex, and PCB in a Lake Ontario pelagic food web. Can. J. Fish. Aquat. Sci. 52: 2660-2674.

NIIMI, A.J. 1996. Evaluation of PCBs and PCDD/Fs retention by aquatic organisms. Sci. Total Environ. 192: 123-150.

MUNKITTRICK, K.R. 1992. A review and evaluation of study design considerations for site-specifically assessing the health of fish populations. Journal of Aquatic Ecosystem Health 1:283-293.

SERVOS, M.R. 1995. Origins of Effluent Chemicals and Toxicity: Recent Research and Future Directions. In Environmental Fate and Effects of Bleached Pulp Mill Effluents, M.R. Servos, J.H. Carey, K.R. Munkittrick, and G.J. Van Der Kraak [eds.] St. Lucie Press, Delray Beach, FL.

The INLAND *Freshwater* Ecosystems

Highlights

- Fresh water, as lakes and rivers, covers 8% of the total Canadian land mass. If wetlands are included, 20% of Canada is covered with fresh water. Excluding the Great Lakes, Canada contains approximately 16% of the total surface area of the world's fresh water.

- Canadian inland waters are important for commercial, subsistence and recreational fishing.

- Contaminants enter Canada's fresh water through industrial and agricultural activity, power production and municipal sewage.

- Long range transport of atmospheric pollutants (LRTAP) contributes to contamination in all parts of Canada. It is the main source of contaminants to Canada's northern fresh waters, where there are only a few major point sources.

- Commonly detected inorganic contaminants in fish and fish habitat are mercury, cadmium, lead and nickel.

- The predominant organic contaminants are polychlorinated biphenyls (PCBs), and the pesticides dichlorodiphenyltrichloroethane (DDT) and toxaphene. The presence of polychlorodibenzo-*p*-dioxins (PCDDs) and polychlorodibenzo-furans (PCDFs) in the aquatic environment may be a result of historical, direct discharges by bleached kraft pulp and paper mills. Polycyclic aromatic hydrocarbons (PAHs) are not of widespread concern in most inland freshwater systems, but may be important at specific locations.

Highlights

■ High concentrations of toxaphene in certain areas have resulted in the closure of some fisheries. Mercury is frequently responsible for consumption advisories for Ontario sport fish. In some areas, restrictions and limits have been placed on consumption of fish containing elevated concentrations of some chemical contaminants.

■ Many metallic and organic contaminants have declined in concentration in air, water, sediments and biota over the last several decades due to the implementation of pollution and regulatory controls. Nevertheless, the presence of persistent anthropogenic organic contaminants in the environment is unnatural and potentially significant.

■ Mercury is an exception to declining trends in that atmospheric deposition to Canada's aquatic ecosystems appears to be increasing with time. In Canada, most metals, including mercury, vary widely in natural concentrations because of highly variable geology. Therefore, contamination by metals must be evaluated in the context of site-specific background concentrations.

Table of Contents

Inland Freshwater Ecosystems

List of Figures

Inland Freshwater Ecosystems

List of Tables

List of Sidebars

Introduction

Fresh water is a renewable yet precious resource. It is essential to life and for many agricultural, commercial and industrial purposes. Canada has an abundance of freshwater resources. Lakes cover 8% of the Canadian land mass (Statistics Canada 1992), with enough water to flood the entire nation to a depth of 2 m (Statistics Canada 1994). If wetlands are included, 20% of Canada is covered in fresh water (Pearse *et al.* 1985).

Canada has twice as much fresh water available per person as any other country (Healey and Wallace 1987). The nation's river flows rank third in the world behind Brazil and Russia. Seven of the world's 20 largest lakes by area, including Superior, Huron, Great Bear, Great Slave, Erie, Winnipeg and Ontario are found wholly or partly within Canada's borders (Beeton 1984).

There are 25 river basin regions in Canada that drain into five different receiving waters, the Atlantic, Pacific, and Arctic Oceans, the Gulf of Mexico, and Hudson Bay (Figure 1; Table 1). The largest drainage basin is that of Hudson Bay. The basin extends around Hudson Bay and across the prairies, contributing approximately 29% of the total Canadian river flow (Pearse *et al.* 1985).

Canada's rivers and lakes supports commercial, recreational, and subsistence fisheries. Fresh waters are utilized also for domestic consumption

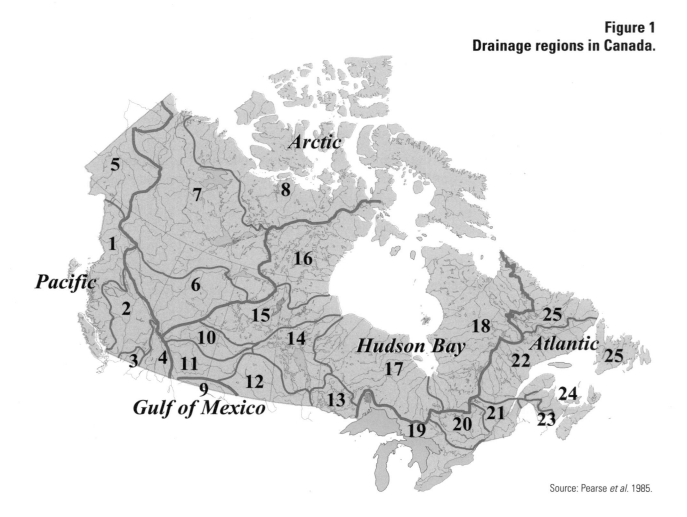

**Figure 1
Drainage regions in Canada.**

Source: Pearse *et al.* 1985.

and urban, industrial, and agricultural uses, as well as for shipping and transportation, and hydroelectric power generation. Canada's abundant freshwater ecosystems serve as aquatic and wildlife habitat, for recreational and aesthetic purposes (Healey and Wallace 1987), and comprise an important part of Canada's cultural and ecological heritage.

Despite the overall abundance of water, there is regional disparity. Quebec, Ontario, Manitoba and Saskatchewan all have well over 10% of their total surface area as fresh water. Prince Edward Island and the Yukon Territory are much poorer in freshwater resources; less than 1% of their total area is fresh water (Table 2). Moreover, water supplies do not always meet the demand because most users of water in the western half of Canada live in the south, along the boundary with the U.S., whereas most of the water supply in western Canada is north of 60° latitude. Numerous persistent contaminants from industrial and municipal effluents, and pesticides from agricultural activity tend to be released into

Table 1   Canadian river drainage basins, drainage areas, human population and mean annual river flows.

OCEAN BASIN Region/River Drainage Basin	Drainage Area (10^3 km²)	Human Population (thousands, 1981)	Total Mean Annual River Flows (m³ .sec⁻¹)
PACIFIC			
1. Pacific Coastal	350	620	16 000
2. Fraser-Lower Mainland	230	1 700	4 000
3. Okanagan-Similkameen	14	190	74
4. Columbia	90	160	2 000
5. Yukon	330	23	2 500
ARCTIC			
6. Peace-Athabasca	490	290	2 900
7. Lower Mackenzie	1 300	43	7 300
8. Arctic-Islands	2 000	13	10 000
GULF OF MEXICO			
9. Missouri	26	14	12
HUDSON BAY			
10. North Saskatchewan	150	1 100	230
11. South Saskatchewan	170	1 200	240
12. Assiniboine-Red	190	1 300	50
13. Winnipeg	110	80	750
14. Lower Saskatchewan- Nelson	360	220	1 900
15. Churchill	300	68	700
16. Keewatin	690	5	3 900
17. Northern Ontario	690	160	6 000
18. Northern Quebec	950	110	17 000
ATLANTIC			
19. Great Lakes	320	7 600	3 100
20. Ottawa	150	1 300	2 000
21. St. Lawrence	120	5 200	2 000
22. North Shore-Gaspé	400	650	8 700
23. St. John-St. Croix	37	390	780
24. Maritime Coastal	110	1 300	3 100
25. Newfoundland-Labrador	380	570	9 300

Source: Pearse *et al.* 1985.

southern waters compromising some consumptive uses, recreational fisheries, and other recreational uses. Not all of Canada's freshwater systems serve human uses equally well.

Canada's freshwater ecosystems naturally vary widely in water quality from highly coloured, organic-laden waters of the northern muskegs, to the white, chloride precipitate-laden playa lake waters in parts of the Prairies, to the pure, sediment-free run-off and lakes of the Precambrian Shield (Healey and Wallace 1987).

This chapter discusses chemical contaminants in Canada's inland waters, including those in the Arctic. Contaminant issues in the four Canadian Great Lakes are discussed in the chapter on the Great Lakes ecosystem. The diversity of aquatic ecosystems within the various freshwater drainage basins of Canada is immense. As well, many contaminant issues are specific to particular types of freshwater ecosystems. It is beyond the scope of this chapter to provide an exhaustive summary of the available information on contaminants associated with fresh waters and their fisheries resources. Therefore, this chapter focuses on selected contaminants of general relevance throughout most or all of the

ecosystems described. More detail on some contaminant issues or on specific localities is provided in case studies.

MAJOR CONTAMINANT ISSUES

There are a number of major contaminants associated with the various freshwater ecosystems in Canada. In northern and eastern areas on the Precambrian Shield, metals are released by non-ferrous mining, smelting and refining activities. These metals enter aquatic ecosystems through emission into the atmosphere followed by deposition, by direct release of effluents into water, or via drainage from tailing ponds. In reservoirs that have been created by hydroelectric development, methylmercury is produced by microbial activity. Contamination is also associated with general industrial development, including agriculture and human population centres. Numerous organochlorine compounds such as polychlorinated biphenyls (PCBs) and the dichlorodiphenyltrichloroethane- (DDT) group compounds are universally present in the fresh water environment. The presence of other organochlorine compounds, such as polychlorodibenzo-*p*-dioxins (PCDDs) and polychlorodibenzo-furans (PCDFs), in the aquatic environment may be the result of direct

Table 2   Provincial or territorial land and freshwater areas.

Province or Territory	Land Area (km²)	Freshwater Area (km²)	Largest Lake	Freshwater by percent
Newfoundland	370 000	34 000	Smallwood Reservoir	9.2
Prince Edward Island	5 700	—[1]		NA
Nova Scotia	53 000	2 700	Bras d'Or Lake	5
New Brunswick	72 000	1 400	Grand Lake	1.9
Quebec	1 400 000	180 000	Lake Mistassini	14
Ontario	890 000	180 000	Lake Nipigon[2]	20
Manitoba	550 000	100 000	Lake Winnipeg	19
Saskatchewan	570 000	82 000	Lake Athabasca	14
Alberta	640 000	17 000	Lake Clair	2.6
British Columbia	930 000	18 000	Williston Lake	1.9
Yukon Territory	480 000	4 500	Kluane Lake	1
Northwest Territories	3 300 000	130 000	Great Bear Lake	4.1

[1] Amount is too small to be expressed.
[2] Excluding the Laurentian Great Lakes.

Source: Statistics Canada 1994.

discharges to waterways by bleached kraft pulp and paper mills. Polycyclic aromatic hydrocarbons (PAHs) are not of general importance in most inland freshwater systems, but locally can be significant.

Long-range transport of atmospheric pollutants (LRTAP) is a major mechanism by which contaminants are distributed to all of Canada's freshwater environments, but particularly those in the north. All classes of contaminants, including organochlorine compounds and PAHs, many of which have never been used in the north, are detected in various abiotic and biotic samples. The presence of these contaminants raises concern because of the dependence of Aboriginal peoples on subsistence fisheries, and the possible intake of significant amounts of contaminants by northern consumers (Muir *et al.* 1996a). This concern is explored further in the Arctic Marine Ecosystem Chapter.

LIMNOLOGICAL FEATURES OF IMPORTANCE TO CONTAMINANT DYNAMICS

Contaminant dynamics in freshwater ecosystems can be affected by many conditions such as lake or river size, trophic status, climate, and water chemistry including acidity, and geological setting. These conditions vary greatly within Canada. Some factors of particular importance to metal accumulation by fish are lake water and sediment chemical characteristics, especially calcium concentration, pH, organic acids and sulphur, as well as temperature (Klaverkamp *et al.* 1991). Canada's inland water bodies range in size from potholes to the 31 000 km^2 Great Bear Lake. Water chemistry varies from some of the most dilute lakes (low concentrations of dissolved salts) in the world to some of the most concentrated (saltiest), and from clear, neutral lakes to brown and acidic ones. Most lakes found in the low-carbonate, granitic Precambrian Shield have low concentrations of salts and nutrients, limited buffering capacity and are known as oligotrophic lakes. Many of the smaller prairie lakes occur in limestone-rich areas, have much higher buffering capacity, higher salt and nutrient concentrations and are termed mesotrophic or eutrophic (Pearse *et al.* 1985).

Rivers provide connections for transfer of chemical contaminants. In contrast, lakes provide storage of contaminants. The principal factors affecting contaminant processes in streams and rivers are gradient, hydrodynamics and temperature. The principal factors affecting contaminant processes in lakes are depth, area, volume and climate (Rosenberg *et al.* 1997). Most storage of contaminants in lakes occurs in the sediments that serve as vast contaminant

Productivity and Contamination

A factor that may affect the bioaccumulation of organic contaminants by biota is plankton biomass. Lakes in Ontario with low plankton biomass (low productivity) had higher concentrations of contaminants in that biomass relative to lakes with high productivity and higher plankton biomass (Taylor et al. 1991). This inverse relationship of contaminant concentration and plankton biomass was strongest for more hydrophobic and persistent compounds (Taylor et al. 1991). Therefore, phosphorus control and increased fish stocking might exacerbate contaminant problems in some lakes because it would decrease productivity and biomass of plankton, resulting in higher contaminant concentrations in plankton biomass, and ultimately higher contaminant concentrations in aquatic vertebrates (Taylor et al. 1991). Higher concentrations of some contaminants are observed in fish from less productive lakes compared with those from more productive lakes (Larsson et al. 1992). Lower contaminant concentrations in more productive lakes were attributed to the higher growth rates of fish, and to higher biological turnover rates, as well to the faster sedimentation of particles to which contaminants were adsorbed (Larsson et al. 1992).

reservoirs (Boudou and Ribeyre 1989). Mention is made elsewhere in this chapter of the use of sediment cores to estimate the contribution of anthropogenic loading to the burdens of contaminants (e.g., metals) in lakes.

BIOLOGICAL RESOURCES

The Canadian freshwater fisheries are diverse and important for commercial, recreational and subsistence purposes. The three types of fisheries may rely on different species in different regions of Canada (Pearse 1988). Generally, commercial and recreational fisheries are more important in southern than in northern Canada. For example, from November 1994 to November 1995, Saskatchewan, Alberta, Manitoba, northwestern Ontario, and the Northwest Territories sold 13 000 000 kg of various species of fish, worth approximately $26 million, through the

Acidification and the Bioavailability of Metals

Different forms of metals vary in their availability to biota. The acidity of water is an important factor affecting this process. Lake acidification due to acid rain increases both methylmercury production (Xun et al. 1987) and fish mercury concentrations (Wiener et al. 1990). Acidification of aquatic systems causes metals such as cadmium to desorb from particulates and become more available for uptake by organisms (Campbell and Evans 1991; Wiener and Stokes 1990; also see Campbell and Stokes 1985). On the other hand, hydrogen ions compete with the metals for uptake at biological surfaces such as gills so that an increase in available metals may not necessarily lead to increased uptake and accumulation by an organism (Bendell Young and Harvey 1988). Increased bioavailability of metals in acidified freshwater systems can result in metal accumulation in terrestrial animals feeding on contaminated aquatic biota (St. Louis et al. 1993).

Freshwater Fish Marketing Corporation (C. Craig personal communication). The contribution by the Northwest Territories was relatively small. For example, total reported experimental and commercial landings in the Northwest Territories were 1 500 000 kg with a landed value of $1.7 million in 1995/96 (DFO 1997). In 1994 and 1995, New Brunswick had catches of freshwater fish valued at $1.8 and $2.1 million, respectively, and Quebec had catches valued at $3.1 and $3.6 million, respectively (DFO unpublished).

In 1990, inland recreational anglers in British Columbia spent $186 million (DFO 1994) on direct expenditures including money spent on bait, licences, lodges, outfitters, camps and transportation (DFO 1994). Recreational fishing in the province of Manitoba resulted in direct expenditures of $78 million in 1990. In 1990, Ontario had the largest recreational fishery in the country, including that in the Great Lakes, with total direct expenditures by fishers of approximately $1.1 billion (DFO 1994). Comparable figures in 1990 for recreational fisheries in Quebec were $490 million; New Brunswick, $24 million; Nova Scotia, $22 million, Newfoundland and Labrador, $52 million; and in Canada as a whole, $2.4 billion (DFO 1994). In 1990, direct expenditure on recreational fisheries in the Northwest Territories was $7.8 million (DFO 1994). In 1995/96, about 13 000 sport fishing licences were sold in the Northwest Territories west area to Northwest Territories residents and to other Canadian residents and nonresidents (DFO 1997). For 1994/95, revenue of approximately $270 000 was generated in the Northwest Territories West Area by the sale of 14 000 licences (DFO 1995).

The freshwater subsistence fishery is important to Canadians living in northern and remote areas, and is a culturally integral part of the lifestyle of the Aboriginals of Canada (DFO 1994). A total of 61 000 kg of fish were caught in a year for subsistence purposes by aboriginal communities in the south-central district of the Northwest Territories (DFO 1995).

Assessment of Chemical Contamination

SOURCES, DISTRIBUTION, TRENDS AND EFFECTS
METALS AND ORGANOMETALLIC COMPOUNDS
MERCURY

Metals are natural constituents of the earth's crust and occur at measurable natural background concentrations in water, sediments and all biota in Canada (Mierle 1992). Background concentrations of metals are highly variable reflecting Canada's vast geological variability. Because of this, contamination must be examined in the context of site-specific natural background levels.

Background mercury concentrations in sediments range from 5 to 310 ng.g^{-1} dry weight determined under the National Geochemical Reconnaissance program with 180 000 sites covering 2.2 million km^2 (Friske and Coker 1995). Highest background concentrations were 23 times the lowest values for streams, and 9 times those for lakes. Lowest mercury concentrations were encountered in sediment underlain with shale, while highest values were obtained in areas overlain with shale. Some of the highest sediment mercury concentrations occurred in relatively isolated areas, while some of the lowest sediment mercury concentrations occurred in highly industrialized southeastern Ontario (Friske and Coker 1995).

Anthropogenic global sources of mercury have been estimated to exceed natural ones by about two to three times and to have increased the atmospheric burden by about 1.5%. A significant source of mercury to the global environment is the deliberate use of mercury in gold mining in the Amazon. As much as 800 to 1 000 t of mercury is estimated to have reached the atmosphere from the 1 200 t of metallic mercury used in the Amazon over the past 10 years (Homero 1989 in Barbosa *et al.* 1995).

Although the major current sources of mercury contamination to Canadian fresh waters are global, reaching Canada via LRTAP, there are substantial Canadian sources. Canadian sources released an estimated 39 t of mercury to the atmosphere in 1990 (Nriagu 1994). More than half of this input was from non-ferrous metal production (Nriagu 1994). Lesser sources were the recycling of metals, power generation, commercial fuel and fuel wood combustion, and municipal and hospital refuse (Nriagu 1994).

In Canada in the past, industrial point sources, especially mercury cell chlor-alkali plants, directly released large quantities of mercury to the freshwater environment that resulted in the contamination of fish and fish habitat (e.g., English-Wabigoon River system, See Rudd *et al.* 1983). With government regulation and voluntary industry controls, the direct point source discharges of mercury to the environment have decreased significantly over the past 20 years.

The creation of reservoirs is another important source of mercury contamination to fish (Morrison and Thérien 1991). Unlike industrial and LRTAP sources, it does not add new mercury to the system. Rather, the mercury that is present in newly flooded land becomes more bioavailable and toxic because of the increased rate of conversion to methylmercury. Reservoir development leads to increased concentrations of methylmercury in water, sediment and biota (e.g., in South Indian Lake, La Grande, and Churchill Falls); (See Case Study: Hydroelectric Developments and Mercury Contamination in Northern Aquatic Environments).

Although there is enormous natural variability in background mercury concentrations in sediments, those near anthropogenically contaminated sites are generally higher in mercury than the expected background concentrations. For example, in St. Lawrence River sediments, concentrations of mercury increased close to urban areas (Roche Ltee. 1995). Sediment cores show a slow, constant increase in mercury concentrations from the early 1900s to the 1980s in lakes in northern Canada and northwestern Ontario (Lockhart *et al.* 1995), attributed to anthropogenic emissions of mercury. Mercury concentrations in the top layers were high in comparison to the concentrations in the bottom layers due to increases in flux to the sediments (Figure 2). Rates of accumulation of mercury in sediments in lakes in a subarctic region of Quebec show rates of mercury accumulation 2.3 times background concentrations since about 1940 (Lucotte *et al.* 1995). Rates of accumulation in sediments in northwestern Ontario and the Northwest Territories are estimated to be 2 to 8 times background rates (Nriagu 1994). Increases

in mercury concentrations have also been observed in sediment cores from Lake Laporte and Cabonga Reservoir in Quebec (Roche Ltee. 1995).

Mercury has been detected in surface water in all locations examined throughout Canada, although concentrations are usually low in locations with no known point source contamination. Trends in mercury concentrations over time are generally not available because of the difficulty of comparing data collected using modern techniques with those from older water sampling techniques. The older techniques often gave inaccurately high values due to inadvertent contamination of samples. Concentrations of mercury in surface water of reservoirs are not usually higher than in areas with no known point source contamination. For example, mean concentrations of total mercury in reservoir surface waters from various sites in Canada were found to range from less than 1 to 2 ng.L^{-1}. In reservoirs that are not newly established Methylmercury concentrations are typically small relative to total mercury (Table 3). Acidification

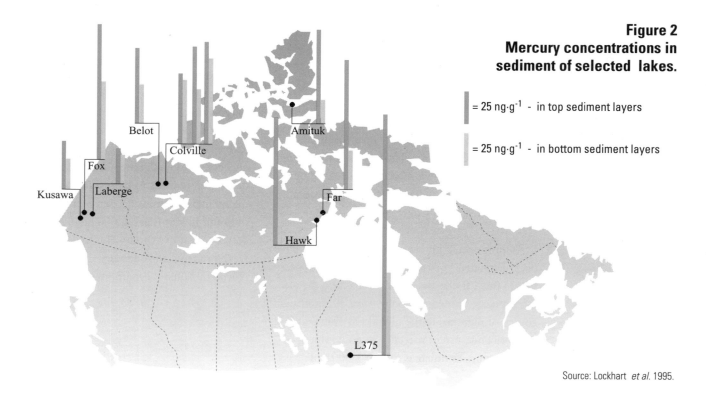

**Figure 2
Mercury concentrations in sediment of selected lakes.**

= 25 ng·g^{-1} - in top sediment layers

= 25 ng·g^{-1} - in bottom sediment layers

Belot
Amituk
Colville
Fox
Laberge
Kusawa
Far
Hawk
L375

Source: Lockhart *et al.* 1995.

of surface waters elevates mercury concentrations in tributaries of the Gulf of St. Lawrence (Van Coillie *et al.* 1984) and other regions of Canada.

Mercury contamination of fish is pervasive in Canada's fresh waters, even in areas with no known point sources of anthropogenic contamination (Table 4). It is due almost entirely to accumulation of methylmercury through the food web (Bloom 1992). Because methylmercury accumulates over the life span of a fish and is expected to be of higher concentration in larger, older fish, some provinces restrict the size of fish that can be consumed from certain lakes. For example, the Ontario Ministry of the Environment and Energy's Guide to Eating Ontario Sport Fish has restrictions for various contaminants based on fish sizes and human consumption frequencies for over 1 600 lakes in Ontario. Mercury is the most prevalent of these contaminants. Mercury concentrations in fish flesh above the 0.5 $\mu g.g^{-1}$ (wet weight) guideline, established by Health Canada for unrestricted commercial sale, have been found in lake trout from some arctic and subarctic lakes, particularly in larger, older fish (Muir *et al.* 1996b).

Elevated concentrations of mercury in fish have been measured in flooded waterways in northern Manitoba (Bodaly and Hecky 1979; Bodaly *et al.* 1984; Jackson 1991) and other reservoirs including those in Quebec, Newfoundland and Ontario. Elevated mercury concentrations due to mercury chlor-alkali production have also been found in fish in northwestern Ontario (Scott and Armstrong 1972; Armstrong and Hamilton 1972; Armstrong and Scott 1979; Rudd and Turner 1983; Rudd *et al.* 1983; Parks *et al.* 1984; Figure 3a).

There are cases where mercury from natural sources can cause mercury concentrations in fish to be well above the recommended guidelines for human consumption. Walleye, lake trout and northern pike collected from uncontaminated Lac Ste. Therese in the Northwest Territories historically have mean concentrations of mercury around 1 $\mu g.g^{-1}$ wet weight, concentrations attributed to water quality and basin morphology (Stephens 1995). Lake trout from a commercial fishery on Kamick Lake, Northwest Territories, had concentrations that ranged from 0.57 to 2.0 $\mu g.g^{-1}$ wet weight. This was attributed to the release of gaseous mercury during the weathering process of sulphide-bearing black slates (Shilts and Coker 1995).

Due to the slow biological loss of mercury from the fish, mercury concentrations in fish in contaminated areas remain elevated for many years, even where the pollution source is controlled. Figure 3a shows that mercury concentrations in northern pike from Clay Lake (part of the Wabigoon/English/Winnipeg River system in northwestern Ontario that was highly contaminated by a chlor-alkali plant during the 1960s) in 1987 were still 20 times greater than the background concentrations of mercury found

Table 3   Mean mercury concentrations in reservoir surface waters from various sites in Canada.

Location	Year Sampled	Methylmercury Concentrations	Total Mercury	Reference
		------------- (ng.L⁻¹) -----------		
Opinaca, Quebec	1992	0.15	2.0	Bodaly[1] (unpublished data)
Ladouceur, Quebec	1992	0.08	1.9	"
LG2, Quebec	1992	0.09	1.5	"
South Indian Lake, Manitoba (inshore)	1994	0.07	0.7	"
South Indian Lake, Manitoba (offshore)	1994	0.06	0.7	"
Sundance Cooling Pond	1995	0.25	1.7	AGRA 1996

1. D. Bodaly, Freshwater Institute, Winnipeg, Canada.

in northern pike from uncontaminated control sites. Mercury concentrations in fish decreased from a high of 9 µg.g^{-1} in 1971 to about 2 µg.g^{-1} in 1980, but there has been no appreciable decline since then. Figure 3b shows the same downward trend in much less-contaminated northern pike from the Souris River, Manitoba, which probably reflects more stringent controls on the use and release of mercury to the environment. Concentrations generally decreased from 1980 to 1987, after which declines were minimal or non-existent.

Other factors being constant, mercury concentrations in fish tend be higher in small rather than in large lakes. This relationship, observed in a series of six lakes of increasing size in northwestern Ontario, was explained by the fact that small lakes tend to be warmer than larger lakes (Bodaly *et al.* 1993). In warmer water, methylmercury production increases (Callister and Winfrey 1986). This relationship may have important implications for the impact of long-term climate changes on the methylation and accumulation of mercury by fish.

Concentrations of methylmercury less than 10 µg.g^{-1} in fish are not suspected of causing acute toxic effects in fish. Nevertheless, sublethal effects, such as behavioral effects and effects on reproduction, may be occurring at concentrations less than 10 µg.g^{-1} (Spry and Wiener 1991). While overt adverse effects of mercury on aquatic biota are not obvious, dramatic effects of mercury are reported in piscivorous birds and mammals, including humans that consume fish contaminated with mercury. Effects can include nervous system derangements, pathological changes in body organs and death (Nriagu 1994). Other effects include decreased reproductive success (Evers *et al.* 1995). Elevated concentrations of

mercury in the liver and other tissues may be the cause of reproductive failures of many fish-eating birds (bald eagle, tern, cormorant, gull and loon) and mammals (otter) in some regions of northern Ontario, Quebec and Nova Scotia (Fimreite 1974; Environment Canada 1991a; Barr 1986).

Figure 3a and 3b
Mercury concentrations in muscle of northern pike in Clay Lake, Ontario, 1971 to 1987 and in the Souris River, Manitoba, 1980 to 1992.

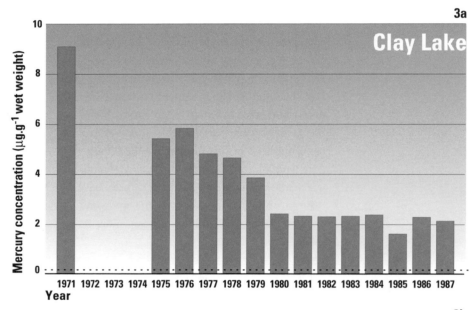

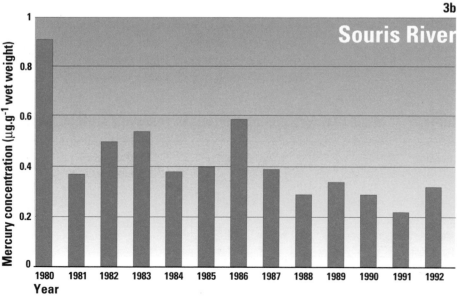

Dotted line in figure 3a indicates background concentration from control lakes.

Source: Parks *et al.* 1994; Green and Beck 1994.

Table 4   Mercury concentrations in fish muscle in inland waters with no known point sources of contamination.

Location	Species	Mean or Range ($\mu g.g^{-1}$, wet weight)	Reference
Cornwallis Island, Northwest Territories	Arctic char	0.05 - 0.42	Muir and Lockhart 1994
Peter Lake, Northwest Territories	Lake trout	0.15 - 1.9	Muir *et al.* 1995b
North Saskatchewan River, Saskatchewan	Goldeye Walleye Sauger Pike	0.56 0.55 0.74 0.28	Merkowsky 1987
Souris River, Manitoba	Walleye Northern pike	0.33 - 0.59 0.22 - 0.41	Green and Beck 1994
Assiniboine River, Manitoba	Walleye Sauger Northern pike	0.14 - 0.77 0.24 - 0.47 0.20 - 1.06	Green and Beck 1995
Tadenac Lake, Ontario	Smallmouth bass Northern pike Lake char	0.58 1.01 0.40	Wren *et al.* 1983
Northwest Ontario Lake Size Series, Ontario	Yellow perch	0.04 - 0.14	Bodaly *et al.* 1993
Moose, Albany, Attawapiskat, Winisk Severn Rivers, Ontario	Northern pike	0.14 - 0.28	McCrea and Fischer 1986
South-central Ontario	Pumpkinseed sunfish	0.04 - 0.33	Wren and MacCrimmon 1983
New Brunswick	Brook trout (whole fish)	0.03 - 0.84	W. Fairchild (personal communication 1994)

CADMIUM, LEAD, NICKEL AND OTHER METALS

CADMIUM

Canada is one of the world's top five industrial producers of cadmium. Sources of cadmium to inland fresh water include both natural and anthropogenic origins. Anthropogenic sources of cadmium exceed natural sources by about 20 fold and include metal production, particularly base metal smelting and refining, stationary fuel consumption (power generation and heating), transportation, solid waste disposal, and sewage application (CEPA 1994a). An estimated 160 t of cadmium are released annually to the Canadian environment, 150 t into the atmosphere and 12 t directly into the aquatic environment (CEPA 1994a). In addition, approximately 340 t of cadmium slag, sludge and solid wastes are estimated to be disposed of on land (CEPA 1994a). According to the recently available estimates, base smelters (primarily lead/zinc) account for 82% (130 t) of the total cadmium released to the Canadian environment (CEPA 1994a).

Concentrations of cadmium in air downwind of smelters are higher than in areas remote from them. Air concentrations near smelters ranged from mean values of 46 ng.m^{-3} in Flin Flon, Manitoba (copper/zinc base metal smelter), 20 to 40 ng.m^{-3} in Trail, British Columbia (lead/zinc smelter), and 10 to 30 ng.m^{-3} at Belledune, New Brunswick (lead smelter), (Bezak 1991; Kenyon 1991, in CEPA 1994a; Murphy 1991) compared with a mean concentration in ambient air in Canada of 2 ng.m^{-3} (Environment Canada 1991b). Surface waters near known point sources of contamination such as smelters and urban areas have higher cadmium concentrations than surface waters in more remote areas (Roche Ltee. 1995). Mean cadmium concentration in surface water near Sudbury, Ontario, (Inco Smelter) was 120 ng.L^{-1} while cadmium concentrations in other central Ontario Lakes with no known point source averaged 11 ng.L^{-1} (Stephenson and Mackie 1988). The mean cadmium concentration in remote Lake 382 at the Experimental Lakes Area in northwestern Ontario was 1.6 ng.L^{-1} in the 1980s (Malley *et al.* 1989). This value is probably several times above natural concentrations prior to regional industrialization which took place after 1850 (see Case Study: A Whole Lake Cadmium Experiment to Observe Fate and Effects).

Cadmium and lead are usually found associated with suspended particles in the water and with bottom sediment. However, in the St. Lawrence River, cadmium is found primarily in the dissolved phase, suggesting that local inputs are important (Roche Ltee. 1995).

Cadmium concentrations in sediment and particulate matter in Canada are higher near point sources such as smelters, mines and urban centres and decline with increasing distance from these sources. For example, Figure 4 indicates that cadmium concentrations in surface sediments decline with increasing distance from the Flin Flon base metal smelter in Manitoba. The same trend has been found at other lakes where point sources dominate inputs. Significant concentrations of metals have also been observed in the Abitibi region of Quebec due to mining activities (Roche Ltee. 1995).

Concentrations of metals in sediments indicate that current loadings are higher than those observed in the pre-industrial past. Present day loadings of cadmium were estimated to be 1.8 times as much as background in sediment from 14 Ontario lakes (Johnson 1987). Northwestern Ontario was estimated to receive three times the loading of cadmium in the 1980s that it did prior to the 1850s (Malley 1996). Cadmium concentrations in sediment also appear to be increasing in the Cabonga Reservoir in Quebec in contrast to decreases observed in most areas of

Figure 4

Cadmium concentrations in the upper 2 cm of sediment cores from lakes versus distance from the Flin Flon, Manitoba smelter, 1985.

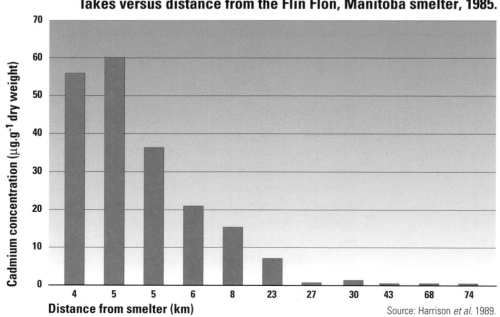

Source: Harrison *et al.* 1989.

the St. Lawrence River (Roche Ltée. 1995). Sediment cores sampled close to point sources show temporal increases in metals coinciding with releases by those sources. For example, Figure 5 shows an increase in cadmium concentrations in more recent layers in sediments taken from Phantom Lake, 5 km from the Flin Flon smelter (Harrison *et al.* 1989), probably reflecting the startup of the smelter in 1930.

In Canada, concentrations of cadmium in freshwater organisms can be elevated above expected background concentrations due to a proximity to smelters (Harrison and Klaverkamp 1990), or to urban, or industrialized areas (Roche Ltee. 1995). Cadmium, zinc, copper, mercury and selenium concentrations in liver of northern pike were highest closest to the Flin Flon, Manitoba, smelter and lower in lakes further away (Harrison and Klaverkamp 1990). Nevertheless, the relationship between point sources and concentrations in fish is not as clear-cut as for sediments. Figure 6 shows that cadmium concentrations in liver from both northern pike and white sucker were highest closest to the Flin

Flon smelter, but lakes in Saskatchewan remote from point sources had fish with variable cadmium concentrations in liver. Cadmium concentrations in kidneys of white suckers were 3 to 7 times higher in fish from lakes near the Flin Flon smelter compared to fish from lakes further away (Klaverkamp *et al.* 1991).

Although concentrations of cadmium and lead in freshwater fish from waters with no known point source of contamination are generally found to be low, there are exceptions. These include cadmium in lake trout from the Great Whale study area in Quebec and cadmium in white sucker from Lake Brûlé in Labrador. Peterson *et al.* (1989) found higher concentrations of cadmium in trout from lakes in northern Nova Scotia than in those from New Brunswick and southern Nova Scotia.

The possibility that loadings of cadmium below the Canadian Water Quality Guideline set in 1979 for the protection of freshwater aquatic life could have adverse effects at an ecosystem level was explored in a small Precambrian Shield lake in

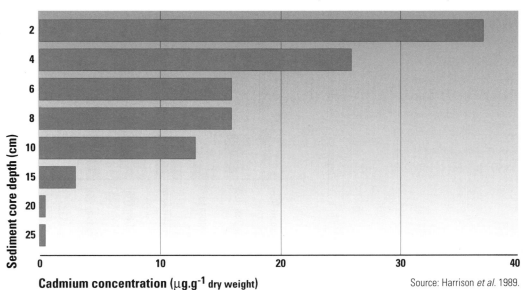

Figure 5
Mean cadmium concentrations in sediment cores from Phantom Lake, 5 km from Flin Flon, Manitoba.

Source: Harrison *et al.* 1989.

northwestern Ontario. Further details are provided in the Case Study: A Whole Lake Experiment to Observe Fate and Effects later in this chapter.

LEAD

Globally, anthropogenic sources of lead are estimated to outweigh natural ones by 200 times (Pacyna 1995). Canada mines about 300 000 t of lead annually. In 1990, anthropogenic emissions of lead into the atmosphere in Canada were estimated at 29 000 t (Nriagu 1994). About half of this was released during the smelting of non-ferrous metals. About 3 000 t was released by motor vehicles, 1 750 t by aircraft, and 1 100 t from waste oil and tires. The phase-out of leaded gasoline since 1970 has had a dramatic impact on emissions to the air, causing the emissions from automobiles to fall from about 19 000 t.yr^{-1} in 1970 to about 3 000 t.yr^{-1} in 1990. Other sources of lead are paints and dyes, plumbing, electrical components and electronics, and plastics and chemicals (Hodson *et al.* 1984). A significant direct source of lead to the aquatic environment is spent lead shot and lost fishing weights (Scheuhammer and Norris 1995).

Deposition of lead in Ontario is highest in the industrialized southern region at 60 compared with 30 g.ha^{-1}.yr^{-1} in the north (Nriagu 1994). During the 1980s, the deposition of lead in the northern areas of Ontario exceeded the pre-1850 flux by 1 500 fold. With

further reductions in lead emissions to 3% of their 1974 rate the current deposition of lead in northern Ontario is still 300 times greater than the pre-1850s rate (Nriagu 1994).

Lead concentrations in sediment and particulate matter, like those of cadmium, are highest near point sources such as smelters, mines and urban centres and decline with increasing distance from these sources. For example, in the St. Lawrence River, lead concentrations in particulate matter decrease gradually from Cornwall to Quebec (Roche Ltee. 1995). As a further example of the relationship between point sources of lead and lead in the environment, profiles in sediment cores collected in 1978 showed increases in lead concentrations at the same time as lead/zinc mining was occurring near the Kootenay Lake area, British Columbia (MacDonald *et al.* 1994).

Figure 6
Cadmium concentrations in liver of northern pike and white sucker versus distance from the Flin Flon, Manitoba smelter.

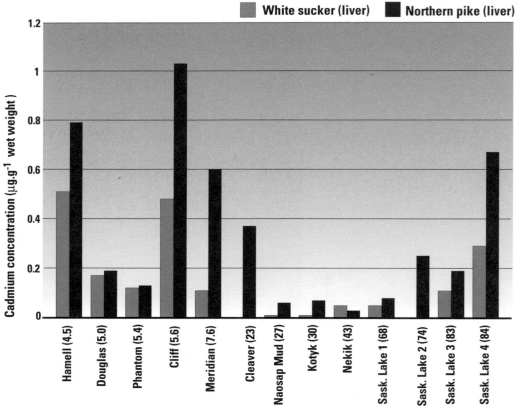

Distance from smelter (km)

Source: Harrison and Klaverkamp 1990.

Aquatic Effects of Mining in Canada

Over the past 25 years the number of Canadian mining operations has varied from 103 to 177. In 1993, the Assessment of the Aquatic Effects of Mining in Canada (AQUAMIN) (Mining Association of Canada and Environment Canada 1996) was initiated prior to updating and strengthening the Metal Mining Liquid Effluent Regulations (MMLER). There had been few programs monitoring the environmental contamination caused by mines. Nevertheless, case studies of 18 mines showed in every case that degradation in water quality from such substances as zinc, cyanide, fluoride, aluminum, nickel, ammonia, nitrate-containing compounds, copper, and acids could be attributed to the mining operation. Downstream water quality may be affected over many kilometres. Although few operations had regular monitoring programs that examined effects on biota as well as on water and sediments, it was clear that older mine sites, particularly those with acid mine drainage, typically had more pronounced effects than newer sites. Metals in mine discharges; effluent-caused acidification; anoxia; and high concentrations of ammonia, calcium, sulfate and iron in waters were all linked to effects on fish. Effects included elimination of fish populations, interference with spawning migrations, contamination of muscle and liver tissues with metals, and increased metallothionein concentrations. Improvements in effluent quality and wastewater management in older facilities and improved design in new facilities have resulted in an overall reduction of adverse effects of mining on fish and fish habitat.

The report of the AQUAMIN program included a recommendation that the revised MMLER require that mine operators develop, conduct and report on a site-specific Environmental Effects Monitoring (EEM) program that monitors key components of aquatic systems. Environment Canada is now establishing a working group to define the EEM objectives and requirements for the mining sector.

Present-day loadings of lead average 17 times background loadings as estimated in sediment from 14 Ontario lakes (Johnson 1987). Lead concentrations in sediment appear to be increasing in the Cabonga Reservoir in Quebec in contrast to decreases observed in most areas of the St. Lawrence River (Roche Ltee. 1995). Generally, the dramatic reduction in emissions of lead from gasoline over the past several decades is reflected in declines in concentrations of lead in air, water and surface sediments (LaZerte 1993; Mudroch 1993). Lead concentrations in fish increased during the decades of leaded gasoline use, and decreased with its ban as a gasoline additive (Green and Beck 1995). The decline of lead contamination in aquatic ecosystems is one of the clearest examples that pollution controls result in a less contaminated environment.

NICKEL

The major anthropogenic source of nickel in Canada is primary base metal production from the smelters in Sudbury, Ontario, and Thompson, Manitoba (CEPA 1994b). In 1988 and 1989, respectively, 27 and 49 t of nickel were discharged in waste water from the smelters in Sudbury. Comparable figures for discharges from the Thompson smelter were 26 and 15 t in 1988 and 1990, respectively. The estimated annual loading of nickel to water in Canada from mining, smelting and refinery operations was 64 t (CEPA 1994b).

Nickel is released into the inland freshwater systems via natural and anthropogenic activities. Nickel reaches water bodies closest to smelters through atmospheric transport, resulting in high inputs to these lakes. Direct waste water discharges also carry nickel to freshwater systems. Concentrations of nickel in uncontaminated surface waters in Canada range from 1 to 10 $\mu g.L^{-1}$ (CEPA 1994b). Léger (1991) found nickel concentrations in surface waters to be consistently less than 2 $\mu g.L^{-1}$ in several thousand samples from fresh waters in Atlantic Canada. Nickel concentrations in water are highest in areas close to smelters such as near Sudbury, Ontario, with concentrations averaging 130 $\mu g.L^{-1}$, and near natural point sources such as the Smoking Hills, Northwest Territories where burning bituminous shale has released large quantities of nickel to the local surface waters (Havas and Hutchinson 1983). Keller *et al.* (1992) concluded that although major reductions in emissions to the atmosphere from the Sudbury smelter initially led to significant decreases of nickel in lakes near the smelter, there has been no evidence of a further decline since 1981.

Background concentrations of nickel in the surficial sediments of lakes on the Canadian Shield and Appalachian regions of Canada are approximately 5 to 50 $\mu g.g^{-1}$. Concentrations in sediments close to point sources are well above these levels (CEPA 1994b), with concentrations reaching over 4 000 $\mu g.g^{-1}$ dry weight in contaminated lakes closest to the Sudbury smelter (Bradley and Morris 1986), or in the Welland River, downstream of a steel manufacturing plant (CEPA 1994b).

Freshwater Mussels

Although freshwater mussels (also called naiads, unionids, or clams) of the families Margaritiferidae and Unionidae are found throughout the world, they reach their greatest diversity in North America. To date 297 taxa (281 species and 16 subspecies) have been identified in Canada and the U.S. Depending on the species, adult mussels range in size from 4 cm to more than 30 cm and occupy a wide range of freshwater habitats. Freshwater mussels are planktivores: they feed by filtering the minute plants and animals suspended in the waters. Mussels, in turn, serve as food for a wide variety of freshwater fish and other vertebrate predators, including aquatic birds and mammals. Traditionally, freshwater mussels have been used by Arboriginals for tools and ornamental objects. Until the advent of plastics, mussels also were important for the manufacture of buttons. Today, the commercial value of freshwater mussels lies in the freshwater pearl industry and the use of their shell for the production of cultured pearls mainly in Japan.

Some species of freshwater mussels live for several decades and thus are considered to be good bio-indicators for monitoring the health of aquatic ecosystems (Stewart and Malley 1997). They are dependant on good water quality and physical habitat for successful growth. They also require an environment that will support populations of host fish onto which the young temporarily attach themselves for development and dispersal. Because of pollution and habitat disruption, many mussel species are quickly disappearing. As filter feeders, mussels have the ability to concentrate the contaminants attached to particles in the water. This ability combined with their stationary nature, makes mussels excellent indicators of point sources of persistent contaminants.

There are very few data on the concentrations of nickel in organs and tissues of freshwater fish in Canada. Generally, concentrations of nickel in fish and aquatic biota appear to be elevated near point sources (CEPA 1994b). For example, nickel concentrations were below detection limits in most fish livers analyzed by Bradley and Morris (1986) from around the Sudbury smelter, except for fish from the most contaminated lakes, where it was detected in liver. Alikhan *et al.* (1990) found decreasing nickel concentrations in crayfish with increasing distance from the Sudbury smelter.

Contamination of freshwater systems by metals is not restricted to that caused by mercury, cadmium, lead, and nickel. Concentrations of copper and zinc are elevated above background concentrations in sediments and biota in the vicinity of non-ferrous metal smelters. Uranium mines in northern Saskatchewan and Ontario discharge arsenic, nickel, molybdenum, cobalt, chromium, selenium and uranium, in addition to radioactive radium-226 and lead-210. For example, concentrations of uranium, molybdenum, lead, salts, and sulfate were elevated in surface waters downstream of the Cluff Lake Mine in Saskatchewan. The Hope Brook gold mine in Newfoundland has contributed to elevated concentrations of copper and cyanide in waters and increasing concentrations of metals in sediments. Increased mercury and arsenic concentrations were found in fish in the vicinity of the Nova Scotia Gold mine (Mining Association of Canada and Environment Canada 1996). Fish exposed to the effluents from diamond mines accumulate aluminum, barium, cobalt, nickel, iron and chromium.

Little is known about metal and metalloid contamination of the arctic fishery and the presence of these contaminants in native diets. With the exception of mercury, concentrations of metals in freshwater fish from the Canadian arctic have received relatively little attention. Extensive surveys have been conducted for a few metals in fish populations near specific sources such as smelters and mines (e.g., Bohn and Fallis 1978; Johansen *et al.* 1991; Harrison and Klaverkamp 1990). Until the late 1980s, information was particularly limited on metal concentrations in subsistence fisheries (Muir *et al.* 1996a; Lockhart *et al.* 1992). The geographical coverage of contaminant measurements is very good in the Yukon, where most major subsistence fisheries have been surveyed, but is much less detailed in the Northwest Territories and northern Quebec. In the Northwest Territories, most measurements have been made in fish from Great Slave Lake and nearby smaller lakes, and the Mackenzie River and Delta. Contaminant analysis of freshwater fishes in the Nunavut region of the Northwest Territories and the Nunavik area of northern Quebec is limited mainly to arctic char and is insufficient to assess spatial trends.

Production of metallothionein, one of the most easily measured biological responses to metals, provides a degree of protection from metal toxicity. Wild rainbow trout from South Buttle Lake exhibited elevated concentrations of hepatic metallothionein in response to metal contamination in the water (Roch and McCarter 1984). In white suckers from Hamell Lake, closest to the Flin Flon, Manitoba, smelter in 1976, McFarlane and Franzin (1980) observed reduced spawning success, reduced larval and egg survival, smaller egg size, and reduced longevity. These fish had higher concentrations of hepatic and renal metallothionein and were more resistant to acute cadmium toxicity than suckers from Thompson Lake, located 20 km from the smelter (Klaverkamp *et al.* 1991). The suckers from Hamell Lake had apparently developed a capacity to survive metal concentrations that would otherwise be toxic. Nevertheless, the protection afforded by metallothionein has limits. In 1986 no white suckers were captured from Hamell Lake despite intensive fishing efforts, indicating that the entire population had become extinct, presumably due to metal toxicity (Klaverkamp *et al.* 1991).

ORGANOMETALLIC COMPOUNDS

Areas of contamination include large commercial shipping harbours and recreational boating marinas associated with the use of tributyltin (TBT) in marine paints. Studies have shown that TBT and its degradation products occur in freshwater sediments (Maguire *et al.* 1996). Although it is likely that freshwater biota in Canada have been harmed by TBT in some locations, there have been no documented cases of TBT effects in freshwater biota (Maguire *et al.* 1996). Suitable biomonitors for TBT in freshwater environments, such as freshwater snails, amphipods or bivalves, have not yet been developed as they have in the marine environment.

ORGANIC CHEMICALS
ORGANOCHLORINE COMPOUNDS
PESTICIDES AND PCBs

Sources of organochlorine compounds to aquatic freshwater ecosystems are diverse and depend greatly on regional activities and the nature of the contaminant. An estimated 19 000 t or about 50 % of Canadian usage of pesticides, including currently approved organochlorine pesticides such as lindane and endosulfan, occurs in the Lake Winnipeg watershed for agricultural, forestry and domestic purposes (Muir *et al.* 1997). Pesticides are transported to freshwater ecosystems via run-off, volatilization, spray drift, and atmospheric transport (Muir 1991).

Formerly used organochlorine compounds, such as the pesticides toxaphene, DDT and mirex as well as PCBs, are still found in freshwater ecosystems due to their persistence, atmospheric

Uranium Development

As with metal mining in general, uranium tailings contain iron sulfide that may generate acid mine drainage. In addition to problems of acid mine drainage and metal pollution, uranium tailings contain a variety of radioactive atoms, including radium-226, thorium-230, lead-210, and polonium-210. Some of these radionuclides become soluble under acidic conditions and thus are leached from the tailings area and distributed more widely. These radionuclides can become incorporated into food chains. The radioactivity in uranium tailings is sufficient for them to be considered low-level radioactive wastes. The Atomic Energy Control Board is responsible for closely regulating the development, operation, closure, and site decommissioning of uranium mines and mills.

Table 5   Organochlorine compound concentrations in surface freshwater systems in Canada.

Location	Total DDT	Lindane	Total PCB (ng.L⁻¹)	Methoxychlor[1]	Reference
Alberta surface waters	<1.0 to 4.0	<1.0 to 50			Anderson 1994
Saskatchewan (Qu'Appelle River)		<1.0 to 5.0			Dunn 1995
Manitoba surface waters		<1.0 to 140		<DL[2]	Currie and Williamson 1995
Northern Ontario (Hudson Bay lowlands)			6 to 16	≤40	McCrea and Fischer 1986
Northwestern Ontario (L375)[3]		0.40 to 1.5		<0.02 to 0.12	Muir and Grift 1995
South central Ontario			0.80 to 4.0		MacDonald and Metcalfe 1991
Northern Quebec (Ungava and Hudson Bay regions)	<0.40	0.80 to 1.1[5]	<9[5]		Langlois 1987
Quebec/Labrador border	<1.00[4]	<1.0 to 2.0	<5	<10	Lockerbie 1987
Arctic (Amituk Lake)		0.30			Bidleman et al. 1995

1   Other organochlorine compounds were detected but were not included in this table.
2   Less than detection limit; <30 ng.L⁻¹ (provincial laboratory); <12 ng.L⁻¹ (federal laboratory).
3   Lake 375 in the Experimental Lakes area, northwestern Ontario.
4   Sum of p,p'-DDE, p,p'-DDD, o,p'-DDT, p,p'-DDT.
5   Mean concentrations.

Table 6   Organochlorine compound concentrations in suspended or bottom sediment from freshwater systems in Canada.

Location	Sample[1]	Total DDT	Lindane	Toxaphene (ng.g⁻¹ dry weight)	Total PCB	α-Endosulfan[2]	Reference
Saskatchewan (Saskatchewan River)	ss		<DL[3] to 4.4				Chacko et al. 1991
Manitoba (Red River)	ss		<DL to 0.80			<DL to 9	Chacko et al. 1991
Northern Ontario	sediment	6.3 to 10	0.02 to 0.20	2.6 to 5.3			Muir et al. 1995a
Quebec/Labrador region	sediment	<1.0 to 3.0	<1.0		<5 to 15	<10	Lockerbie 1987
South-central Ontario	sediment				9.5 to 66		MacDonald and Metcalfe 1991
Southern Quebec (St. Lawrence River)	sediment				10 to 530		Kaiser et al. 1990
Arctic	sediment	0.09 to 5.1	0.01 to 0.6	0.01 to 17			Muir et al. 1995a

1   ss =suspended sediment, sediment=sediment slices.
2   Other organochlorine compounds were detected but were not included in this table.
3   <detection limit.

transport from areas where the pesticides are still used, and the cycling of these chemicals in the environment. LRTAP is the major source of these chemicals to most freshwater areas of Canada; however, the Great Lakes appear to be the major source of mirex and PCBs to the St. Lawrence River (Roche Ltee. 1995). New additions of PCBs to the atmosphere occur as they are released from previously used electrical equipment and hydraulic fluids.

Organochlorine pesticides have been detected in areas with high historical and existing pesticide use, including agricultural areas in Alberta

Figure 7
Organochlorine compound concentrations in sediment from freshwater lakes along a north-south transect of Canada.

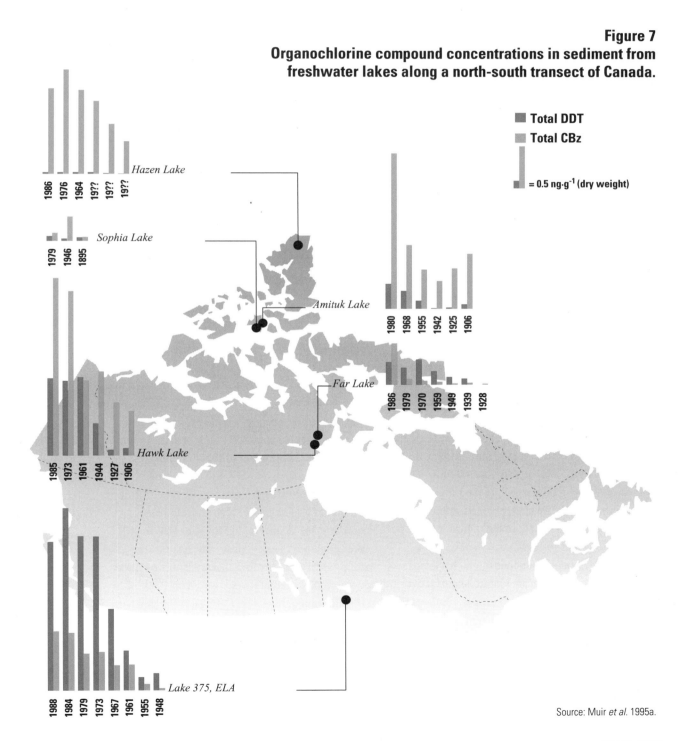

Source: Muir *et al.* 1995a.

and Manitoba (Anderson 1994; Currie and Williamson 1995), as well as in areas where pesticide use has been virtually non-existent (Barrie *et al.* 1992), such as northern Ontario (McCrea and Fischer 1986; Muir and Grift 1995), northern Quebec (Langlois 1987), and the Arctic (Bidleman *et al.* 1995) (Table 5). Because most of these pesticides are hydrophobic, concentrations in water are usually quite low. Historically used pesticides such as total DDT in surface water are usually only slightly above the detection limit in most areas in Canada. Lindane, γ–hexachlorocyclohexane (γ–HCH), methoxychlor, and γ-endosulfan, currently used in Canada for a variety of pest control applications, are found more often than total DDT in surface waters in Canada; nevertheless, detection frequencies are still quite low. The highest concentrations shown in Table 5 were reported from a long-term database on Manitoba surface waters; the lowest concentration was detected in an arctic freshwater lake.

Overall, there is little information on long-term temporal trends of organochlorine pesticides in fresh water other than in the Great Lakes. Generally α–HCH is decreasing in fresh water in various sites across western Canada such as the Assiniboine, Dauphin, Pembina, Red, Roseau, Saskatchewan, Souris, Winnipeg, and Hayes Rivers and Pipestone Creek in Manitoba (Currie and Williamson 1995), and the Qu'Appelle, Carrot, North Saskatchewan, Churchill and Red Deer Rivers in the Prairie provinces (Dunn 1995). This decrease may be a result of the reduction in the amount of α–HCH present as an impurity in the HCH insecticide used globally from 60 to 70% prior to 1972 to less than 1% now. From 1974 to 1994, lindane concentrations have remained fairly constant in Manitoba surface waters, although some rivers showed increases while others had decreasing concentrations (Currie and Williamson 1995).

Concentrations of total PCBs in surface waters throughout some freshwater systems of Canada are quite low, ranging from less than detection

limits (these vary among studies) to 16 ng.L^{-1} (Table 5). It is more common to find PCB residues in the sediments and suspended sediments than in water. Table 6 shows the range of concentrations of PCBs in sediment sampled from various Canadian sites. Highest concentrations of PCBs (530 ng.g^{-1}) were detected in sediment from the St. Lawrence River in southern Quebec.

Total DDT, total PCBs and toxaphene are the major organochlorine compounds found in sediments from various sites. Other organochlorine compounds such as chlordane, nonachlor, heptachlor epoxide, hexachlorobenzene (HCB) and dieldrin are also detected in sediments across Canada. These compounds are detected less frequently and usually at lower concentrations than total DDT, total PCBs or toxaphene.

The major organochlorine compounds present in all sediment samples collected from northwestern Ontario to Hazen Lake in the Northwest Territories were DDT-related compounds, HCHs, chlorobenzenes (CBz), toxaphene and PCBs. Toxaphene was the major pesticide in sediments of most of the arctic lakes (63°N; Muir *et al.* 1995a). Concentrations of total DDT in surface sediments in these lakes decrease with increasing northerly latitude. In contrast, concentrations of total CBz were generally higher in the surface sediments of the arctic lakes at latitudes greater than 63°N than in the lakes located in northwestern Ontario (Figure 7; Muir *et al.* 1995a). The figure also shows that, generally, the deeper sediment slices of most lakes contained lower concentrations of total DDT and total CBz, indicating that transport of these compounds to remote areas is recent. Peak deposition occurred in the 1980s in the northern areas and in the 1960s and 1970s in the southern areas (Muir *et al.* 1995a). Other pesticides such as toxaphene, chlordanes, dieldrin and HCHs did not show consistent latitudinal gradients but were most concentrated in lakes at 63°N and 75°N. These findings support the cold condensation

Table 7   Organochlorine compound mean concentrations in various fish species from freshwater systems in Canada.

Location	Species/tissue	Total PCBs	α-HCH	Total DDT	Toxaphene	Mirex[1]	Reference
				(ng.g⁻¹ wet weight or lipid weight basis)			
Manitoba (southern)	Burbot/liver (lipid)	1 900	54	620	810	10	Muir *et al.* 1990
(northern)	Burbot/liver (lipid)	940	39	460	1 500	17	
Northern Ontario	Northern pike/whole fish (wet weight)	10 to 60	<DL to 3				McCrea and Fischer 1986
Northern Ontario	White sucker/whole fish (wet weight)	10 to 90	<DL to 3				McCrea and Fischer 1986
Northwestern Ontario	Burbot/liver (lipid)	870 to 1 300	31 to 43	1 000 to 1 500	1 700 to 2 300	14 to 17	Muir *et al.* 1990
Central Ontario (four lakes)	Lake trout/muscle (lipid)	1 600 to 2 500					MacDonald and Metcalfe 1991
	Yellow perch/muscle (lipid)	3 100 to 4 300					
	Smallmouth bass/muscle (lipid)	1 200 to 2 900					
Quebec/Labrador border	Northern pike/muscle (wet weight)	<5 to 68	<1 to 2	<1 to 120			Lockerbie 1987
Northern Quebec	Lake whitefish/muscle (wet weight)	2.1[2]					Langlois and Langis 1995
	Northern pike/muscle (wet weight)	4.1[2]					
Yamaska River Noire River, Quebec	Redhorse sucker/muscle (wet weight)	1.5 to 6		900 to 3 900			Metcalfe-Smith *et al.* 1995
	Redhorse sucker/liver (wet weight)	50 to 260					
NWT and northern Quebec	Lake trout/muscle (wet weight)	9 to 47	0.6 to 3		2 to 20		Muir *et al.* 1996b
New Brunswick, Nova Scotia	Various/muscle (wet weight)	<DL	<DL			<DL	Prouse and Uthe 1994
Arctic	Burbot/liver (lipid)	300 to 560	15 to 25	51 to 160	930 to 1 700	3.7 to 8.6	Muir *et al.* 1990

1   Other organochlorine compounds were detected but were not included in this table.
2   Median values.

hypothesis (See the Arctic Marine Ecosystem Chapter) that more volatile organochlorine compounds are transported to the arctic in gaseous forms and then condensed or removed from the atmosphere in the colder arctic temperatures and deposited there (Muir *et al.* 1995a). This hypothesis was supported also by analysis of PCBs in sediment from 11 remote lakes from 49° N to 82° N latitude. Fluxes and burdens of total PCBs declined with increasing latitude (Muir *et al.* 1996a). Highest concentrations of total PCBs were observed deeper in the core in the most lakes (1960 to 1970s) except for those in the high arctic where the maxima were in the surface layers. This suggests that the onset of PCB deposition occurred earlier in the mid-latitude and sub-arctic lakes (1930s to 1940s) than in the highest latitude lakes (1950s to 1960s) (Muir *et al.* 1996b).

Environmental Effects Monitoring Program for the Pulp and Paper Industry

In 1992, the revised Pulp and Paper Effluent Regulations of the Fisheries Act required pulp and paper mills and off-site treatment facilities to initiate an Environmental Effects Monitoring (EEM) program to observe the effects of pulp and paper effluent on the aquatic environment. The objective of the EEM program is to assess the adequacy of national regulations to protect fish, fish habitat and the use of fisheries resources. The EEM studies were to be conducted on a 3-year cycle. The first reports were presented in April 1996. The design of the first cycle included an adult fish survey, a benthic invertebrate survey, water and sediment quality analysis, the sublethal toxicity testing of mill effluent, tissue analysis for chlorinated PCDDs and PCDFs and tainting of fish tissue. The results of the first cycle were used to design a second cycle.

Organochlorine compounds accumulate to varying degrees in fish in freshwater systems. In areas that are more industrialized or closer to more developed regions, concentrations are greater in the fish populations than in more remote regions (e.g., in southern Ontario lakes compared to northern Ontario lakes) (Rasmussen *et al.* 1990). MacDonald and Metcalfe (1991) found that concentrations of total and individual PCB congeners in biota of lakes receiving point source inputs of PCBs were significantly higher than those in the lakes receiving only atmospherically transported PCBs. Total PCBs are some of the major organochlorine compounds found in lipids of various fish species in Canada (Table 7). Toxaphene was usually the next most concentrated organochlorine found, followed by total DDT. In walleye, lake whitefish and burbot from the commercial fishery of Lake Winnipeg, PCBs, DDT and toxaphene were the major contaminants. Mirex and α–HCH were usually the least concentrated in the lipid of various fish species of Canada. Except for γ–HCH, these compounds are no longer used in Canada but because of their persistence in the environment, they are still present and sometimes accumulate in the aquatic environment and biota.

In general, Muir *et al.* (1990) found a decline in concentrations of total PCBs, penta- to nonachlorobiphenyls, total DDT, dieldrin, mirex and lindane in burbot liver with increasing northerly latitude in a transect of lakes from northwestern Ontario to the Northwest Territories from 1985 to 1986 (Figure 8). No similar distribution was observed for other compounds such as toxaphene (Figure 8), tri- and tetrachlorobiphenyls and HCH concentrations in burbot liver (Muir *et al.* 1990). The more volatile compounds did not increase in concentrations in the burbot liver with increasing latitude as the cold condensation theory would predict; however, the less volatile compounds did show decreasing concentrations with increasing latitude. This may indicate that the less volatile compounds may not be transported to the more northern areas to the same extent as more volatile compounds.

Data from the Ontario Ministry of the Environment from 1978 to 1981 indicate that bioaccumulation of PCBs in the top fish predators in some Ontario lakes, such as lake trout, increased with increasing length of the food chain (Rasmussen *et al.* 1990). Lake trout from the shortest food chains had the lowest levels, 1.7 and 3.1 μg.g^{-1} lipid in northern and southern lakes, respectively, while trout from the longest food chains had 6.0 and 18 μg.g^{-1} lipid in northern and southern lakes, respectively. Bentzen *et al.* (1996) point out that the relationship between accumulation of PCBs (and DDT) with food chain length in Ontario lakes may be due to food chain length *per se* or to the fact that longer food chain results in increased lipid content. The presence of the opossum shrimp appears to enhance accumulation of these contaminants at the top of the food chain (Bentzen *et al.* 1996).

Few data are available by which to discern temporal trends of concentrations of organochlorine compound contaminants in freshwater fish outside of the Great Lakes region. Lake whitefish collected in 1992 from Lake Laberge, Northwest Territories, had markedly lower concentrations of total DDT, CBz, HCH and dieldrin, but higher concentrations of toxaphene, total PCBs, and chlordane than round whitefish collected in 1974 (Figure 9a; Muir *et al.* 1996a). Although it appears that some concentrations changed over the 18-year period, other factors such as lipid content, age, diet and other biological differences between round and lake whitefish can influence results (Muir *et al.* 1996a). Nevertheless, lipid content of the two groups of fish could not account for the differences and mean ages of the fish were the same. Although the time period is relatively short,

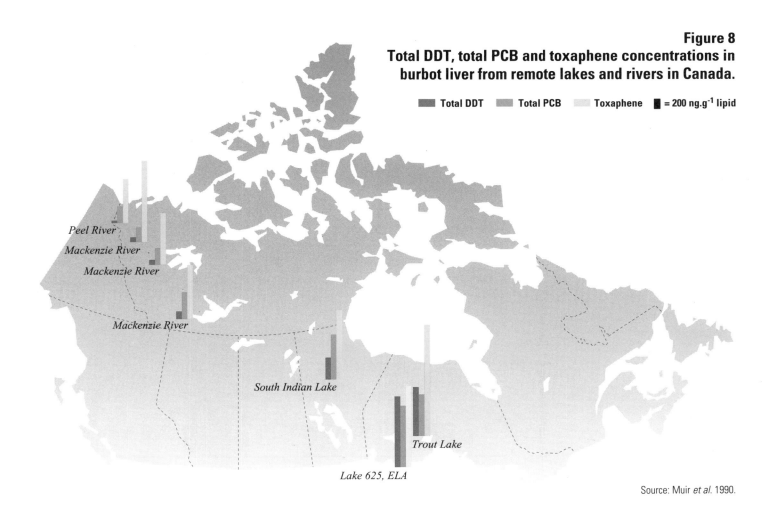

Figure 8
Total DDT, total PCB and toxaphene concentrations in burbot liver from remote lakes and rivers in Canada.

■ Total DDT   ■ Total PCB   ▨ Toxaphene   ■ = 200 ng.g^{-1} lipid

Peel River

Mackenzie River

Mackenzie River

Mackenzie River

South Indian Lake

Trout Lake

Lake 625, ELA

Source: Muir *et al.* 1990.

Figure 9a
Organochlorine compound concentrations in whitefish from Lake Laberge, 1974 and 1992.

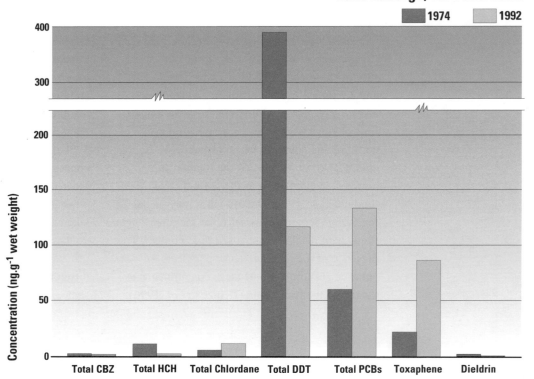

Figure 9b
Organochlorine compound concentrations in burbot liver from the Mackenzie River near Fort Good Hope, Northwest Territories, 1986 to 1994.

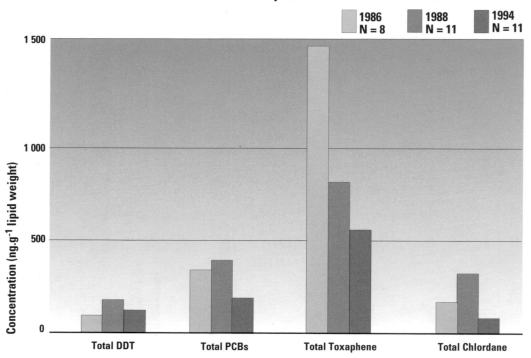

Source: Muir *et al.* 1996b.

concentrations of toxaphene appeared to have significantly decreased in burbot liver from the Mackenzie River at Fort Good Hope from 1986 to 1994 (Figure 9b; Muir *et al.* 1996a). Concentrations of total DDT, total PCBs, and total chlordane were not significantly lower in 1994 (Figure 9b).

Total PCB and DDT concentrations in Lake Winnipeg burbot liver did not decline over the period 1986 to 1995, but there was a 10-fold decline in total DDT concentrations in walleye from 1970 to 1995 (Muir *et al.* 1997). Concentrations of total PCBs in sauger, goldeye, walleye and freshwater drum from three sites and suckers from one site on the Red River in Manitoba decreased significantly between 1978 and 1983; however, total DDT showed no trend over the same time period (Beck 1986).

It is expected that concentrations of organochlorine compounds in organisms will decrease in the St. Lawrence River over time in response to decreases in concentrations in water and sediment observed over the past 25 years due to improved industrial, municipal and commercial practices (Roche Ltee. 1995). From 1988 to 1993, 49 of 50 industrial plants considered to be most important in polluting the St. Lawrence River, cut their discharges by 74%. The metallurgical and organic chemicals sectors made the most progress. Nevertheless, significant effects still occur from discharges of toxic effluents to the river containing oils and greases (32% of pollutants from 1988 to 1993), metals (29%), other metals (8.8%), PCBs (1.4%), and PAHs (0.4%).

General improvements in the level of contamination in the St. Lawrence River are also expected due to the decrease in contaminant concentrations in the upstream Great Lakes.

Although organochlorine pesticides and PCBs are present in various fish species from freshwater systems across Canada, overall impacts of these contaminants on fish, aquatic ecosystems, and human health are difficult to assess. Nevertheless, in some areas, high concentrations of certain organochlorine compounds have caused the closure of some fisheries. For example, the important subsistence and commercial fishery in Lake Laberge was closed due to high concentrations of toxaphene in fish (See the Case Study: Contaminant Cycling in Freshwater Lakes).

PCDDs AND PCDFs

Kraft pulp and paper mills employing chlorine bleaching processes and releasing effluents to freshwater systems in Canada contribute the bulk of the PCDDs and PCDFs to the freshwater environment. Mills not employing bleached kraft processes also contribute small amounts (Servos *et al.* 1994).

A national program was launched in 1988 to measure PCDDs and PCDFs in sediment and fish near Canada's bleached kraft mills, including 31 that discharged into inland waters (Whittle *et al.* 1993; Figure 10). The national program and other recent studies throughout Canada found PCDDs and PCDFs in many abiotic and biotic compartments of the freshwater environment, albeit sometimes in trace amounts (Servos *et al.* 1994; Owens *et al.* 1994; Clement *et al.* 1989; Merriman *et al.* 1991; Pastershank and Muir 1995; Whittle *et al.* 1993). Because PCDDs and PCDFs are hydrophobic, they partition from water into suspended and bottom sediments very rapidly (Servos *et al.* 1989), therefore, dissolved concentrations in the water column are usually very low (i.e., less than 0.1 to less than 0.4 pg.L[-1]). Pastershank and Muir (1995) found

I-TEFs / TEQ

International toxicity equivalent factors (I-TEFs) have been assigned to 17 of the most hazardous PCDDs and PCDFs (Safe and Phil 1990). Because PCDD and PCDF congeners are normally found in combination in samples from contaminated areas, I-TEFs allow the total toxicity of a mixture of PCDDs or PCDFs to be expressed as a single toxic equivalent (TEQ) value. The most toxic congener, 2,3,7,8-TCDD, has a I-TEF value of 1.0, the remaining 16 PCDD, PCDF and PCB congeners were assigned I-TEF values according to their toxicities relative to 2,3,7,8-TCDD. The I-TEF for 2,3,7,8-TCDF, for example, is 0.1, meaning that it is approximately one-tenth as toxic as 2,3,7,8-TCDD. The I-TEFs for the lower chlorinated PCDD and PCDF congeners such as dichlorodibenzodioxin, have not been assigned an I-TEF value and are assumed to be zero. I-TEFs have been adopted by the scientific and regulatory community in eight countries including Canada, the United States, the United Kingdom, Norway, The Netherlands, Denmark, Italy and the Federal Republic of Germany.

Source: Safe and Phil 1990.

that concentrations in centrifuged water were generally below the detection limit (< 0.1 to < 0.4 pg.L^{-1}) on the Athabasca River 1 km downstream of the Weldwood Pulp and Paper mill at Hinton, Alberta. This was thought to be due to the river's dilution capacity and the association of the more highly chlorinated compounds with suspended material in the water column. Generally, PCDD and PCDF concentrations were highest in sediment, water, suspended sediments, invertebrates and fish at sampling sites within 50 km downstream of the town of Hinton and the mill outlet, and returned to normal background concentrations in fish at sites more than 120 km downstream from the mill

effluent source (Pastershank and Muir 1995). Decreases of contamination with increasing distance downstream have also been found at other mills (Owens *et al.* 1994). The rate of decrease in concentrations downstream of a mill varies for each river depending upon flow rates, type of sediment, and channel characteristics of the receiving rivers.

The national program found that, overall, the sediment samples upstream of the 31 inland pulp mills were not contaminated by PCDDs and PCDFs, unless they were contaminated from mills still further upstream (Trudel 1991). However, sediments downstream of the mills

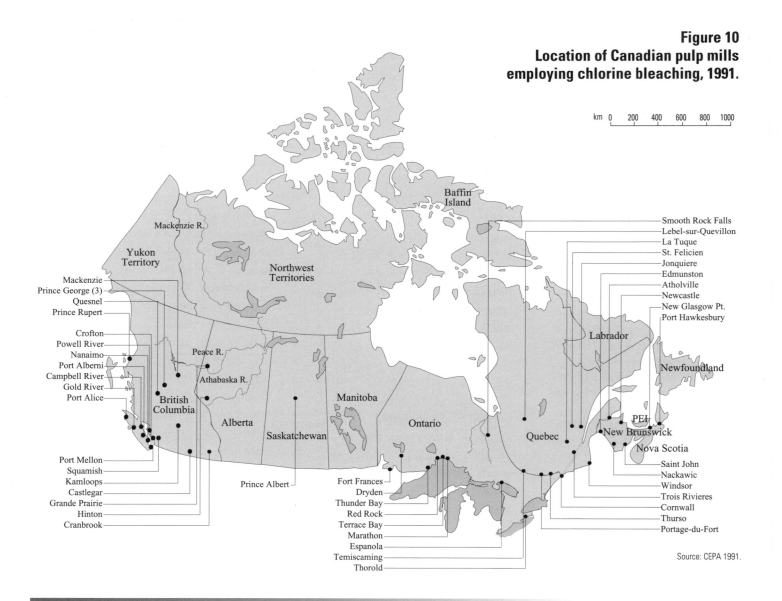

Figure 10
Location of Canadian pulp mills employing chlorine bleaching, 1991.

Source: CEPA 1991.

were contaminated, mainly with tetrachlorinated PCDFs and octachlorinated PCDDs (Trudel 1991). The most toxic PCDD congener 2,3,7,8,-tetrachlorodibenzo-dioxin (2,3,7,8-TCDD), was absent in most sediments collected, but a PCDF congener, 2,3,7,8-tetrachlorodibenzo-furan (2,3,7,8-TCDF) was found at the majority of the freshwater sampling sites (Trudel 1991). Table 8 shows the minimum and maximum concentrations of PCDD and PCDF (toxic equivalents) TEQs at each of the bleached kraft mills; areas where concentrations exceeded 60 pg.g^{-1} TEQs are denoted by an asterisk. Of the total 31 inland mills sampled, 10 were found to have high concentrations in the sediments (Trudel 1991). Four of these mills were located in British Columbia, five in Ontario, and one in Quebec. None of the other mills in Quebec, the Prairie region, New Brunswick or Nova Scotia showed high concentrations of sediment contamination.

PCDD and PCDF congeners are found not only in the vicinity of mills but have also been detected in various arctic lakes, away from any point sources that may contribute these contaminants. Concentrations of 2,3,7,8-TCDD and 2,3,7,8 TCDF ranged from less than the detection limit value up to 28 pg.g^{-1} wet weight in burbot liver (Muir *et al.* 1996a). The concentrations of these compounds were consistently lower in muscle and whole fish than in the liver. Concentrations in fish muscle from arctic lakes ranged from less than the detection limit to 3 pg.g^{-1} wet weight in Arctic char, lake trout and whitefish (Muir *et al.* 1996a). The most commonly detected PCDD and PCDF congener in fish from arctic lakes was octachlorodibenzo-dioxin (OCDD). This was present in muscle of lake trout from Lake Laberge and Kusawa Lake at 7.5 and 13 pg.g^{-1} wet weight, respectively (Muir *et al.* 1996a).

Emissions of PCDDs and PCDFs to freshwater systems at all inland mills have decreased in the past few years in response to new regulations that specify that 2,3,7,8-TCDD TEQ concentrations must be less than 15 pg.L^{-1} (CCME 1995). Pulp and paper mills are reducing the emissions of PCDDs and PCDFs to the aquatic environments by substituting chlorine dioxide for molecular chlorine in the bleaching process.

The Northern River Basin (NRB) study provides information on the PCDD and PCDF concentrations in northern Alberta rivers in the early 1990s when mills were undergoing processing changes (Pastershank and Muir 1995). The major PCDD and PCDF congeners detected in muscle of mountain whitefish and northern pike collected in the Athabasca River downstream of Hinton in 1992 were 2,3,7,8-TCDD and 2,3,7,8-TCDF. The concentrations detected exceeded the draft guideline (CCME 1995) for protection of fish-eating wildlife (1.1 pg.g^{-1}), and exceeded the guideline (CCME 1995) for the protection of aquatic life (50 pg.g^{-1} lipid basis) (Pastershank and Muir 1995). Other PCDDs and PCDFs that were detected in the sediments were not found in the mountain whitefish and northern pike, suggesting that they were eliminated or metabolized by the fish. Walleye and goldeye had the lowest 2,3,7,8,-TCDD and 2,3,7,8-TCDF TEQs of all fish species sampled on the Athabasca River in 1992. Lower accumulation in these fish may be explained by such factors as different food webs and uptake parameters, and their location, which at 300 and 630 km downstream was much greater than for the northern pike, whitefish, and longnose suckers (less than 200 km downstream). The Weldwood of Canada Ltd. mill at Hinton on the Athabasca River did not shift from 45 % to 100% chlorine dioxide until 1993. The Weyerhauser Canada Ltd. mill at Grande Prairie on the Wapiti River progressed from 25% to 70% to 100% chlorine dioxide substitution in 1989, 1991, and 1992, respectively. Owens *et al.* (1994) found a steep decline in TCDD and TCDF concentrations in mountain whitefish following process changes at the bleached kraft mill at Grande Prairie, especially after the shift to 100% chlorine substitution.

Table 8   Minimum and maximum PCDD and PCDF concentrations in sediments around bleached kraft pulp mills across Canada.

Mill Location	Name of Mill	Concentrations as toxic equivalents (TEQs) pg.g^{-1} Minimum	Maximum
British Columbia			
Thompson River	Weyerhaeuser Canada Ltd.*	16	330
Kitimat River	Eurocan Pulp and Paper Ltd.[2]	16	70[1]
Williston Lake	Fletcher Challenge Canada Ltd.*	16	230
Fraser River	Northwood Pulp and Timber Ltd.	16	43
Fraser River	Canfor Prince George Pulp & Paper, Prince George Pulp and Intercontinental Pulp Ltd.	16	22
Fraser River	Cariboo Pulp and Paper Co.	17	35
Kootenay River	Crestbrook Forest Industries Ltd.*	16	240
Williston Lake	Finlay Forest Industries Ltd.[2]	16	45
Fraser River	Quesnel River Pulp Co. Ltd.[2]	16	40
Columbia River	Celgar Pulp Company*	16	80
Prairie Provinces			
Wapiti River	Proctor and Gamble Cellulose Ltd.	16	19
Athabasca River	Weldwood of Canada Ltd.	16	16
North Saskatchewan River	Weyerhaeuser Canada Ltd.	16	16
Saskatchewan River	Manfor Ltd. (Repap)[2]	16	32[1]
Ontario			
St. Lawrence River	Domtar Inc.*	36	110[1]
Wabigoon River	C.P. Forest Products Ltd.*	19	210
Spanish River	E.B. Eddy Forest Products Ltd.	18	23
Rainy River	Boise Cascade Canada	16	17
Lake Superior, Marathon, Ont.	James River Marathon Ltd.*	16	66
Lake Superior, Nipigon Bay	Domtar Inc.*	16	270
Old Welland Canal, Twelve Mile Creek, Welland River	Fraser Inc.	46	46[1]
Lake Superior, Jackfish Bay	Kimberley Clark of Canada Ltd.*	17	163
Kaministiugia River	C.P. Forest Products Ltd.	17	20[1]
Matagami River	Malette Kraft Pulp and Power	17	19
Quebec			
Ottawa River	Tembec Inc.	17	17
Saguenay River	Cascades Inc.	16	16
Saint Maurice River	Les Produits Forestiers Canadiens Pacifique Ltée.	26	26
Quevillion River	Produits des Pâtes et Papiers Domtar*	64	64
Ottawa River	Stone Consolidated Inc.	21	21
Mistassini River	Donohue St.-Felicien Inc.	17	17
Ottawa River	Industries James MacLaren Inc.	17	17
Saint Maurice River	Stone Consolidated Inc.	20	46[1]
Saint François River	Les Papiers Fins Domtar	19	19
New Brunswick/Nova Scotia			
Saint John River	St. Anne-Nackawic Co. Ltd.*	16	44
Saint John River	Fraser Inc.	17	20

[1]   Uncertain evaluation due to high detection limit.

[2]   Control mill.

*   Mills where concentrations above 60 pg·g^{-1} TEQ were found in sediments. Other sites may also have maximum concentration above this, but due to high detection limits were not marked.

Source: Trudel 1991.

In 1990, Health and Welfare Canada recommended consumption restrictions to safeguard the health of those who consume mountain whitefish, burbot and bull trout due to elevated PCDD and PCDF concentrations in the Athabasca River near Hinton, Alberta (Government of Canada News Release 1990-66). Restrictions were also placed on the consumption of mountain whitefish from the Wapiti River near Grande Prairie, Alberta.

Exposure to pulp and paper mill effluents directly affects fish in inland waters (Munkittrick *et al.* 1994; Servos *et al.* 1994; Gagnon *et al.* 1994). However, these effects may not be related solely to the organochlorine compounds in the effluent since fish from water with treated effluent and effluent from mills employing newer chlorine dioxide processes also show these effects (See Case Study: Impact of Pulp and Paper Mill Discharges at Jackfish Bay, Lake Superior in the Great Lakes Ecosystem Chapter). Effects can include increased condition factors, decreased gonad size, increased liver size, induction of hepatic mixed function oxidase (MFO) or ethoxyresorufin-*O*-deethylase, (EROD) enzyme, and reduced circulating concentrations of sex steroids (Servos *et al.* 1994). The most common biological response in fish across many sites has been the induction of MFO activity (Servos *et al.* 1994). Other effects, such as altered immune response (Andersson *et al.* 1988; Larsson *et al.* 1988), differential growth, and impaired reproduction (Larsson *et al.* 1988) have also been observed. Nevertheless, many responses at the physiological and population level are not consistent, not well understood and appear to be site specific (Gagnon *et al.* 1994). For example, in the St. Maurice River, Quebec, white suckers downstream of a major bleached kraft mill had decreased serum testosterone concentrations in males, and increased concentrations in females, and reduced gonad size in both sexes. In addition, MFO activity decreased with distance from the mill (Hodson *et al.* 1992). Gagnon *et al.* (1994), however, found decreased 11-ketotestosterone but not serum testosterone in male white suckers

downstream of the mill. Female white suckers downstream of the mill had decreased testosterone and 17β-estradiol levels. Decreases in gonad size were not observed in either sex by Gagnon *et al.* (1994). A clear relationship between the physiological effects observed by Gagnon *et al.* (1994) and impaired reproduction as measured by gonad weight and fecundity was not apparent in white suckers. This was in contrast to reduced gonad size observed by Hodson *et al.* (1992) in both sexes. Pulp mill effluents can have a severe impact on the biotransformation of steroid hormones in fish and on the reproductive success of exposed populations. The identity of the compounds responsible for reproductive hormonal effects is unknown (Munkittrick *et al.* 1997).

OTHER AROMATIC COMPOUNDS

Natural seepage of PAH-containing hydrocarbons into the Mackenzie River is estimated to be as high as 7 900 L per day at Fort McMurray in Alberta to less than 16 L per day at the Liard Plateau in the Northwest Territories (Carey *et al.* 1990). The movement of PAHs into freshwater environments can also occur via atmospheric transport to distant areas such as northern Canada (Lockhart *et al.* 1992). This process also plays a very important role in the contamination of more developed areas such as the St. Lawrence River (Roche Ltee. 1995).

In general, data on PAHs in water, sediment and fish in most freshwater environments throughout Canada are limited. Concentrations of total PAH at non-contaminated sites such as the North Saskatchewan River, Lake 375 in northwestern Ontario, upstream sites on Rainy River, Ontario, and Hawk Lake, Northwest Territories, were much lower in water, suspended sediments, and sediments than those in naturally-impacted or anthropogenically-contaminated areas such as downstream sites on the Rainy River, Ontario; the sites downstream of areas draining wood preserving operations in New Brunswick and Nova Scotia; and the Mackenzie River, Northwest Territories (Table 9). In areas

Table 9   PAH concentrations in water, suspended sediment or bottom sediment from freshwater systems in Canada.

| Sampling Location | Sample | Concentration ($\mu g \cdot g^{-1}$ dry weight in sediments or $ng \cdot L^{-1}$ in water) | | | | | Reference |
		Pyrene	Phenanthrene	B[a]P	Fluoranthene[1]	Total PAHs	
North Saskatchewan River, Alberta to Saskatchewan	water	<DL[4]	<DL	<DL	<DL		Ongley et al. 1988
	ss[2]	<DL to 27.0	<DL to 85		<DL to 67.8		
	sed[3]	<DL to 8.67	<DL to 65				
Lake 375, Ontario	sed					530[6]	Lockhart et al. 1993
Rainy River, Ontario upstream of mills	water[5]	1.5 to 4.0	0.8 to 4.2		1.1 to 3.3	16 to 42	Merriman et al. 1991
Rainy River, Ontario upstream of mills	ss	<DL	<DL		<DL	<DL	Merriman et al. 1991
Yukon River basin	water	0.46 to 7.71	0.39 to 2.0	<DL to 0.15	0.15 to 1.10		Alaee et al. 1994
Hawk Lake, Northwest Territories	sed					130[6]	Lockhart et al. 1993
Rainy River, Ontario downstream of mills	water	2.3 to 16	2.1 to 13		1.9 to 9.3	80 to 600	Merriman et al. 1991
Rainy River, Ontario downstream of mills	ss	<DL to 118	<DL to 54		<DL to 134	<DL to 310	Merriman et al. 1991
Lake Saint Pierre, Quebec (portion of the St. Lawrence River)	sed			<100 to 500	<100 to 1 200	<100 to 2 700	Langlois and Sloterdijk 1989
Chemical Brook, Newcastle, New Brunswick[7]	sed	120 to 1 700 000	<DL to 5 600 000	390 to 190 000		3 600 to 11 000 000	Kieley et al. 1986
Salmon River, in Truro, Nova Scotia[7]	sed	250 to 1 300 000	280 to 1 900 000	56 to 150 000		1 500 to 6 300 000	Kieley et al. 1986
Mackenzie River, Northwest Territories	water	<DL	<DL	<DL	<DL	110 to 1 800	Carey et al. 1990
	ss		<DL to 37	<DL to 53	<DL to 11	52 to 420	

1   Other PAHs may have been detected but were not included in this table.
2   Suspended sediment.
3   Bottom sediment.
4   Less than detection limit.
5   Centrifuged water.
6   Concentrations obtained from graph.
7   Downstream of wood preserving plant.

Table 10   Mean PAH concentrations in fish from freshwater systems.

Location	Species/sample	Pyrene	Naphthalene	Phenanthrene	Fluoranthene	Fluorene[1]	Total PAHs	Reference
				$(ng \cdot g^{-1}$ wet weight)				
Northwest Territories, Slave River area	Arctic grayling/muscle	0.02 to 0.07	2.4 to 2.5	0.25 to 0.28	0.02 to 0.06	0.04 to 0.09	5.8 to 6.2	Muir *et al.* 1996b
	Burbot/muscle	0.05 to 0.06	1.8 to 2.1	0.16 to 0.21	0.01 to 0.03	<0.01 to 0.01	4.4 to 5.2	
	Lake trout/muscle	0.02 to 0.09	1.9 to 4.7	0.25 to 0.36	0.02 to 0.07	0.04 to 0.17	4.7 to 7.3	
	Northern pike/muscle	0.04 to 0.12	2.2 to 2.8	0.23 to 0.33	0.02 to 0.14	<0.01 to 0.03	3.0 to 6.7	
	Round whitefish/muscle	0.03 to 0.14	2.1 to 3.6	0.19 to 0.30	0.01 to 0.07	0.03 to 0.10	5.0 to 7.3	
	Walleye/muscle	0.05 to 0.07	1.9 to 5.6	0.25 to 0.36	0.03 to 0.06	<0.01	4.2 to 6.9	
Mackenzie River, Northwest Territories	Burbot/liver	<DL to 22.7	<DL to 137	<DL to 6.26	<DL to 4.67	<DL to 180		Lockhart *et al.* 1989
Lake Saint Pierre (portion of the St. Lawrence River)	Various/whole fish	<50 to 1 200						Langlois and Sloterdijk 1989
Rainy River, Ontario	Yellow perch/muscle		<DL			<DL to 120		Merriman *et al.* 1991
	Smallmouth bass/muscle		<DL			<DL to 75		
	Log perch/muscle		39			71.9		

1   Other PAHs may have been detected but were not included in this table.

that have high environmental concentrations of PAHs, concentrations appear to decrease in water and sediment downstream of the source (Kieley *et al.* 1986; Merriman *et al.* 1991). Nevertheless, concentrations of PAHs in water of the St. Lawrence River showed increases from Cornwall to Quebec City; the opposite of other contaminant gradients observed in the river (Roche Ltee. 1995).

PAHs in fish are of particular concern in locations where they are found at high concentrations as a result of industrial loadings. Because fish metabolize and excrete aromatic hydrocarbons, concentrations of total PAH in muscle of fish from uncontaminated sites such as in the Slave River area are usually low (Table 10). In the Mackenzie River, the most frequently detected compounds in burbot liver were naphthalene, phenanthrene, chrysene/benzo[a]anthracene, fluorene and fluoranthene (Lockhart *et al.* 1989; Table 10; Muir *et al.* 1996a). Concentrations of PAHs in muscle were generally lower than in liver (Lockhart *et al.* 1989). The higher molecular weight PAHs, such as benzo[a]anthracene, benzofluoranthene and benzo[a]pyrene (B[a]P), were generally undetectable in muscle and liver from the Mackenzie River burbot (Lockhart *et al.* 1989). Given that PAHs are metabolized, the concentrations of PAHs in organisms will be lower than the concentration of exposure. To determine if fish are exposed, it is often more reliable to look for these substances in other environmental compartments and to observe biochemical changes in fish.

No spatial patterns were apparent in PAH concentrations in fish collected from various ecosystems. Concentrations of PAHs in fish from freshwater systems that drain hydrocarbon-bearing oil sands were similar to those in fish from isolated lakes that have no identifiable point sources (Muir *et al.* 1996a). This implies that the PAHs found in fish from the Slave River region of the Northwest Territories are of natural origin or the result of LRTAP. Bile from some fish species analyzed from this region contained much higher total PAH concentrations than muscle (Muir *et al.* 1996a). This again reflects the ability of fish to metabolize and excrete aromatic hydrocarbons and helps explain the low concentrations found in muscle.

A number of studies have linked concentrations of PAHs in sediments to the concentrations of aromatic compounds in bile, the hepatic activities of contaminant-induced enzymes, and the presence of hepatic neoplasms in bottom fish (Krahn *et al.* 1986; Varanasi *et al.* 1987; Stein *et al.* 1990; Johnson *et al.* 1993). Cancer-related biological effects associated with individual PAHs are shown in Table 11. In areas with low or undetectable concentrations of other organic compounds but high concentrations of PAHs, lesions or tumors have been found in or on fish (Krahn *et al.* 1986; Varanasi *et al.* 1987). Consequently, it is believed that PAHs alone are sufficient to induce biological effects such as lesions, including gonadal tumors (Dickman and Steele 1986), oral and skin papillomas, and dermal and liver tumors (Baumann *et al.* 1987).

Table 11   Cancer–related biological activities associated with selected individual PAHs.

Compound	Biological Activity
Benzo[a]anthracene	Tumor initiator
Chrysene	Tumor initiator
Indeno[1,2,3-c,d]pyrene	Tumor initiator
Benzo[a]pyrene	Complete carcinogen, tumor initiator
Benzo[b]fluoranthene	Complete carcinogen, tumor initiator
Dibenzo[a,h]anthracene	Complete carcinogen, tumor initiator
Fluoranthene	Co-carcinogen with B[a]P
Pyrene	Co-carcinogen with B[a]P
Benzo[g,h,i]perylene	Co-carcinogen with B[a]P

Source: Gammage 1983.

Conclusions

Sources of contaminants to freshwater ecosystems in Canada include industrial, and agricultural activities and municipal effluents. Agricultural uses are the main source of pesticides. LRTAP brings contaminants to aquatic ecosystems from sites far removed from their origin both within Canada and outside of Canada.

Increased controls on direct discharges to water and air from point sources over the last 20 years and the banning of many pesticides and PCBs in favour of more biodegradable products has led to decreased contamination of water, sediment and biota in many Canadian aquatic ecosystems. Nonetheless, some of these persistent chemicals are still being used outside of Canada, and Canadian ecosytems contain large quantities available for recycling. Therefore, the data suggest that the trends towards lower concentrations of contaminants seen over the past decades may have slowed or ended. Concentrations may decline, remain stable, or increase over time depending upon the chemical, its use and chemical behaviour.

Compared with point sources, LRTAP is becoming an increasingly significant factor in delivering contaminants to freshwater ecosystems and is especially important for volatile, persistent contaminants that are released to the atmosphere outside of Canada.

Concentrations of naturally occurring substances in the environment such as metals (i.e. mercury, cadmium, lead and nickel) and PAHs in petroleum hydrocarbons can become elevated through anthropogenic activities. Complicating the assessment of contamination by substances that have both natural and anthropogenic sources is the fact that concentrations of these substances may naturally vary widely in water, sediment and biota. This variation may be due to geological variability or natural processes such as petroleum hydrocarbon seepage that can result in elevated local concentrations. Other complications include the vast variability in types of water bodies and water chemistry in Canada, rates of flow or water turnover, climate and other influences. Human activities significantly influence the movement and concentrations of chemical contaminants in water bodies. These activities and their resulting effects include over-harvesting of fish or other resources, destruction of habitat, flooding, diversion of water, water level manipulation, eutrophication and acidification. Thus, detailed assessments of contamination must be specific to a particular site, locality, or type of aquatic environment.

Despite the complexity and variability in ecosystem types throughout Canada's inland fresh waters, some patterns of current contamination emerge. Mercury, cadmium, lead, PCBs, DDT and other organochlorine pesticides (including some not registered for use in Canada, e.g., toxaphene), PCDDs and PCDFs are significant contaminants in many freshwater ecosystems. High concentrations of toxaphene and mercury in certain areas have resulted in fishery closures and consumption restrictions.

Elevated concentrations of mercury in fish due to reservoir development remain a major contaminant issue in Canadian fresh waters. Depending upon the local conditions of the particular reservoir, mercury concentrations in fish can remain elevated for many years. Recent findings are showing that climate warming may increase the concentration of mercury in fish.

Most contaminant monitoring in freshwater fish has focused on measurements of metals in edible portions (e.g., muscle) of fish. However, many contaminants also accumulate in liver, kidney, gills and other body organs. To determine if deleterious effects are occurring in fish, emphasis needs to be placed on analysis of tissues in addition to muscle, including liver, kidney and gills.

There is a large amount of evidence that the organisms respond to exposure to a range of contaminant types. Some organisms may adapt to these exposures. Biological responses such as MFO and metallothionein induction have been observed in some fish species. Although there are few data on direct toxic effects to fish, chronic exposure to these compounds may affect the overall health of the biota, consequently increasing susceptibility to other stresses. Further studies are necessary to distinguish contaminant effects that may be occurring from the background of high variability in physiological and life history responses due to seasonal, regional, latitudinal and site-specific differences among the diverse freshwater habitats in Canada.

Physiological changes including decreased gonad size, increased liver size, and MFO induction have been observed in fish downstream from industrial sites such as pulp and paper mills. It is becoming clear, however, that some or all of the adverse effects of pulp and paper mill effluents on the hormonal status and reproduction of fish downstream of the mills are not due solely to the presence of organochlorine compounds. Notwithstanding the importance of controlling highly toxic PCDDs and PCDFs in fisheries resources, it appears that other substances in the effluent are more closely linked to the fish reproductive effects. Natural toxins produced by

trees for protection from their insect and fungal pests, and then concentrated during the pulping process, possibly play a role.

Emerging issues in freshwater ecosystems include increasing urbanization, with an associated increase in effluents containing a wide range of domestic and industrial compounds, increasing pharmaceutical use by the population, and the introduction of new drugs and chemicals. Some compounds of concern are a class of nonionic surfactants (nonylphenol polyethoxylates) that have been found to be persistent in marine sediments and likely behave similarly in freshwater sediments.

There is growing concern that increasing penetration of UV-B radiation into clear, oligotrophic lakes may influence contaminant dynamics, concentrations and effects. Three widespread environmental problems are all acting to increase exposure of aquatic ecosystems to UV-B radiation. There is an increase in the amount of UV-B radiation reaching Canada's surface as a result of stratospheric ozone depletion. Penetration of UV-B radiation is increased by climate warming and acidification (Schindler *et al.* 1996). These stresses reduce the concentration of dissolved organic carbon in the water which normally acts as a shield against the radiation. As summarized by

Table 12 Mercury concentrations in food chains of north-central Quebec lakes and recently created hydroelectric power reservoirs in northern Manitoba and Quebec.

Trophic Level	Mercury concentration (μg.g^{-1} wet weight)		
	North-central Quebec reference Lakes	Southern Indian and Notigi (Manitoba)	LG-2 (Quebec)
Periphyton	0.130 - 0.190	—	0.15 - 1.3
Benthic insects	0.067 - 0.330	—	0.14 - 1.7
Plankton	0.010 - 0.070	0.02 - 0.10	0.060 - .14
Fish (muscle)			
Long nose sucker	0.060 - 0.320	—	0.410 - 0.67
Lake whitefish	0.070 - 0.300	0.17 - 0.71	0.480 - .57
Northern pike	0.250 - 0.900	0.35 - 2.4	1.30 - 3.0
Walleye	0.320 - 1.300	0.23 - 2.6	1.90 - 2.8

Source: Chaire en environnement 1993, Schetagne 1990, Brouard *et al.* 1990, Jackson 1988, Bodaly *et al.* 1984, all cited in Tremblay *et al.* 1993.

Schindler *et al.* (1996), such radiation decreases primary production in surface waters and alters chemical processes that determine the fate and effects of metals and other contaminants in water.

Case Study:

HYDROELECTRIC DEVELOPMENTS AND MERCURY CONTAMINATION IN NORTHERN AQUATIC ENVIRONMENTS

Increases in mercury contamination have been observed in organisms found in most reservoirs created for hydroelectric development. Bacterial action in the flooded lands of the reservoir can encourage the methylation of mercury to produce methylmercury, the form toxic to humans and other vertebrate consumers of fish.

Methylmercury in the flooded soils can then be mobilized and incorporated into the food chain through mechanical erosion caused by dynamic processes such as waves and ice movement (Louchouarn *et al.* 1993); growth of periphyton which accumulates mercury; and sediment burrowing of benthic organisms, including aquatic insects that resuspend mercury adsorbed to sediments (Benoit and Lucotte 1994).

Mercury concentrations in all organisms from major reservoirs of northern Manitoba and Quebec were higher than in reference lakes of north-central Quebec (Table 12). Top predator species such as northern pike and walleye are particularly susceptible to mercury contamination in reservoirs. Most fish caught in new reservoirs have mercury concentrations that exceed the consumption limit of 0.2 $\mu g.g^{-1}$ wet weight recommended by Health Canada for people who frequently consume fish (Bodaly *et al.* 1984; Boucher *et al.* 1985).

Mercury contamination is also a problem in river segments downstream of new reservoirs because of mercury export from the reservoirs. All fish species examined in the La Grande River downstream of the LG-2 reservoir in Quebec exhibited higher mercury concentrations in muscle than those examined from the reservoir itself (Messier and Roy 1987; Brouard *et al.* 1990; Figure 11). For non-piscivorous species, especially lake whitefish, this was largely but not entirely due to significant changes in diet (Johnson *et al.* 1991). This

Figure 11
Mercury concentrations in muscle of fish in the LG-2 reservoir and in the downstream La Grande River, Quebec.

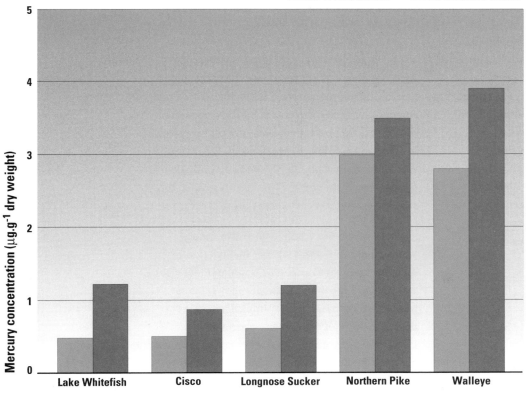

Source: Brouard *et al.* 1990.

species usually feeds on zooplankton and benthos but was found to feed actively on fish or fish parts coming through the turbines of the LG-2 generating station (Brouard *et al.* 1992).

Further downstream in estuaries and the coastal environment, the effects of mercury export from reservoirs appear to be limited to the freshwater plume of affected rivers. In James Bay and Hudson Bay, there is no evidence of increased mercury contamination related to hydroelectric developments in marine organisms (Brouard *et al.* 1990).

The decrease in mercury concentrations in fish in hydroelectric reservoirs over time appears to depend on the extent of flooding and the fish species involved (Bodaly and Johnston 1992). Mercury concentrations in non-piscivorous fish began to decline after 3 years post-impoundment in Southern Indian Lake and after 9 years in LG-2 reservoirs, whereas concentrations in piscivorous species were still stable or increasing at least 12 years after impoundment (Brouard *et al.* 1990; Strange *et al.* 1991; Morrison and Thérien 1995). This same trend, over a longer period of time, was observed in the Smallwood Reservoir in Labrador. Non-piscivores returned to background concentrations of mercury after 16 years, while mercury concentrations in piscivores remained elevated after 21 years (Anderson *et al.* 1995). Continued monitoring in reservoirs will be required to determine the length of time for methylmercury to return to pre-impoundment concentrations, but the available information suggests that the problem of mercury contamination in hydroelectric reservoirs will persist for decades (Bodaly and Johnston 1992).

To better understand how to ameliorate or minimize methylmercury production and accumulation by fish, a small wetland and pond ecosystem at the Experimental Lakes Area (ELA) in northwestern Ontario was subjected to experimental flooding over several years.

This flooding increased the depth of the pond by 1.3 m, the surface area by a factor of 3, and the water volume by a factor of 6 (Kelly *et al.* 1997). Each winter the reservoir was drained to pre-flood levels to simulate winter drawdown practiced in many northern hydroelectric reservoirs.

After flooding, the methylmercury concentrations in the flooded pond were up to 20 times higher than they were in water in the pre-flooded pond (Figure 12). Overall, the flooded wetland produced 35 times more methylmercury than before flooding (Kelly *et al.* 1997). After flooding, small fish (finescale dace) held in enclosures in the pond accumulated about 4 to 5 times more methylmercury than they did under pre-flood conditions. The route of uptake was through food rather than directly from the water. After flooding, predatory insects had 2.5 to 3 times and zooplankton had 4 to 5 times the pre-flood body burdens of methylmercury (Hall *et al.* 1997). Freshwater unionid mussels from a nearby lake introduced in cages increased their methylmercury body burdens by 2.4 times in 3 months (Malley *et al.* 1996).

The increase in waterborne methylmercury with flooding was caused by mechanisms that increased the production of methylmercury. First flooding increased the microbial decomposition of the flooded vegetation. This process produced low-oxygen conditions which promoted mercury methylation. Further, the flooded environment was warmer than before. This rise in temperature also stimulated mercury methylating organisms.

This experiment demonstrated that natural wetlands are important sites of methylation of mercury and that flooding of a wetland increases the already high natural rate of methylmercury production by more than 30-fold. The decomposition of the flooded vegetation leads to increased methylmercury concentrations in water, the food chain, and eventually, fish.

The decomposition of the flooded vegetation, including peat, may take centuries, but monitoring in boreal reservoirs indicates that concentrations of methylmercury in fish may return to normal 10 to 50 years after flooding. An important conclusion of this work is that wetland areas should be avoided when considering site placement of a reservoir for flood control, water conservation, or hydroelectric development purposes (Rudd *et al.* 1997). The best location for a reservoir is a canyon with as little upland soil as possible.

Case Study:
A WHOLE LAKE CADMIUM ADDITION TO OBSERVE FATE AND EFFECTS

Experimental additions of low concentration of cadmium were made to a small lake in the Experimental Lakes Area (ELA) in northwestern Ontario (Malley 1996) to observe fates and effects of cadmium in the water not exceeding the Canadian Water Quality Guideline (CWQG) deemed safe for protection of aquatic life in softwater lakes (0.2 μg.L^{-1} cadmium) (Reeder *et al.* 1979). After one year

Figure 12

Methylmercury concentrations in water flowing into and out of a small pond at the Experimental Lakes Area in northwestern Ontario, prior to, during and after flooding.

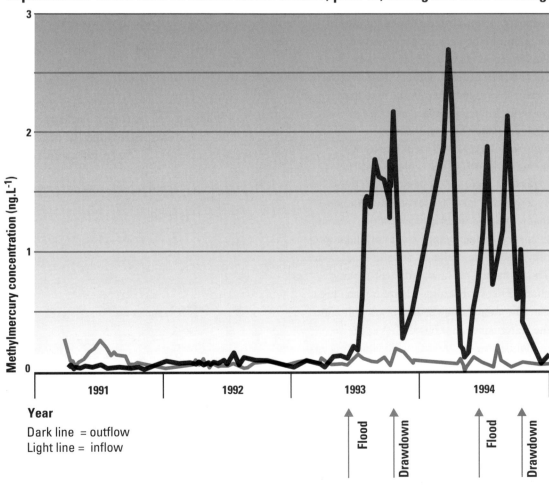

Year

Dark line = outflow
Light line = inflow

Source: Kelly *et al.* 1997.

of addition, all components of the food web were substantially contaminated with cadmium. After six years of additions, lake trout (Figure 13), white sucker, mussels and crayfish (Malley 1997) were still accumulating cadmium in liver and kidney and other body organs, reaching up to 50 times above background concentrations. Almost 95% of the added cadmium remained in the lake adsorbed to the sediments. Sediment concentrations of cadmium in certain areas of the lake approached or exceeded the severe effect sediment quality Ontario guideline of 10 $\mu g.g^{-1}$ dry weight (Stephenson *et al.* 1996).

Fish, mussels and crayfish tissues had elevated concentrations of metallothionein which binds metals such as cadmium, thus reducing adverse effects. The additions of cadmium to the lake were stopped before the large organisms reached a steady state concentration of cadmium at the CWQG. Nevertheless, cadmium in the lake trout posterior kidney approached 15 $\mu g.g^{-1}$ wet weight, a value known to cause histopathological lesions in laboratory mammals and fish and which would be expected to harm individual fish in the experimental lake. The experiment indicates that regulating water concentrations of some persistent contaminants is not sufficient to prevent contamination of aquatic ecosystems. The contaminants have sediment-seeking characteristics that allow them to reach high concentrations in the sediment even though the water concentrations do not exceed guidelines. In this experiment, sediments at first were a sink for the contaminant. Later, the sediments became a source of contaminants. When additions of cadmium ceased, cadmium concentrations in water fell rapidly (Lawrence *et al.* 1996), but the sediments will continue to supply cadmium to the water and prevent the water concentrations from returning to background levels for several decades at least.

Figure 13
Trends in mean cadmium concentrations in posterior kidney of lake trout from ELA Lake 382, 1987 to 1992.

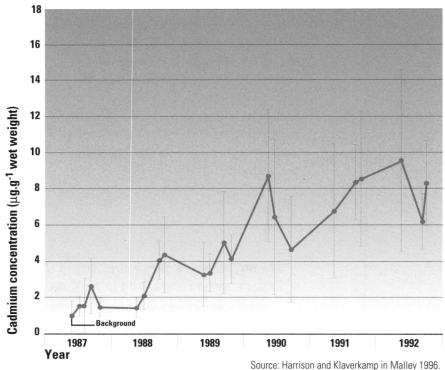

Source: Harrison and Klaverkamp in Malley 1996.

Figure 14
Mean δ^{15} nitrogen isotope measurements and toxaphene concentrations in organisms at various levels in the food chain in Lake Laberge, Yukon Territory.

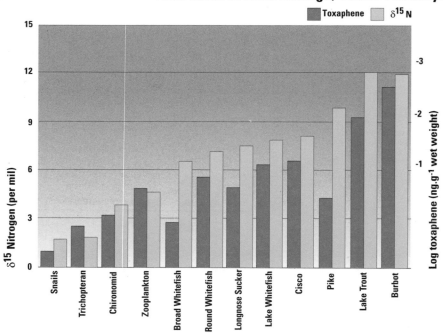

Tissues for nitrogen analysis were whole animals for snails, insects and zooplankton and muscle for fish. Tissues for toxaphene analysis were the same except for whole round whitefish and cisco and burbot liver.

Source: Kidd *et al.* 1995 a,b.

Case Study:

CONTAMINANT CYCLING IN FRESH WATER: THE LAKE LABERGE STORY

High concentrations of toxaphene in fish from Lake Laberge, Yukon Territory, resulted in the closure of an important commercial, sport and native subsistence fishery in 1991 (Muir *et al.* 1996a). On May 8 1991, Yukon's Chief Medical Officer of Health warned against the consumption of burbot liver, and recommended a limitation on the consumption of lake trout flesh due to high concentrations of toxaphene. The flesh of burbot, northern pike, and whitefish and whitefish eggs from Lake Laberge were deemed as safe to eat by the Yukon Government. Concentrations of toxaphene and other organochlorine compounds such as DDT and PCBs in Lake Laberge fish were up to 125 times higher than those found in the same species in other regional lakes (Kidd *et al.* 1993; Muir and Lockhart 1994). Analysis of sediments from Lake Laberge indicated that this lake does not receive greater loadings of toxaphene than other regional lakes. It was hypothesized that different food web structure in Lake Laberge resulted in elevated concentrations of atmospherically-deposited contaminants in the fish (Muir *et al.* 1996a). To find out if the food web structure of Lake Laberge was different from that of the other lakes the food web of Lake Laberge and two reference lakes (Fox Lake and Kusawa Lake) were characterized using stable nitrogen isotope ratios ($\delta^{15}N$) and fish stomach content analyses (Kidd *et al.* 1995a). The increases in $\delta^{15}N$ in the trophic levels is presented in Figure 14, along with the general increase in toxaphene concentrations up the food chain. The food chain in Lake Laberge was found to have three levels of consumers with lake trout and burbot at the top. Organisms in the lowest level were lowest in $\delta^{15}N$, those in the second level were intermediate, and the top predators, burbot and lake trout, had the highest amounts.

The fish community structure in Lake Laberge is thus different from that in the reference lakes, with few lake trout (high in lipid) and lake whitefish and many burbot and longnose sucker (Muir *et al.* 1996a). The lake trout and burbot in Lake Laberge feed at a higher trophic level than those species in the reference lakes, a factor which has been implicated elsewhere in higher contaminant concentrations in lake trout (Rasmussen *et al.* 1990). Thus, the high concentrations of toxaphene in lake trout and burbot in Lake Laberge are attributed to their position at the top of an exceptionally long food chain.

Case Study:

RAPID RELEASE OF CONTAMINANTS IN MELTING SNOW AND ICE: FLIN FLON, MANITOBA

Although emissions from smelters, and the subsequent deposition of acidic substances and metals, occur year-round, winter transport to aquatic systems is restricted by snow and ice on the lake surface. In the spring the metals that have accumulated over the winter enter the water in a "slug discharge" that may result in the release of high concentrations of metals and low pH. This in turn can result in acute adverse effects on aquatic life as has been documented in the vicinity of Flin Flon, Manitoba (Franzin and McFarlane 1981). The slug discharge may pose a particular risk because early spring is also a time for egg development and spawning.

References

°AGRA. 1996. Environmental considerations for augmenting Wabamun Lake water levels. AGRA Earth & Environmental Limited. Calgary, Alberta.

ALAEE, M., D. Gregor, and M. Palmer. 1994. A study of the sources and fate of organochlorine contaminants in the Yukon River basin, p. 152-156. *In* J.L. Murray, and R.G. Shearer [eds.] Environmental Studies no.72. Indian and Northern Affairs Canada, Ottawa.

ALIKHAN, M. A., G. Bagatto, and S. Zia. 1990. The crayfish as a "biological indicator" of aquatic contamination by heavy metals. Wat. Res. 24: 1069-76.

ANDERSON, A-M. 1994. Overview of pesticide data for Alberta surface waters. CAESA Water Quality Monitoring Committee. Appendix A 4. Alberta Environmental Protection. 92 p.

ANDERSON, M.R., D.A. Scruton, U.P. Williams, and J.F. Payne. 1995. Mercury in fish in the Smallwood Reservoir, Labrador, twenty one years after impoundment. Water, Air Soil Poll. 80: 927-930.

ANDERSSON, T., L. Förlin, J. Härdig, and C. Larsson. 1988. Biochemical and physiological disturbances in fish inhabiting coastal waters polluted with bleached kraft pulp mill effluents. Mar. Environ. Res. 24: 233-36.

ARMSTRONG, F. A. J., and A. L. Hamilton. 1972. Pathways of mercury in a polluted northwestern Ontario lake, p. 131-156. *In* P. C. Singer [ed.] Trace metals and metal-organic interactions in natural waters. Ann Arbor Science Publishers, Ann Arbor, Mich.

ARMSTRONG, F. A. J., and D. P. Scott. 1979. Decrease in mercury content of fishes in Ball Lake, Ontario, since imposition of controls on mercury discharges. J. Fish. Res. Bd. Can. 36: 670-672.

BARBOSA, A.C., A.A. Boischio, G.A. East, I. Ferrari, A. Gonçalves, P.R.M. Silva, and T.M.E. da Cruz. 1995. Mercury concentration in the Brazilian Amazon. Environmental and Occupational Aspects. Water Air Soil Pollut. 80: 109-121.

BARR, J.F. 1986. Population dynamics of the common loon associated with mercury-contaminated waters in northwestern Ontario. Can. Wildlife Service, Occasional Paper No. 56.

BARRIE, L. A., D. Gregor, B. Hargrave, R. Lake, D. C. G. Muir, R. Shearer, B. Tracey, and T. Bidleman. 1992. Arctic contaminants: Sources, occurrence and pathways. Sci. Tot. Environ. 122: 1-74.

BAUMANN, P. C., W. D. Smith, and W. K. Parland. 1987. Tumor frequencies and contaminant concentrations in brown bullheads from an industrialized river and a recreational lake. Trans. Am. Fish. Soc. 116: 79-86.

BECK, A. E. 1986. PCB and pesticide residues in fish from the Red River, Manitoba: 1978-1983. Man. Dept. Environ. Workplace Safety Health. Environ. Management Division. Water Standards and Studies Report No. 86-2. 89 p.

BEETON, A.M. 1984. The world's great lakes. J. Great Lakes Res. 10: 106-113.

BENDELL Young, L., and H. H. Harvey. 1988. Metals in Chironomidae larvae and adults in relationship to lake pH and lake oxygen deficiency. Verh. Internat. Verin. Limnol. 23: 246-51.

BENOIT, C., and M. Lucotte. 1994. Enquête sur le mercure du Moyen-Nord québécois. Interface 15: 26-35.

BENTZEN, E., D.R.S. Lean, W.D. Taylor, and D. Mackay. 1996. Role of food web structure on lipid and bioaccumulation of organic contaminants by lake trout (*Salvelinus namaycush*). Can. J. Fish. Aquat. Sci. 53: 2397-2407.

BEZAK, D. 1991. Heavy metals in air: Flin Flon, December, 1988-March 1991. Air Management Section, Manitoba Department of Environment. Report No. 91-02. 22 p.

BIDLEMAN, T. F., R. L. Falconer, and M. D. Walla. 1995. Toxaphene and other organochlorine compounds in air and water at Resolute Bay, N.W.T., Canada. Sci. Tot. Environ. 160/161: 55-63.

BLOOM, N. S. 1992. On the chemical form of mercury in edible fish and marine invertebrate tissue. Can. J. Fish. Aquat. Sci. 49: 1010-1017.

BODALY, R. A., and R. E. Hecky. 1979. Post-impoundment increases in fish mercury levels in the Southern Indian Lake reservoir, Manitoba. Fish. Mar. Serv. MS Rep. No. 1531: 15 p.

BODALY, R. A., and T. A. Johnston. 1992. The mercury problem in hydro-electric reservoirs with predictions of mercury burdens in fish in the proposed Grand Baleine Complex, Quebec. James Bay Publication Series, hydro-electric development: Environmental impacts. Paper No. 3. 15 p.

BODALY, R. A., R. E. Hecky, and R. J. P. Fudge. 1984. Increases in fish mercury levels in lakes flooded by the Churchill River diversion, Northern Manitoba. Can. J. Fish. Aquat. Sci. 41: 682-91.

BODALY, R. A., J. W. M. Rudd, R. J. P. Fudge, and C.A. Kelly 1993. Mercury concentrations in fish related to size of remote Canadian shield lakes. Can. J. Fish. Aquat. Sci. 50: 980-987.

BOHN, A., and B. W. Fallis. 1978. Metal concentrations (As, Cd, Cu, Pb, and Zn) in Shorthorn Sculpins, *Myoxocephalus scorpius* (Linnaeus), and Arctic Char, *Salvelinus alpinus* (Linnaeus), from the vicinity of Strathcona Sound, Northwest Territories. Wat. Res. 12: 659-63.

BOUCHER, R., R. Schetagne, and E. Magnin. 1985. Teneur en mercure des poissons des réservoirs La Grande 2 et Opinaca (Quebec, Canada) avant et après la mise en eau. Rev. Fr. Sci. Eau 4: 193-206.

BOUDOU, A., and F. Ribeyre. 1989. Fundamental concepts in aquatic ecotoxicology. Chapter 3. *In* A. Boudou, and F. Ribeyre, [eds.] Aquatic Ecotoxicology: Fundamental Concepts and Methodologies. Volume 1: 35-75. CRC Press, Inc., Boca Raton, FL.

BRADLEY, R. W., and J. R. Morris. 1986. Heavy metals in fish from a series of metal-contaminated lakes near Sudbury, Ontario. Water Air Soil Poll. 27: 341-54.

BROUARD, D., C. Demers, R. Lalumiere, R. Schetagne, and R. Verdon. 1990. Summary report: Evolution of mercury levels in fish of the La Grande hydroelectric complex, Quebec (1978-1989). Groupe Environnement Shooner Inc. and Vice-Presidence Environnement Hydro-Quebec, Montreal, Quebec. 97 p. + annexes.

BROUARD, D., J.-F. Doyon, and R. Schetagne. 1992. Amplification of mercury concentrations in lake whitefish *(Coregonus clupeaformis)* downstream from the La Grande 2 Reservoir, James Bay, Quebec. Presented to the International Conference on Mercury as a Global Pollutant, sponsored by Electric Power Research Institute, US Environmental Protection Agency. Monterey, California, May 31-June 4, 1992.

CALLISTER, S. M., and M. R. Winfrey. 1986. Microbial methylation of mercury in upper Wisconsin river sediments. Water Air Soil Pollut. 29: 453-65.

CAMPBELL, J., and P. M. Stokes. 1985. Acidification and toxicity of metals to aquatic biota. Can. J. Fish. Aquat. Sci. 42: 2034-49.

CAMPBELL, J., and R. D. Evans. 1991. Cadmium concentrations in the freshwater mussel *(Elliptio complanata)* and their relationship to water chemistry. Arch. Environ. Contam. Toxicol. 20: 125-31.

CAREY, J. H., E. D. Ongley, and E. Nagy. 1990. Hydrocarbon transport in the Mackenzie River, Canada. Sci. Tot. Environ. 97/98: 69-88.

CCME (Canadian Council Ministers of the Environment). 1995. Canadian Environmental Quality Guidelines for Polychlorinated Dibenzo-p-dioxins and Polychlorinated Dibenzofurans. CCME Summary Version. January 1995. Ottawa, Ontario.

CEPA (Canadian Environmental Protection Act). 1991. Effluents from Pulp Mills Using Bleaching. Priority Substances List Assessment Report No. 2. Minister of Supply and Services Canada, Ottawa, Ontario.

CEPA (Canadian Environmental Protection Act). 1994a. Cadmium and its compounds. Priority Substances List Assessment Report. Minister of Supply and Services, Ottawa, Ontatio.

CEPA (Canadian Environmental Protection Act). 1994b. Nickel and its compounds. Priority Substances List Assessment Report. Minister of Supply and Services, Ottawa, Ontario.

CHACKO, V. T., H. H. Vaughan, D. J. Munro, and R. N. Woychuk. 1991. Results of multimedia analysis, large volume extraction and biotoxicity tests in nine Prairie rivers of western Canada; and their role in water quality monitoring. Environment Canada, Western and Northern Region. Winnipeg, Manitoba.57 pages.

CHAIRE en Environnement. 1993. Sources et devenir du mercure dans les réservoirs hydroélectriques. Université du Québec B Montréal, Chaire de recherche en environnement HYDRO-QUÉBEC/CRSNG/UQAM, Montréal, Québec. Rapport annuel 1992-1993.

CLEMENT, R. E., S. A. Suter, E. Reiner, D. McCurvin, and D. Hollinger. 1989. Concentrations of chlorinated dibenzo-p-dioxins and dibenzofurans in effluents and centrifuged particulates from Ontario pulp and paper mills. Chemosphere 19: 649-54.

CURRIE, R. S., and D. A. Williamson. 1995. An assessment of pesticide residues in surface waters of Manitoba, Canada. Water Quality Management Section. Manitoba Environment Report No. 95-08.

DFO (Department of Fisheries and Oceans). 1994. 1990 Survey of Recreational Fishing in Canada. Economic and Commercial Analysis Report No. 148: 156 p

DFO (Department of Fisheries and Oceans). 1995. Annual Summary of Fish and Marine Mammal Harvest Data for the Northwest Territories. Volume 7, 1994-1995: xiii + 81 p.

DFO (Department of Fisheries and Oceans). 1997. Annual Summary of Fish and Marine Mammal Harvest Data for the Northwest Territories, Volume 8, 1995-1996: xii + 80 p.

DICKMAN, M. D., and P. O. Steele. 1986. Gonadal neoplasms in wild carp-goldfish hybrids from the Welland River near Niagara Falls, Canada. Hydrobiologia 134: 257-63.

DUNN, G. W. 1995. Trends in water quality variables at the Saskatchewan/Manitoba boundary. Prairie Provinces Water Board Report No. 124 (Draft).

ENVIRONMENT Canada. 1991a. Toxic Chemicals in the Great Lakes and Associated Effects. Communications Directorate Ontario Region, Toronto, Ontario.

ENVIRONMENT Canada. 1991b. Measurement program for toxic contaminants in Canadian urban air, update and summary report. River Road Environmental Technology Centre, Pollution Measurement Division, Conservation and Protection, Environment Canada. Report No. PMD 91-2. Ottawa, Ontario.

EVERS, D., J.D. Kaplan, M.W. Meyer, J. Counard, P.S. Reaman, E. Braselton, R.W. Bachmann, J.R. Jones, R.H. Peters, and D.M. Soballe. 1995. Mercury exposure in feathers of the Common Loon (*Gavia immer*). Lake and Reservoir Management 11: 137.

FIMREITE, N. 1974. Mercury contamination of aquatic birds in northwestern Ontario. J. Wild. Manag. 38: 120-131.

FRANZIN, W. G., and G. A. McFarlane. 1981. Elevated Zn, Cd and Pb concentrations in waters of ice-covered lakes near a base metal smelter during the snow melt, April 1977. Environ. Poll. (Series B) 2: 11-19.

FRISKE, P. W. B., and W. B. Coker. 1995. The importance of geological controls on the natural distribution of mercury in lake and stream sediments across Canada. Water Air Soil Poll. 80: 1047-51.

GAGNON, M. M., J. J. Dodson, P. V. Hodson, G. Van Der Kraak, and J. H. Carey. 1994. Seasonal effects of bleached kraft mill effluent on reproductive parameters of white sucker *(Catostomus commersoni)* populations of the St. Maurice River, Quebec, Canada. Can. J. Fish. Aquat. Sci. 51: 337-347.

GAMMAGE, R. B. 1983. Polycyclic aromatic hydrocarbons in work atmospheres, p. 653-707. *In* A. Bjorseth [ed.] Handbook of polycyclic aromatic hydrocarbons. Marcel Dekker Inc., New York.

GOVERNMENT of Canada. 1990. Advisory to limit consumption of some freshwater fish in limited areas of Alberta. News Release #1990-66.

GREEN, D. J., and A. E. Beck. 1994. Mercury and other metal residues in fish from the Souris River, Manitoba, Canada, 1979-1992. Water Quality Management Section, Manitoba Environment. Report No. 94-10.

GREEN, D. J., and A. E. Beck. 1995. Mercury and other metals residues in fish from the Assiniboine River, Manitoba, Canada, 1978-1992. Water Quality Management Section, Manitoba Environment. Report No. 95-07.

HALL, B. D., R. A. Bodaly, R. J. P. Fudge, J. W. M. Rudd, and D.H. Rosenberg. 1997. Food as the dominant pathway of methylmercury uptake by fish. Water Air Soil Poll. 100: 13-24.

HARRISON, S. E., and J. F. Klaverkamp. 1990. Metal contamination in liver and muscle of northern pike (*Esox lucius*) and white sucker (*Catostomus commersoni*) and in sediments from lakes near the smelter at Flin Flon, Manitoba. Environ. Toxicol. Chem. 9: 941-56.

HARRISON, S. E., M. D. Dutton, R. V. Hunt, J. F Klaverkamp, A. Lutz, W. A. Macdonald, H. S. Majewski, and L. J. Wesson. 1989. Metal concentrations in fish and sediment from lakes near Flin Flon, Manitoba. Can. Data Rep. Fish. Aquat. Sci. No. 747. 74 p.

HAVAS, M., and T. C. Hutchinson. 1983. The Smoking Hills: natural acidification of an aquatic ecosystem. Nature 301: 23-27.

HEALEY, M.C., and R.R. Wallace. 1987. Canadian aquatic resources. Can. Bull. Fish. Aquat. Sci. 215: 533 p.

HODSON, P. V., M. McWhirter, K. Ralph, B. Gray, D. Thivierge, J. H. Carey, G. Van Der Kraak, D. M. Whittle, and M.-C. Levesque. 1992. Effects of bleached kraft mill effluent on fish in the St. Maurice River, Quebec. Environ. Toxicol. Chem. 11: 1635-51.

HODSON, P.V., D.M. Whittle, P.T.S. Wong, U. Borgmann, R.L. Thomas, Y.K. Chau, J.O. Nriagu, and D.J. Hallett. 1984. Lead contamination of the Great Lakes and its potential effects on aquatic biota, p. 335-369. *In* J.O. Nriagu, and M. S. Simmons [eds.] Toxic Contaminants in the Great Lakes. Wiley Series in Advances in Environmental Science and Technology, No. 14.

JACKSON, T. A. 1988. The mercury problem in recently formed reservoirs of northern Manitoba (Canada): effects of impoundment and other factors on the production of methyl mercury by microorganisms in sediments. Can. J. Fish. Aquat. Sci. 45: 97-121.

JACKSON, T. A. 1991. Biological and environmental control of mercury accumulation by fish in lakes and reservoirs of northern Manitoba, Canada. Can. J. Fish. Aquat. Sci. 48: 2449-70.

JOHANSEN, P., M. M. Hansen, G. Asmund, and N. Palle Bo. 1991. Marine organisms as indicators of heavy metal pollution-experience from 16 years of monitoring at a lead zinc mine in Greenland. Chem. Ecol. 5: 35-55.

JOHNSON, L. L., C. M. Stehr, O. P Olson, M. S. Myers, S. M. Pierce, C. A. Wigren, B. B. McCain, and U. Varanasi. 1993. Chemical contaminants and hepatic lesions in winter flounder (*Pleuronectes americanus*) from the northeast coast of the United States. Environ. Sci. Technol. 27: 2759-71.

JOHNSON, M. G. 1987. Trace element loadings to sediments of fourteen Ontario lakes and correlation with concentrations in fish. Can. J. Fish. Aquat. Sci. 44: 3-13.

JOHNSON, T. A., R. A. Bodaly, and J. A. Mathias. 1991. Predicting fish mercury levels from physical characteristics of boreal reservoirs. Can. J. Fish. Aquat. Sci. 48:1468-75.

KAISER, K. L. E., B. G. Oliver, M. N. Charlton, K. D. Nicol, and M. E. Comba. 1990. Polychlorinated biphenyls in St. Lawrence River sediments. Sci. Tot. Environ. 97/98: 495-506.

KELLER, W., J. R. Pitblado, and J. Carbone. 1992. Chemical responses of acidic lakes in the Sudbury, Ontario, area to reduced smelter emmissions, 1981-89. Can. J. Fish. Aquat. Sci. 49 (Suppl. 1): 25-32.

KELLY, C. A., J. W. M Rudd., R. A. Bodaly, N. P. Roulet, V. L. St. Louis, A. Heyes, T. R. Moore, S. Schiff, R. Ararena, K.J. Scott, B. Dyck, R. Harris, B. Warner, and G. Edwards. 1997. Increases in fluxes of greenhouse gases and methyl mercury following flooding of an experimental reservoir. Environ. Sci. Technol. 31:1334-1344.

KIDD, K. A., J. E. Eamer, and D. C. G. Muir. 1993. Spatial variability of chlorinated bornanes (toxaphene) in fishes from Yukon Lakes. Chemosphere 27: 1975-86.

KIDD, K. A., D. W. Schindler, R. H. Hesslein, and D. C. G. Muir. 1995a. Correlation between stable nitrogen isotope ratios and concentrations of organochlorines in biota from a freshwater food web. Sci. Tot. Environ. 160/161: 381-90.

KIDD, K.A., D.W. Schindler, D.C.G. Muir, W.L. Lockhart, R.H. Hesslein. 1995b. High concentrations of toxaphene in fishes from a subarctic lake. Science 269: 240-242.

KIELEY, K. M., R. A. F. Matheson, and P. A. Hennigar. 1986. Polynuclear aromatic hydrocarbons in the vicinity of two Atlantic Region wood preserving operations. Environment Canada, Environmental Protection Service, Atlantic Region, Dartmouth, N.S. Surveillance Report EPS-5-AR-86-3. 28 p.

KLAVERKAMP, J.F., M.D. Dutton, H.S. Majewski, R.V. Hunt, and L.J. Wesson. 1991. Evaluating the effectiveness of metal pollution controls in a smelter by using metallothionein and other biochemical responses in fish, p. 33-64. *In* M.C. Newman, and A.W. McIntosh [eds.] Metal ecotoxicology concepts and applications. Lewis Publishers, Chelsea, MI.

KRAHN, M. M., L. D. Rhodes, M. S. Myers, L. K. Moore, W. D. MacLeod Jr., and D. C. Malins. 1986. Associations between metabolites or aromatic compounds in bile and the occurrence of hepatic lesions in English sole (*Papophrys vetulus*) from Puget Sound, Washington. Arch. Environ. Contam. toxicol. 15: 61-67.

LANGLOIS, C. 1987. Étude préliminaire de la qualité des eaux de surface de 15 cours d'eau majeurs du nouveau-Québec. Water Poll. Res. J. Canada 22: 530-544.

LANGLOIS, C., and H. Sloterdijk. 1989. Contamination du lac Saint-Pierre (Saint Laurent) par certains pollutants organiques et inorganiques. Revue des Sci. de l'eau 2: 659-79.

LANGLOIS, C., and R. Langis. 1995. Presence of airborne contaminants in the wildlife of northern Quebec. Sci. Tot. Environ. 160/161: 391-402.

LARSSON, C., T. Andersson, L. Förlin, and J. Härdig. 1988. Physiological disturbances in fish exposed to bleached kraft mill effluents. Wat. Sci. Technol. 20: 67-76.

LARSSON, P., L. Collvin, L. Okla, and G. Meyer. 1992. Lake productivity and water chemistry as governors of the uptake of persistent pollutants in fish. Environ. Sci. Technol. 26: 346-52.

LAWRENCE, S.G., M.H. Holoka, R.V. Hunt, and R.H. Hesslein. 1996. Multi-year experimental additions of cadmium to a lake epilimnion and resulting water column cadmium concentrations. Can. J. Fish. Aquat. Sci. 53: 1876-1887.

LAZERTE, B.D. 1993. Recent reductions in lead deposition and surface water concentrations in central Ontario, Canada, pp 504-507. *In* R.J. Allan, and J.O. Nriagu [eds.] Heavy Metals in the Environment. CEP Consultants, Edinburgh, Vol. 2.

LÉGER, D.A. 1991. Data Summary Report on Nickel in Atlantic Canada (1973-1990). Environment Canada, Inland Water Directorate, Water Quality Branch, IWD-AR-WQB-91-60, Moncton, New Brunswick.

LOCKERBIE, D. M. 1987. Cemical characterization of water, sediment and biota from five Labrador-Quebec transboundary basins. Inland Waters Directorate, Atlantic Region, Water Quality Branch. IW/L-AR-WQB-120. vi + 139 p.

LOCKHART, W. L., D. A. Metner, D.A.J. Murray, R. W. Danell, B. N. Billeck, C. L. Baron, D. C. G. Muir, and K. Chang-Kue. 1989. Studies to determine whether the condition of fish from the lower Mackenzie River is related to hydrocarbon exposure. Environmental Studies No. 61. Department of Indian Affairs and Northern Development, Ottawa, Canada. 84 p.

LOCKHART, W. L., P. Wilkinson, B. N. Billeck, G. J. Brunskill, R. V. Hunt, and R. Wagemann. 1993. Polycyclic aromatic hydrocarbons and mercury in sediments from two islolated lakes in central and northern Canada. Wat. Sci. Tech. 28: (8-9): 43-52.

LOCKHART, W. L., P. Wilkinson, B. N. Billeck, R. V. Hunt, R. Wagemann, and G.J. Brunskill. 1995. Current and historical inputs of mercury to high latitude lakes in Canada and to Hudson Bay. Water Air Soil Poll. 80: 603-10.

LOCKHART, W. L., R. Wagemann, B. Tracey, D. Sutherland, and D. J. Thomas. 1992. Presence and implications of chemical contaminants in the freshwaters of the Canadian Arctic. Sci. Tot. Environ. 122: 165-243.

LOUCHOUARN, P., M. Lucotte, A. Mucci, and P. Pichet. 1993. Geochemistry of mercury in two hydroelectric reservoirs in Quebec, Canada. Can. J. Fish. Aquat. Sci. 50: 269-81.

LUCOTTE, M., A. Mucci, C. Hillaire-Marcel, P. Pitchet, and A. Grondin. 1995. Anthropogenic mercury enrichment in remote lakes of northern Quebec (Canada). Water Air Soil Poll. 80: 467-476.

MACDONALD, C. R., and C. D. Metcalfe. 1991. Concentration and distribution of PCB congeners in isolated Ontario Lakes contaminated by atmospheric deposition. Can. J. Fish. Aquat. Sci. 48: 371-81.

MACDONALD, R. W., E. C. Carmack, and C. H. Pharo. 1994. Sediment records of man's activities in the Kootenay Lake drainage basin. Water Poll. Res. J. Canada 29: 103-16.

MAGUIRE, R.J., Y.K. Chau, and J.A.J. Thompson. 1996. Proceedings of the workshop on Organotin Compounds in the Canadian Aquatic Environment. Sidney, British Columbia, February 19-20, 1996. NWRI contribution No. 96-153.

MALLEY, D. F. 1996. Cadmium whole lake experiment at the Experimental Lakes Area: An anachronism? Can. J. Fish. Aquat. Sci. 53: 1862-1870.

MALLEY, D. F., A. R. Stewart, and B. D. Hall. 1996. Uptake of methyl mercury by the floater mussel, *Pyganodon grandis* (Bivalvia, Unionidae) caged in a flooded wetland. Environ. Toxicol. Chem. 15: 928-936. (Experimental Lakes Area Reservoir Project Contribution No. 13).

MALLEY, D.F. 1997. Multi-generational exposure of the crayfish *Orconectes virilis* to low levels of cadmium in a whole lake experiment: Metal concentrations and population parameters. Book of Abstracts: Department of Fisheries and Oceans Green Plan Toxic Chemicals Program Wrap-up Conference. Government Conference Centre, Ottawa, Ontario, January 28-31, 1997. Can. Tech. Rep. Fish. Aquat. Sci. 2163: 40.

MALLEY, D.F., P.S.S. Chang, and R.H. Hesslein. 1989. Whole lake addition of cadmium-109: radiotracer accumulation in the mussel population in the first season. Sci. Tot. Environ. 87/88: 397-417.

MCCREA, R. C., and J. D. Fischer. 1986. Heavy metal and oraganochlorine contaminants in the five major Ontario rivers of the Hudson Bay lowland. Wat. Poll. Res. J. Can. 21: 225-34.

MCFARLANE, G. A., and W. G. Franzin. 1980. An examination of Cd, Cu, and Hg concentrations in livers of northern pike, *Esox lucius*, and white sucker, *Catostomus commersoni*, from five lakes near a base metal smelter at Flin Flon, Manitoba. Can. J. Fish. Aquat. Sci. 37: 1573-78.

MERKOWSKY, J. J. 1987. Biological survey of the North Saskatchewan River. Saskatchewan Parks, Recreation and Culture. Fisheries Technical Report 87-4. 268 p.

MERRIMAN, J. C., D. H. J. Anthony, J. A. Kraft, and R. J. Wilkinson. 1991. Rainy River water quality in the vicinity of bleached kraft mills. Chemosphere 23: 1605-15.

MESSIER, D., and D. Roy. 1987. Concentrations en mercure chez les poissons au complexe hydroélectrique de La Grande Rivière (Québec). Nat. Can. 114: 357-68.

METCALFE-SMITH, J. L, R. J. Maguire, and S. P. Batchelor. 1995. Polychlorinated biphenyl congeners and chlorinated pesticides in fish from the Yamaska River basin, Quebec. Water Qual. Res. J. Canada 30: 179-204.

MIERLE, G. 1992. What limits the rate of methyl mercury uptake via the gills of fish? Proceedings of the eighteenth annual Aquatic Toxicity Workshop. Can. Tech. Rep. Fish. Aquat. Sci. No. 1863: 65-66.

MINING ASSOCIATION OF CANADA and Environment Canada. 1996. Assessment of the Aquatic Effects of Mining in Canada: AQUAMIN. Final Report. 127p.

MORRISON, K.A., and N. Thérien, 1991. Experimental evaluation of mercury release from flooded vegetation and soils. Water Air Soil Pollut. 56: 607-619.

MORRISON, K.A., and N. Thérien, 1995. Changes in mercury levels in lake whitefish (*Coregonus clupeaformis*) and northern pike (*Esox lucius*) in the LG-2 reservoir since flooding. Water Air Soil Pollut. 80: 819-828.

MUDROCH, A. 1993. Lake Ontario sediments in monitoring pollution. Environ. Monit. Assess. 28: 117-129.

MUIR, D. C. G. 1991. Dissipation and transformation in water and sediment, p. 1-87. *In* R. Grover, and A. J. Cessna [eds.] Environmental Chemistry of Herbicides, Vol. II. CRC Press, Boca Raton, FL.

MUIR, D. C. G., and N. P. Grift. 1995. Fate of herbicides and organochlorine insecticides in lake waters, p. 141-156. *In* N. N. Ragsdale, P. C. Kearney, and J. R. Plimmer [eds.] 8th International Congress of Pesticide Chemistry: Options 2000. Washington, DC.

MUIR, D. C. G., and W. L. Lockhart. 1992. Contaminant trends in freshwater biota, p. 121-125. *In* J. L. Murray, and R. G. Shearer [ed.] Synopsis of research conducted under the 1991/1992 Northern Contaminants Program. Environmental Studies Report, No. 68. Indian and Northern Affairs Canada, Ottawa.

MUIR, D. C. G., and W. L. Lockhart. 1994. Contaminant trends in freshwater and marine fish, p. 264-270. *In* J. L. Murray, and R. G. Shearer [eds.] Synopsis of research conducted under the 1992/1993 Northern Contaminants Program. Environmental Studies Report, No. 72. Indian and Northern Affairs Canada, Ottawa.

MUIR, D. C. G., C. A. Ford, N. P. Grift, D. A. Metner, and W. L. Lockhart. 1990. Geographic variation of chlorinated hydrocarbons in Burbot (*Lota lota*) from remote lakes and rivers in Canada. Arch. Environ. Contam. Toxicol. 19: 530-542.

MUIR, D. C. G., N. P. Grift, W. L. Lockhart, P. Wilkinson, B. N. Billeck, and G. J. Brunskill. 1995a. Spatial trends and historical profiles of organochlorine pesticides in Arctic lake sediments. Sci. Tot. Environ. 160/161: 447-57.

MUIR, D. C. G., K. Kidd, and W. L. Lockhart. 1995b. Analysis of food web and sediment samples from Peter Lake for toxaphene, heavy metals, and bone collagen. Report to Department of Indian Affairs and Northern Development, Water Resources Division, Yellowknife, NWT. 10 p.

MUIR, D., B. Braune, B. DeMarch, R. Norstrom, R. Wagemann, M. Gamberg, K. Poole, R. Addison, D. Bright, M. Dodd, W. Duschenko, J. Eamer, M. Evans, B. Elkin, S. Grundy, B. Hargrave, C. Herbert, R. Johnstone, K. Kidd, B. Koenig, L. Lockhart, J. Payne, J. Peddle, and K. Reimer. 1996a. Chapter 3. Ecosystem Uptake and Effects. *In* J. Jensen, K. Adare, and R. Shearer, [eds.] Canadian Arctic Contaminants Assessment Report, Indian and Northern Affairs Canada, Ottawa 1997.

MUIR, D.C.G., A. Omelchenko, N.P. Grift, D.A. Savoie, W.L. Lockhart, P. Wilkinson, and G.J. Brunskill. 1996b. Spatial trends and historical deposition of polychlorinated biphenyls in Canadian midlatitude and Arctic lake sediments. Environ. Sci. Technol. 30: 3609-3617.

MUIR, D.C.G, K.A. Kidd, K. Koczanski, A. Yarechewski, D. Cobb, and R.H. Hesslein. 1997. Bioaccumulation of persistent organochlorines and current use pesticides in the Lake Winnipeg food web. Book of Abstracts: Department of Fisheries and Oceans Green Plan Toxic Chemicals Program Wrap-up Conference. Government Conference Centre, Ottawa, Ontario, January 28-31, 1997. Can. Tech. Rep. Fish. Aquat. Sci. 2163: 19-20.

MUNKITTRICK, K. R., G. J. Van Der Kraak, M. E. McMaster, C. B. Portt, M. R. Van Den Heuvel, and M. R. Servos. 1994. Survey of receiving-water environmental impacts associated with discharges from pulp mills: 2. Gonad size, liver size, hepatic EROD activity and plasma sex steroid levels in white sucker. Environ. Toxicol. Chem. 13:1089-101.

MUNKITTRICK, K.R., L.H. McCarthy, M.E. McMaster, and G.J. Van Der Kraak. 1997. Assessment of threshold exposure levels for biological effects of pulp mill effluents. Book of Abstracts: Department of Fisheries and Oceans Green Plan Toxic Chemicals Program Wrap-up Conference. Government Conference Centre, Ottawa, Ontario, January 28-31, 1997. Can. Tech. Rep. Fish. Aquat. Sci. 2163: 41-42.

MURPHY, M. 1991. Air and water quality summary for the Belledune environmental monitoring committee. 1990. New Brunswick Department of the Environment.

NRIAGU, J.O. 1994. Origin, long-range transport, atmospheric deposition and associated effects of heavy metals in the Canadian environment. A report prepared for Atmospheric Environment Service, Environment Canada, Downsview ON, 30 December 1994.

ONGLEY, E. D., D. A. Birkholz, J. H. Carey, and M. R. Samoiloff. 1988. Is water a relevant sampling medium for toxic chemicals? An alternative environmental sensing strategy. J. Environ. Qual. 17: 391-401.

OWENS, J. W., S. M. Swanson, and D. A. Birkholz. 1994. Bioaccumulation of 2,3,7,8-tetrachlorodibenzo-*p*-dioxin, 2,3,7,8-tetrachlorodibenzofuran and extractable organic chlorine at a bleached-kraft mill site in a northern Canadian river system. Environ. Toxicol. Chem. 13: 343-54.

PACYNA, J.M. 1995. The origin of Arctic air pollutants: Lessons learned and future research. Sci. Total Environ. 160/161: 39-53.

PARKS, J. W., J. A. Sutton, and J. D. Hollinger. 1984. Mercury contamination of the Wabigoon/English/Winnipeg River system, p. 1-353. *In* Mercury pollution in the Wabigoon-English River System of northwestern Ontario, and possible remedial measures. Government of Ontario and Government of Canada, Minister of Supply and Services Canada.

PARKS, J. W., P. C. Craig, and G. W. Ozburn. 1994. Relationships between mercury concentrations in walleye (*Stizostedion vitreum*) and northern pike (*Esox lucius*): implications for modelling and biomonitoring. Can. J. Fish. Aquat. Sci. 51:2090-2104.

PASTERSHANK, G.M., and D.C.G. Muir. 1995. Contaminants in environmental samples: PCDDs and PCDFs downstream of bleached kraft mills, Peace and Athabasca rivers, 1992. Northern River Basins Study Project Report No. 44. Edmonton, Alberta.

PEARSE, P.H. 1988. Rising to the Challenge: A New Policy for Canada's Freshwater Fisheries. Canadian Wildlife Federation.

PEARSE, P.H., F. Bertrand, and J. W. MacLaren. 1985. Water Resources, Uses, and Pressures, p. 21-60. *In* Currents of Change. Inquiry on Federal Water Policy. Final Report. 222 p.

PETERSON, R.H., A. Sreedharan, and S. Ray. 1989. Accumulation of trace metals in three species of fish from lakes in New Brunswick and Nova Scotia (Canada): influence of pH and other chemical parameters. Water Poll. Res. J. Can. 24: 101-17.

PROUSE, N. J., and J. F. Uthe. 1994. Concentrations of pesticides and other industrial chemicals in some sports fish species from a few sites in New Brunswick and Nova Scotia. Canadian Technical Report of Fisheries and Aquatic Sciences 1981: v + 39 p.

RASMUSSEN, J.B., D.J. Rowan, D. R. S. Lean, and J. H. Carey. 1990. Food chain structure in Ontario lakes determines PCB levels in lake trout (*Salvelinus namaycush*) and other pelagic fish. Can. J. Fish. Aquat. Sci. 47: 2030-2038.

REEDER, S.W., A. Demayo, and M.C. Taylor. 1979. Guidelines for surface water quality. Vol 1. Inorganic chemical substances: Cadmium. Environment Canada, Inland Water Directorate, Water Quality Branch, Ottawa, Canada. viii + 19 p.

ROCH, M., and J.A. McCarter. 1984. Hepatic metallothionein production and resistance to heavy metals by rainbow trout (*Salmo gairdneri*): II. Held in a series of contaminated lakes. Comp. Biochem. Physiol. 77C: 77-82.

ROCHE Ltée. 1995. Synthése des connaissances acquises sur la contamination des milieux d'eau douce dans le bassin hydrographique de l'estuaire et du golfe du Saint-Laurent. Sainte-Foy, Québec. Report Prepared for Peches et Oceans Canada, Instit. Maurice-Lamontagne. 37 p.

ROSENBERG, D.M., T.B. Reynoldson, K.E. Day, and V.H. Resh. 1997. Role of abiotic factors in structuring benthic invertebrate communities in freshwater ecosystems. *In* C.G. Ingersoll, T. Dillon, and G.R. Biddinger [eds.] Ecological Risk Assessment of Contaminated Sediments. Proceedings of the Pellston Workshop on Sediment Ecological Risk Assessment, 23-28 April 1995, Pacific Grove, CA. SETAC Special Publication Series. SETAC Press, Pensacola, Florida.

RUDD, J.W.M., and M A. Turner. 1983. The English-Wabigoon River system: II. Suppression of mercury and selenium bioaccumulation by suspended and bottom sediments. Can. J. Fish. Aquat. Sci. 40: 2218-27.

RUDD, J. W. M., M. A. Turner, A. Furutani, A. L. Swick, and B. E. Townsend. 1983. The English-Wabigoon River system: I. A synthesis of recent research with view towards mercury amelioration. Can. J. Fish. Aquat. Sci. 40: 2206-17.

RUDD, J.W.M., C.A. Kelly, R.A. Bodaly, N.R. Roule, V.L. St. Louis, M. Paterson, D. Rosenberg, B. Hall, A. Heyes, T.R. Moore, R. Aravena, B. Dyck, R. Harris, S. Schiff, B. Warner, and G. Edwards. 1997. Increases in fluxes of methyl mercury and greenhouse gases following flooding of an experimental reservoir. Book of Abstracts: Department of Fisheries and Oceans Green Plan Toxic Chemicals Program Wrap-up Conference. Government Conference Centre, Ottawa, Ontario, January 28-31, 1997. Can. Tech. Rep. Fish. Aquat. Sci. 2163: 21-22.

SAFE, S., and D. Phil. 1990. Polychlorinated biphenyls (PCBs), dibenzo-*p*-dioxins (PCDDs), dibenzofurans (PCDFs) and related compounds: environmental and mechanistic considerations which support the development of toxic equivalency factors (TEFs). Crit. Rev. Toxicity. 21:51-88.

SCHETAGNE, R. 1990. Recherches exploratories sur le mercure, Région La Grande 2, 1988. Service recherche en environnement et santé publique, Vice-Présidence Environnement, Montréal, Québec.

SCHEUHAMMER, A.M., and S.L. Norris. 1995. A review of the environmental impacts of lead shotshell ammunition and lead fishing weights in Canada. Occas. Pap. Can. Wildl. Serv. 88: 52 p.

SCHINDLER, D.W., P. J. Curtis, B.R. Parker, and M.P. Stainton. 1996. Consequences of climate warming and lake acidification for UV-B penetration in North American boreal lakes. Nature 379:705-708.

SCOTT, D. P., and F. A. J. Armstrong. 1972. Mercury concentration in relation to size in several species of freshwater fishes from Manitoba and north-western Ontario. J. Fish. Res. Bd. Can. 29: 1685-90.

SERVOS, M.R., D.C.G. Muir, D.M. Whittle, D.B. Sergeant, and G.R.B. Webster. 1989. Bioavailability of octachlorodibenzo-*p*-dioxin in aquatic ecosystems. Chemosphere 19: 969-972.

SERVOS, M. R., S. Y. Huestis, D. M. Whittle, G. J. Van Der Kraak, and K. R. Munkittrick. 1994. Survey of receiving-water environmetal impacts associated with discharges from pulp mills: 3. Polychlorinated dioxins and furans in muscle and liver of white suckers (*Catostomus commersoni*). Environ. Toxicol. Chem. 13: 1103-15.

SHILTS, W.W., and W.B. Coker. 1995. Mercury anomalies in lake water and in commercially harvested fish, Kaminak Lake area, District of Keewatin, Canada. Water Air Soil Pollut. 80: 881-884.

SPRY, D. J., and J. G. Wiener. 1991. Metal bioavailability and toxicity to fish in low-alkalinity lakes: a critical review. Environ. Pollut. 71: 243-304.

ST. LOUIS, V.L., L. Breebaart, J.C. Barlow, and J.F. Klaverkamp. 1993. Metal accumulation and metallothionein concentrations in tree swallow nestlings near acidified lakes. Environ. Toxicol. Chem. 12: 1203-1207

STATISTICS Canada. 1992. Canada Year Book 125th Anniversary 1992. Statisitics Canada, Ottawa, Ontario.

STATISTICS Canada. 1994. Canada Year Book 1994. Statistics Canada, Ottawa, Ontario.

STEIN, J. E., W. L. Reichert, M. Nishimoto, and U. Varanasi. 1990. Overviews of studies on liver carcinogenesis in English sole from Puget Sound; evidence for a xenobiotic chemical etiology II: Biochemical studies. Sci. Tot. Environ. 94: 51-69.

STEPHENS, G.R. 1995. Mercury concentrations in fish in a remote Canadian Arctic lake. Water Air Soil Pollut. 80: 633-636.

STEPHENSON, M., and G. L. Mackie. 1988. Total cadmium concentrations in the water and littoral sediments of central Ontario lakes. Water Air Soil Poll. 38: 121-36.

STEPHENSON, M.L. Bendell Young, G.A. Bird, G.J. Brunskill, P.J. Curtis, W.L. Fairchild, M.H. Holoka, R.V. Hunt, S.G. Lawrence, M.F. Motycka,W.J. Schwartz, M.A. Turner, and P. Wilkinson. 1996. Sedimentation of experimentally-added cadmium and 109 Cd in Lake 382, Experimental Lakes Area, Canada. Can. J. Fish. Aquat. Sci. 53:1888-1902.

STEWART, R., and D.F. Malley. 1997. Technical evaluation of molluscs as a biomonitoring tool for the Canadian mining industry. Aquatic Effects Technology Evaluation Program Project 2.3.1 CANMET, Natural Resources Canada, Ottawa.

STRANGE, N. E., R. A. Bodaly, and R. J. P. Fudge. 1991. Mercury concentrations in fish in Southern Indian Lake and Asset Lake, Manitoba, 1975-1988: the effect of lake impoundment and the Churchill River diversion. Can. Tech. Rep. Fish. Aquat. Sci. 1824: iv + 61 p.

TAYLOR, W. D., J. H. Carey, D. R. S. Lean, and D. J. McQueen. 1991. Organochlorine concentrations in the plankton of lakes in southern Ontario and their relationship to plankton biomass. Can. J. Fish. Aquat. Sci. 48: 1960-1966.

TREMBLAY, A., M. Lucotte, and C. Hillaire-Marcel. 1993. Mercury in the environment and in hydroelectric reservoirs. Great Whale Environment Assessment: Background Paper No. 2. Great Whale Public Review Office. 169 p.

TRUDEL, L. 1991. Dioxins and furans in bottom sediments near the 47 Canadian pulp and paper mills using chlorine bleaching. Water Quality Branch, Inland Water Directorate, Environment Canada. viii + 228 p.

VAN-COILLIE, R.D. Brouard, M.Lachance, and Y. Vigneault. 1984. Ecotoxicological possibilities of acid precipitations for salmon in 4 rivers draining the north coast of the St. Lawrence River. Ann. Limnol. 20: 215-227.

VARANASI, U., J. E. Stein, M. Nishimoto, W. L. Reichert, and T. K. Collier. 1987. Chemical carcinogensis in feral fish; uptake, activation, and detoxication of organic xenobiotics. Environ. Health Perspect. 71: 155-70.

WHITTLE, D. M., C. Mageau, R. K. Duncan, D. B. Sergeant, M. D. Nassichuk, J. Morrisson, and J. Piuze. 1993. Canadian national dioxin sampling program: dioxins and furans in biota near 46 pulp and paper mills using the chlorine bleaching process. Chemosphere 27: 279-86.

WIENER, J. G., and P. M. Stokes. 1990. Enhanced bioaccumulation of mercury, cadmium and lead in low-alkalinity waters: an emerging regional environmental problem. Environ. Toxicol. Chem. 9:821-823.

WIENER, J. G., W. F. Fitzgerald, C. J. Watras, and R. G. Rada. 1990. Partitioning and bioavailability of mercury in an experimentally acidified Wisconsin lake. Environ. Toxicol. Chem. 9: 909-18.

WREN, C. D., and H, R. MacCrimmon. 1983. Mercury levels in sunfish, *Leponyss gibbosus*, relative to lake pH and other environmental variables of Precambrian Shield lakes. Can. J. Fish. Aquat. Sci. 40: 1737-1744.

WREN, C. D., H. R. MacCrimmon, and B. R. Loescher. 1983. Examination of bioaccumulation and biomagnification of metals in a Precambrian Shield lake. Water Air Soil Poll. 19: 277-91.

XUN, L., N. E. R. Campbell, and J. W. M. Rudd. 1987. Measurements of specific rates of net methyl mercury production in the water column and surface sediments of acidified and circumneutral lakes. Can. J. Fish. Aquat. Sci. 44: 750-757.

Personal Communications

CRAIG, C. n.d.

FAIRCHILD, W., Gulf Fisheries Centre, Moncton, N.B., n.d.

The ARCTIC *Marine* Ecosystem

Highlights

■ The Arctic marine ecosystem includes 173 000 km of coastline. The ecosystem has a continental shelf equivalent in area to the combined extent of that in Canada's Atlantic and Pacific Oceans.

■ The Arctic marine ecosystem has many unique attributes including relatively simple but long food chains, low species diversity, low productivity, long-lived species with low reproduction rates and multi-year and seasonal ice.

■ The Arctic is unique in Canada because its waters have been harvested chiefly to meet the subsistence and cultural needs of the Inuit. The marine and freshwater subsistence fisheries are of greater value than commercial or sport fisheries. The estimated annual *per capita* country food harvest for fish and marine and terrestrial mammals in certain Inuit communities is 270 kg, of which about 60 kg is whole fish.

■ The commercial fishery of the Canadian Arctic is relatively small with arctic char worth about $1.2 million annually and Greenland halibut (turbot) worth about $1 million annually.

■ Organochlorine compounds, especially toxaphene, polychlorinated biphenyls (PCBs), dichlorodiphenyltrichloroethane (DDT) and chlordane, are the major contaminants in Arctic biota. Toxaphene and chlordane have never been used in the Arctic but were found in nearly every species of Arctic marine fish and mammal that has been examined. Other contaminants in the Arctic include mercury, metals from mine development, and petroleum hydrocarbons from oil spills. The Arctic marine environment is no longer pristine. Having become a sink, it is now also becoming a source of some contaminants.

Highlights

■ Long-range atmospheric transport (LRTAP) is the main route of delivery of contaminants to the Arctic marine ecosystem. Secondary routes are via ocean currents and rivers. Local sources are also responsible for some contamination.

■ Several lines of evidence suggest that concentrations of mercury are increasing in the Arctic, including in biota, probably as a result of increasing global anthropogenic mobilization of mercury. Since natural background concentrations of these metals in organisms vary through-out the Arctic, it is often difficult to distinguish between the proportions of mercury from natural versus anthropogenic sources. Cadmium is very high in some marine mammals, although its source is probably natural.

■ Although Arctic biota contain a range of organic and inorganic contaminants, there has been little to no research on the biological effects of contaminants on populations of Arctic marine fish and mammals in their natural environment.

Table of Contents

Arctic Marine Ecosystem

List of Figures

List of Tables

List of Sidebars

Introduction

The Arctic and its islands include 173 000 km of coastline, twice that of the Pacific and Atlantic coastlines combined, and over 1 million km^2 of continental shelf. This is equivalent to the combined extent of the shelves in Atlantic and Pacific waters within Canada's 200-mile economic zone. The Canadian Arctic stretches from the Beaufort Sea in the west to Baffin Bay and Davis Strait in the east, and from north of Ellesmere Island to Hudson Bay in the south (Figure 1). In addition to the Arctic archipelago, the Beaufort Sea and Hudson Bay, Canadian Arctic marine waters include the Foxe Basin, James Bay and Hudson Strait. Although most of the Arctic marine areas are above 60° latitude, James Bay extends southward to between 51° and 52°. The Arctic Ocean is the youngest of the world's oceans and the least studied (Alexander 1995).

Many early contaminant studies regarded the Arctic as a pristine site for comparison with more contaminated southern locations, an assumption that has recently been shown to be incorrect (Barrie *et al.* 1997). Although there were isolated studies of organic contaminants and metals in Arctic ecosystems in the 1970s, systematic studies did not begin until the 1990s, much later than those undertaken for other Canadian aquatic ecosystems (Barrie *et al.* 1992; Muir *et al.* 1992; Thomas *et al.* 1992; Lockhart *et al.* 1992; Kinloch *et al.* 1992). The Northern Contaminants Program, particularly the Ecosystem Uptake and Effects sub-program of the Arctic Environmental Strategy (1991 to 1997), reported by Barrie *et al.* (1997), Muir *et al.* (1996), and Gilman *et al.* (1997) has been a source of much of the information included in this chapter.

Recent interest in Arctic ecosystems arose from studies indicating that contaminants are present in the human diet in the North (Kinloch *et al.* 1992) and end up in mothers' milk (DeWailly *et al.* 1989). This led to further study of the contamination of the organisms that comprise "country foods" upon which most aboriginal communities depend for their diet, and the aquatic ecosystems of which these organisms are a part. In contrast to most Canadians in other regions whose diet comes from a variety of sources, the people of the Arctic marine coast, primarily the Inuit living above the tree line, depend greatly on aquatic resources as primary food sources. For the

Figure 1
Circumpolar view of the Arctic marine region.

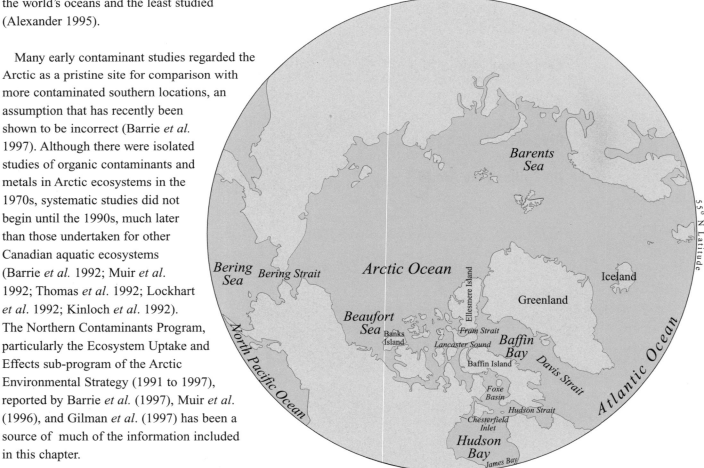

Inuit, country foods provide health, well-being and identity (Egende 1995). However, there are concerns that chronic exposure to chemical contaminants such as polychlorinated biphenyls (PCBs), other organochlorine compounds and metals in the diet may impair the health of the Inuit and the health of top predators such as small-toothed whales and polar bears.

MAJOR CONTAMINANT ISSUES

Organochlorine compounds, especially toxaphene and chlordane, are major contaminants in Arctic biota. Toxaphene and chlordane have never been used in the Arctic but are found in nearly every species of Arctic marine fish and mammal examined (Muir *et al.* 1992). Other organic contaminants in the Arctic include PCBs, dichlorodiphenyltrichloroethanes (DDTs), hexachlorocyclohexane (HCH), hexachlorobenzene (HCB), polycyclic aromatic hydrocarbons (PAHs), and petroleum (aliphatic) hydrocarbons.

Concentrations of mercury appear to be increasing in Arctic biota (Wagemann *et al.* 1996). Since there are varying natural background concentrations of these metals in organisms in various parts of the Arctic, it is often impossible to distinguish between the proportion of mercury contributed by natural versus anthropogenic sources. Cadmium is very high in some marine mammals, but the source of the cadmium is probably natural. Anthropogenically-contributed cadmium is not believed to be an important contaminant in the Arctic marine ecosystem.

OCEANOGRAPHIC FEATURES OF IMPORTANCE TO CONTAMINANT DYNAMICS

Contaminants enter the Arctic Ocean through inflowing ocean currents, deposition from the atmosphere, northward flowing rivers, direct run-off from the land, and direct disposal into the ocean (Macdonald and Bewers 1996). The primary controller of contaminant distribution in the Ocean is the vertical stratification into ocean layers that reflects several sources of water masses, and processes on the continental shelf (Macdonald and Bewers 1996). The surface layer of the Ocean, or polar mixed layer, 30 to 50 m deep, floats upon the denser salt water. It is highly influenced by the brine released during ice formation in winter, by melting ice in summer, and by freshwater run-off via the rivers. These processes establish horizontal density gradients that drive the surface ocean currents (Barrie *et al.* 1992). Below the polar mixed layer are thick layers of water originating from the Pacific or Atlantic Oceans (Macdonald and Bewers 1996). Most of the marine primary production occurs in the polar mixed layer and within the ice. This polar mixed layer collects contaminants from the atmosphere, from run-off, or from ice-melt water. These contaminants may enter the food chain in these biologically-active surface waters and potentially have direct adverse effects.

A dominant feature of the Arctic Ocean is the presence of multi-year and seasonal ice that influences all of its biological regimes. Ice provides a surface on which atmospheric contaminants fall, and into which contaminated sediments from the coastal shelf can be incorporated. As ice moves off the shelves and circulates around the central Arctic Ocean, it is a vehicle for transporting contaminants from the atmosphere and the rivers over long distances across the sea. It is speculated that annual freeze-melt cycles cause non-volatile contaminants to become more concentrated at the ice surface. The period of melting of snow and ice with associated release of contained contaminants coincides with the period of maximum primary productivity, providing enhanced opportunity for contaminants to enter the food chain (Macdonald and Bewers 1996). In addition to these seasonal ice edges, in the high Arctic seas and bays there are extensive areas of permanently open water, (polynyas) that do not freeze over or freeze lightly. These open areas tend to be at fixed geographical locations indicating they are controlled by recurring physical processes (Lewis *et al.* 1996).

Large quantities of sea water flow into the Arctic Ocean from the North Pacific Ocean via the Bering Strait, and from the Atlantic Ocean via the Fram Strait and Barents Sea. Most important for contaminant transfer are the currents in the top 200 m of the oceans. These surface currents are a relatively slow means of transport of contaminants to the Arctic, taking years to decades to transport water from industrialized, temperate coastal regions to the Arctic Ocean (Barrie *et al.* 1997). Whether contaminants are dissolved or adsorbed to particles determines how they are transported by oceanographic processes. Dissolved contaminants tend to cycle around the central Arctic basins in the surface waters with a period of several years to one to two decades. Adsorbed contaminants tend to move around in the intermediate depth waters where they may remain for several decades. They may also move to deep water where they may remain for several centuries (Schlosser *et al.* 1995). In addition to contaminants from temperate, coastal areas, these waters receive chemicals such as HCH which are deposited from the atmosphere (Macdonald and Bewers 1996).

Despite some transport by ocean currents, the main route of entry for most chemical contaminants to the Arctic Ocean is long range atmospheric transport (LRTAP) (Barrie 1986; Hargrave *et al.* 1988; Bidleman *et al.* 1989; Patton *et al.* 1989; Barrie *et al.* 1992; Olsson 1995; Poole *et al.* 1995; Fellin *et al.* 1996). Air pollution in the Arctic was first noted almost 40 years ago when pilots observed a relatively thick haze over the region in the winter months (Mitchell 1956). Recent interest in contaminant

"Brown Snow"

Brown-coloured snow is regularly deposited in the central Canadian Arctic during April and May. One such event, documented in 1988 near Chesterfield Inlet (Welch et al. 1991), deposited thousands of tonnes of fine particulates that included PAHs, PCBs, DDT-related compounds, toxaphene, the herbicide trifluralin, and the insecticides methoxychlor, endosulfan and hexachlorocyclohexane. A trace of trajectories of air movement showed that the air traveled from the Soviet Arctic, across the Arctic Ocean to the western Canadian Arctic and Chesterfield Inlet. This indicated an Asian source for the contaminants, most likely western China. That conclusion was supported by the mineral composition of clay in the particulates and the presence of soot particles, which are residues of coal burning. The industrial use of coal is extensive in China and eastern Russia.

Cold Condensation Theory

"Cold condensation" is also referred to as global distillation, cold finger, cold trap effect, global fractionization or the grasshopper effect. The term refers to a repeated cycle of volatilization of contaminants from a warmer region, followed by condensation or deposition, revolatilization and transport to a colder region where they may accumulate (Dunbar 1973; Wania and Mackay 1993; Gregor and Gummer 1989). This process is believed to be responsible for at least some of the movement of contaminants through the Earth's atmosphere from the warmer climates to the polar regions. Some chemicals never complete the journey. Those that do may fail to condense, so that cold condensation at the poles occurs with varying effectiveness. In the Arctic, selective enrichment of some contaminants, especially organochlorine compounds, may also occur during this process (Mackay and Wania 1995; Muir et al. 1996). Through cold condensation, the Arctic has become a sink for contaminants, such as gaseous toxaphene (Bidleman 1994b).

Once these chemicals, which include organochlorine compounds and probably mercury, reach the Arctic, they may enter terrestrial, freshwater, and marine ecosystems and become bioavailable to marine fish and mammals.

With the dramatic decline in usage of some organochlorine compounds, such as China's reduction of forms of HCH in the past decade (Barrie et al. 1997), the atmospheric concentrations of HCH have dropped by as much as 8-fold. Because the atmospheric contributions are low, the Arctic Ocean becomes oversaturated with HCH at certain times of the year. In this process revolatilization of HCH occurs, and the Arctic becomes a source of contaminants to the atmosphere. The Arctic Ocean can also be a source of HCH to the Atlantic Ocean as Arctic waters flow out through the archipelago (Barrie et al. 1997).

dynamics in the Arctic has centered on the "cold condensation theory" (See Sidebar on Cold Condensation Theory).

Arctic Cod (*Boreogadus saida*)

A small relative of Atlantic cod and one of several species of cod-like fish classified as gadoids, Arctic cod play a key role in the Arctic marine foodchain because they are abundant and are consumed by most marine vertebrates. Although not harvested commercially or for subsistence use, they are the basic component of the diet of at least 12 species of Arctic marine mammals, 20 species of seabirds and 4 species of fish. Circumpolar in their distribution, Arctic cod have been found farther north than any other fish. They inhabit Arctic seas off northern Russia, Alaska, Canada and Greenland and are commonly found close to shore among ice flows and also offshore at depths greater than 900 m. Temperatures of 0° to 4°C are believed to be optimal for the survival of Arctic cod, although they have been found in water colder than 0°C. Arctic cod are the main consumers of plankton in the Arctic seas. Unlike Atlantic cod which are groundfish, Arctic cod rarely feed on bottom dwelling organisms. Instead they rely mainly on those organisms making up the drifting plankton population in the upper layers of the water column. Arctic cod have a short life span. Maturing at about 3 years of age when about 20 cm long, they live only for about 6 years. In northern Canadian waters, spawning occurs in late autumn and winter under ice cover.

Arctic cod are of great importance in Arctic food webs because they are abundant and are the basic component of the diets of virtually all other marine vertebrates including seabirds, seals, narwhal, and beluga. Thus, Arctic cod occupy a central role in the movement of energy and contaminants from lower to upper trophic levels, and ultimately to humans. There is very limited information on the concentrations of organochlorine compounds including PCBs and PAHs in Arctic cod. In a review by Muir et al. (1992), muscle of Arctic cod collected at three eastern Arctic locations in the mid 1980s had toxaphene concentrations 5 to 10-fold higher than those for DDT or PCB residues. There is very little information on the concentrations of metals, including mercury and cadmium, in Arctic cod. Macdonald and Sprague (1988) determined that although Arctic cod had one of the lowest average whole-body cadmium concentrations of lower trophic level Arctic marine organisms, they appeared to be high in cadmium concentrations compared to equivalent organisms from the Baltic Sea and Gulf of Bothnia. Even though knowledge of contaminant concentrations in Arctic cod is limited, their position in the Arctic food web is pivotal in the biomagnification of organochlorine compounds and heavy metals from prey to predator to humans.

As summarized by Macdonald and Bewers (1996), in winter, air flows are mainly from the European and Asian continents into the Arctic, producing arctic haze in spring. This has been linked to direct inputs of contaminants from the heavily industrialized areas of eastern and northern Europe and Asia. In summer, the air flows are reversed. As well, episodic events can transport materials from other parts of the world. In contrast to slower ocean currents, atmospheric air masses rapidly bring volatile and semi-volatile compounds and aerosols containing particle-reactive lead, PCBs and PAHs to the Arctic. These contaminants are deposited on snow, ice, sea water and northern river drainage basins (See Sidebar on Brown Snow).

Rivers bring about 3 300 km^3 of freshwater to the Arctic Ocean annually. This flow occurs mainly via four major rivers — the Yenisey, the Ob, the Lena, and the smallest of the four, the Mackenzie (Aagaard and Carmack 1989). Nevertheless, the amount of freshwater flowing into the Arctic Ocean is approximately 1% of the inflow of seawater entering via Fram Strait. As indicated above, Canadian and Russian rivers are important oceanographically as they contribute to the polar-mixed layer. The Mackenzie River is the only major North American river to flow through the Arctic Ocean basin region. The bulk of the Yukon River flow mixes with Pacific Ocean waters before entering the Arctic Ocean. The Churchill, Nelson, Severn and Great Whale Rivers drain the Hudson Bay ocean basin region. Also draining this ocean basin region, are the Attawapiskat, Albany and La Grande Rivers that flow into James Bay. Canadian freshwater drainage basins into the Arctic are described in Figure 1 and Table 1 of the Inland Freshwater Ecosystems Chapter.

The fate of contaminants brought to the Arctic by rivers depends on how the rivers enter the sea and on contaminant properties, such as their tendency to be dissolved or adsorbed to particles. Particles with their contaminant loads may be trapped in coastal sediments, subject to resuspension, directly entrained into ice, or entrained into ice after sedimentation. Dissolved contaminants enter the Arctic Ocean and are diluted (Macdonald and Bewers 1996).

BIOLOGICAL RESOURCES

Productivity is generally low in the Arctic primarily because the sunlight needed for growth is limited. Ice algae bloom in the spring as light increases seasonally and snow begins to melt, but the major phytoplankton bloom, responsible for most of the aquatic primary production, occurs in June and July with the ice melt. Although spring and summer days are very long in the Arctic, the presence of seasonal ice cover prevents most of the light from reaching the water column from about the time of summer equinox until early July. Thus, about half of the growing season is unavailable to most of the primary producers. Cloud cover tends to increase after the spring, further reducing available light. The ice edge, whether at a polynya or at seasonally melting ice is an area where marine mammals and seabirds congregate to feed. It is a site where contaminants that are present can move quickly from the environment into primary producers, and other biota, including higher predators.

Nutrient inputs to the Arctic marine ecosystem are from aerial deposition, upwelling of nutrient-rich water from southern water masses, and riverine inputs. Although Arctic productivity is generally low, there are areas of unusually high productivity such as Lancaster Sound where upwelling under the polar ice and near the ice edge provides a rich source of nutrients. High productivity occurs in the Northwater Polynya in Smith Sound and the northern part of Baffin Bay where upwelling brings nutrients to upper waters and keeps the ice thin and unconsolidated on a large area of sea. Here, chlorophyll has been recorded at the very high concentration of 510 mg.m^{-2} (Lewis *et al.* 1996).

Arctic food webs have unique characteristics that influence the extent and manner in which contaminants accumulate. The Arctic is relatively low in species diversity. Because of this low diversity, some food chains may be very simple, although most Arctic food chains are longer than in the temperate regions. A generalized Arctic marine food web based on that in Lancaster Sound is shown in Figure 2. In Arctic marine food webs, bioaccumulation of organic contaminants and metals in surface waters can begin with ice algae, phytoplankton, or kelp that take up dissolved contaminants from the water. Nutrient inputs to Arctic waters tend to carry particulate and dissolved organic matter, and with them, contaminants. Nutrients consumed by microorganisms or zooplankton are excreted to be used again by primary producers. Persistent contaminants tend to remain and accumulate in lower food chain organisms and up the food chain. Invertebrates feed on the ice algae, bacteria, protozoa, and microscopic and larger algae, and in turn are consumed by marine fish. Seabirds and marine mammals, including seals and toothed whales, may be at a higher trophic level. Polar bears, humans and sled dogs are on the top level of the Arctic marine food chain, however, humans and dogs may be higher if they consume polar bear. A difference between the Arctic and temperate food chains is that there may be several levels of carnivory; for example, the polar bear is a third-level carnivore in the food chain. Humans in the Arctic are at a higher place in the food chain than humans in southern latitudes, who tend to consume relatively more plant material. Therefore, people in the Arctic are subject to more levels of biomagnification of contaminants through the food chain.

Productivity is highly cyclic due to extreme seasonal fluctuations in temperature, light levels and nutrient inputs. As a result, many marine mammals migrate for overwintering, feeding, or breeding and accumulate contaminants from outside the Arctic. Another consequence of this cyclic behaviour is that contaminants can move through several levels of a food chain very rapidly during periods of high biological activity in the summer.

Compared to temperate regions, the Arctic has slower-growing and longer-lived cold-blooded biota, such as the fish. Slow-growing organisms can be exposed to bioaccumulating contaminants for a long period of time before being consumed by the next level in the food chain. Arctic warm-blooded marine mammals also tend to be long-lived, because of their large size, not because they grow slowly. These animals are also exposed to bioaccumulating contaminants for extended periods of time before being consumed. Fish and marine mammals have adapted to the cold temperatures of the Arctic by accumulating high lipid reserves. Large-bodied marine mammals have high maintenance energy requirements,

especially in winter when the stored energy-rich lipids are utilized for survival. The high lipid body content increases the potential for accumulation and concentration of contaminants, especially of lipophilic compounds such as many organo-chlorine compounds (Barrie *et al.* 1992).

These features of Arctic food webs and physiology enhance the likelihood of biomagnification. Contaminants absorbed originally by algae are passed up through the food chain. At each level they are only slowly excreted and metabolized and accumulate primarily in lipid reserves. Most of these contaminants are obtained from the diet which becomes increasingly

**Figure 2
Simplified Arctic marine food web.**

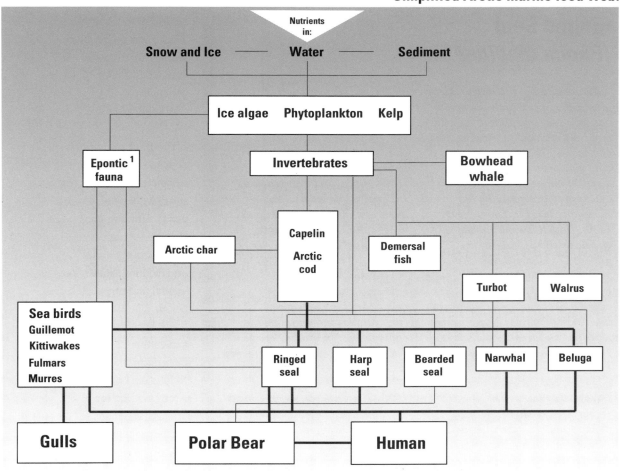

[1] Inhabiting the under ice surface
Heavy lines highlight energy flow mediated by arctic cod.

Source: Welch *et al.* 1992; Muir *et al.* 1996.

contaminated at each trophic level. Critical to the high amount of biomagnification in Arctic food chains is that growth efficiency is often low, that is, little of the food consumed is converted to growth of the animal and the persistent contaminants in the food are concentrated in a body mass that is growing slowly. Although some Arctic invertebrates have high growth efficiency (Welch *et al.* 1996), many others such as short clams at Resolute Bay have low productivity (growth) and they can accumulate large amounts of contaminants. The warm-blooded mammals have greater energy requirements for the same body weight than fish and incidentally tend to consume more contaminants. As well, marine mammal growth efficiencies are low. As a result, biomagnification of PCBs, organochlorine compounds, and mercury in Arctic biota is greatest in marine mammals, although it is also high for some fish (Muir *et al.* 1992).

As predators in the Arctic marine food web, marine mammals are of special interest for monitoring spatial and temporal trends and biological effects of contaminants. The ringed seal is used for marine contaminant monitoring because it is by far the most abundant and widely distributed resident Arctic pinniped. These seals consume mainly Arctic cod and crustaceans (amphipods, mysids and euphausids). Higher in the food chain, polar bears can also be used as a marine contaminant monitor. They preferentially consume ringed seal blubber and skin and lesser amounts of bearded seals, beluga, and walrus. Sled dogs are also potential monitors for contaminants since they typically eat seal, walrus, whale meat and fat, and occasionally polar bear. Thus, their potential for biomagnification is high (H.E. Welch personal communication).

The Inuit traditionally depend on the Arctic marine environment for subsistence and cultural needs. Hunting and fishing are important to their identity, social organization, status, and mental and physical health. The traditional lifestyle of the Inuit makes them not only an integral part of local food chains of the Arctic but the top predators. Because northern native people rely on marine mammals and fish for the majority of their protein and caloric

Ringed Seal
(Phoca hispida)

The smallest of the marine Arctic pinnipeds, ringed seals are one of only two seal species in the world that have adapted to living in land-fast sea ice. The ringed seal is circumpolar in its Arctic distribution, occupying areas of ice-covered oceans and a few freshwater lakes in Europe and Baffin Island. Named for the characteristic pale rings on its greyish or yellowish coat, the ringed seal grows to about 1.5 m in length and 90 kg in weight. A coastal inhabitant, the ringed seal lives near the Arctic ice pack and feeds on crustaceans, molluscs and some fish, such as Arctic cod. Females bear a single white-coated pup each year in a den dug into the snow. Born in March or April, young seals remain with their mothers until about mid June when they migrate into the open ocean. After maturing at age 4 years of age they return to the ice-covered fjords and bays where they remain

for the rest of their lives. A ringed seal can live from 35 to 40 years. A common species, the ringed seal was once important to Arctic coastal peoples as a source of leather, oil and meat. In some Inuit communities, Arctic ringed seals still play a vital role as an important traditional food source which has both cultural and recreational significance.

Like many other marine mammals, the ringed seal accumulates appreciable concentrations of fat-soluble contaminants in its blubber. DFO scientists have studied contaminant distributions in the Arctic ringed seal extensively since the early 1970s. For example, Arctic ringed seals have been used to study spatial and temporal trends of compounds such as the DDT-group of insecticides and PCBs, which are introduced into the Arctic through LRTAP.

intake, contaminants in their food resources are a human health concern (Gilman *et al.* 1997); (See Case Study on Contaminants in Country Foods).

In keeping with the traditional dependence of the Inuit on aquatic resources for subsistence, the marine and freshwater subsistence fisheries are of greater value than commercial or sport fisheries (Lanno *et al.* 1994). About 60 kg of whole fish per capita, are harvested annually in the subsistence fisheries (Berkes 1990). The estimated annual per capita country food harvest for fish and marine and terrestrial mammals in the Inuit communities of Keewatin, Kitikmeot and Baffin regions ranges from 110 to 600 kg, with a mean of 270 kg (Wong 1985). In the Arctic region, 410 beluga, 350 narwhal and 380 walrus were reported harvested during the 1994/95 season, entirely from the eastern Arctic region (DFO 1995), although these data may be incomplete. Ringed seals are also important in the harvest.

The commercial fishery of the Canadian Arctic is relatively small with arctic char valued at about $1.2 million annually and Greenland halibut (turbot) worth about $1 million annually (Welch 1995). In 1994/95, the Northwest Territories commercial fisheries sold approximately 72 000 kg of sea-run arctic char for about $190 000 (DFO 1995). During the same period, approximately 280 000 kg of turbot was harvested that had a landed value of about $590 000 (DFO 1995). Most of this turbot was harvested from the Pangnirtung district of South Baffin Island in the eastern Arctic. There is no commercial fishery in James Bay or Hudson Bay; however, there is a subsistence fishery utilized mainly by aboriginal peoples (Morin and Dodson 1986). Arctic cod is the only marine fish species that is relatively abundant in many Arctic areas. Ringed seal is the most common and most important seal in the Canadian Arctic and is used for food and clothing by the northern populations.

Assessment of Chemical Contamination

SOURCES

Atmospheric transport is a route of entry of mercury to the Arctic (Pacyna and Keeler 1995). A large amount of the mercury is of anthropogenic origin. The main global sources of anthropogenic emissions to the atmosphere are the combustion of fossil fuels to produce electricity and heat, the incineration of refuse, and industrial manufacturing (Pacyna and Keeler 1995). Evidence suggests that up to 50% of various metallic atmospheric contaminants in the Arctic during winter and summer are from fuel burning (Pacyna and Keeler 1995). Emissions from sources in Europe and Asia (the Urals, the Norilsk area, and the industrial regions in central Europe) contribute more than half of the atmospheric contamination measured in the Arctic. Europe seems to contribute less contamination than Russia to the Canadian Arctic (Pacyna 1995).

Reservoirs, known to promote the production of methylmercury, are potential sources of mercury to the Arctic. These have been developed for hydroelectric power along freshwater rivers that drain into James Bay and Hudson Bay (i.e. the Churchill River and the Nelson River in Manitoba; La Grande River in Quebec); (See Case Study in the Inland Freshwater Ecosystems Chapter: Hydroelectric Developments and Mercury Contamination in Northern Aquatic Environments).

Although most of the anthropogenic inputs of metals to the Arctic come from distant sources, there are numerous natural and anthropogenic local point sources in the Arctic. Point sources of metals, such as mercury and cadmium, include mineral outcrops, mining activities, and abandoned mines (Thomas *et al.* 1992). Metals are released from the Smoking Hills at Cape Bathurst, Northwest Territories, an area of spontaneous burning bituminous shales (Thomas *et al.* 1992). Anthropogenic point sources include oil and gas exploration and production (over 1 200 wells have been drilled in the Arctic), disposal of drilling fluids and sump materials (Thomas *et al.* 1992), and combustion of fossil fuel for heating (particularly for mercury) (Ottar *et al.* 1986).

The major source of organochlorine compounds to the Arctic is the atmosphere. The transportation cycle begins with release of contaminants in warmer southern regions, transport to a colder region, condensation or deposition and revolatilization (Wania and Mackay 1993; Gregor and Gummer 1989; See Sidebar on Cold Condensation Theory). The presence of widely-used organochlorine compounds (especially toxaphene and chlordane) and their metabolites in fish and marine mammals in the Arctic may be indicative of contaminants from North American, Asian, or European sources transported to the Arctic via LRTAP. Asian and European LRTAP sources include soil dust and polluted air masses (Welch *et al.* 1991; Wania and Mackay 1993; See Sidebar on Brown Snow). North American LRTAP sources include the use of toxaphene as a pesticide on agricultural crops in the southern U.S. during the 1960s, 1970s and 1980s (Bidleman *et al.* 1989; Stern *et al.* 1992). It is estimated that 90 to 95% of PCBs that were present in soils in Asia, Europe and North America when concentrations peaked in the early 1970s have revolatilized since the 1970s and become available for LRTAP, probably to the Arctic (Gregor 1994). Thus, no decrease in deposition of PCBs to the Arctic is expected in the short term. HCHs are the most abundant pesticides in arctic air and water. Barrie *et al.* (1997) provided estimated annual uses of α-HCH and γ- HCH by country in both 1980 and 1990. In 1980, China used more than 10 fold as much of each of these as the next largest user, India. In 1990, Chinese HCH use declined to a minute

fraction of its previous level and India, the former Soviet Union, and France were the largest users.

There are very few point sources of the above organic contaminants in the Arctic since they have never been manufactured or systematically used there. Two identifiable sources are abandoned Distant Early Warning line (DEW-line) sites and some northern airports (Thomas *et al.* 1992). Direct disposal of PCBs into soils occurred at DEW-line sites in the Canadian Arctic that resulted in contamination of the nearby marine environment (Bright *et al.* 1995a). Oil exploration and production offshore on the Canadian Beaufort Shelf may also contaminate the Arctic marine environment. (Thomas *et al.* 1986).

Anthropogenic point sources of PAHs include waste oils, aviation fuel, and diesel fuel that have been found in abandoned DEW line sites and at some northern airports (Thomas *et al.* 1992). Petroleum hydrocarbons are also released into the Arctic environment from local oil spills and landfills and waste sites that contain abandoned fuel drums. Natural releases of hydrocarbons and smoke containing sulphur dioxide and sulfuric acid aerosols occur from the Smoking Hills at Cape Bathurst, Northwest Territories, an area of spontaneously burning bituminous shale (Thomas *et al.* 1992). Natural hydrocarbon releases to the Mackenzie River occur at Norman Wells (Wilson *et al.* 1973).

Some contaminants in water or ice in the Arctic marine environment are transported directly by ocean currents via the Bering Strait from the Pacific Ocean and Fram Strait from the Atlantic Ocean (Alexander 1995). Contaminants may also be transported via freshwater river transport (Barrie *et al.* 1992). The Mackenzie River, and the rivers of Asia and Europe that flow northward into the Arctic drain approximately 110 km^2 of northern Asia, northern Europe and North America (Barrie *et al.* 1992). These rivers may transport contaminants to the Arctic Ocean that originate from point sources, atmospheric deposition and reservoirs. It is estimated that the Mackenzie River carries about 100 kg of total HCH, 30 kg of total DDT, 370 kg of total PCBs, 5 kg of total chlordane, and 530 kg of lead to the Arctic Ocean annually (Barrie *et al.* 1997). All the other Canadian rivers entering the Arctic Ocean and Hudson Bay introduce lesser quantities. Russian rivers are much greater sources of organochlorine compounds to the Arctic Ocean than the Canadian ones, bringing 15 000 to 30 000 kg of total HCH annually compared with 150 kg by all the Canadian rivers (Barrie *et al.* 1997).

DISTRIBUTION, TRENDS AND EFFECTS
METALS

Present loading of mercury to the Arctic is considerably higher than in the past (Wagemann *et al.* 1995). Pacyna and Keeler (1995) estimate that 60 to 80 t of mercury annually emitted in Asia, Europe and North America is deposited in the Arctic. Judging from increases in the uppermost layers of sediments, mercury deposition in the Arctic appears to have increased more in recent times than other metals, perhaps by a factor of two (Macdonald and Bewers 1996). This is seen in sediment cores taken from Hudson Bay in 1992 and 1993 where mercury concentrations are as high as 34 ng.g^{-1} in the top slice, compared with 15 to 22 ng.g^{-1} deep in the core (Figure 3; Lockhart *et al.* 1995). High sedimentation rates (up to 1 400 g.m^{-2}.yr^{-1}) resulted in mercury fluxes up to 47 µg.m^{-2}.yr^{-1} (Lockhart *et al.* 1995). The observation of increasing mercury in recent sediment layers is evidence that there are anthropogenic sources in addition to the natural ones (Barrie *et al.* 1997). The evaluation of mercury concentrations in Arctic marine biota and attribution to natural and anthropogenic sources is complex because natural background concentrations can vary considerably (Wagemann *et al.* 1995). Determining the relative contributions from natural and anthropogenic sources to mercury concentrations in biota at particular sites may not be possible.

Information on mercury concentrations in Arctic marine fish is very limited. Rarely is the most toxic form, methylmercury, measured. Table 1, with data from 1976 to 1994 on the concentrations of metals in muscle shows that the concentrations of mercury are generally low, below the 0.5 µg.g^{-1} wet weight, established by Health Canada for the protection of human consumers of fish. Much of the information comes from studies carried out in the 1970s and 1980s prior to and following mining developments (Bohn and McElroy 1976; Bohn and Fallis 1978; Fallis 1982). Little or no metal contaminant information exists on Arctic marine fish from Hudson Bay, Hudson Strait or areas south and west of Lancaster Sound (Muir *et. al.* 1996). Methylmercury is found at higher concentrations in liver and kidney than in muscle. Data on methylmercury in these organs or in the whole body are needed for assessment of biological effects of mercury on aquatic organisms (See the Inland Freshwater Ecosystems Chapter).

Marine mammals had considerably higher mercury concentrations in liver than in muscle, with the lowest concentrations in muktuk (the fat and skin layer of the animal, Table 2). Ringed seals and beluga from the western Arctic had significantly higher concentrations of mercury in liver than those from the eastern Arctic (Figure 4; Wagemann *et al.* 1995). Average mercury concentrations in muscle of beluga from the western Arctic were more than twice the Health Canada guideline for fish muscle (Muir *et al.* 1996). Total mercury concentrations in beluga liver were 18 times and in beluga kidney were 4 times higher than the concentrations in muscle (Muir *et al.* 1996). Concentrations of mercury in polar bear liver from the western Arctic were much higher than from the eastern Arctic and appeared to be highest in areas bordering the Beaufort Sea (Braune *et al.* 1991). This geographical pattern was also reported for mercury concentrations in liver of ringed seal and beluga. Data on geographic distribution reflect the natural geological variability of mercury present in sediments and water (Wagemann *et al.* 1995). Sedimentary geological formations of the western Arctic tend to have higher natural concentrations of mercury than those of the central and eastern Arctic. Nevertheless, it cannot be ruled out that the west-east spatial differences are attributable in whole or in part to geographical differences in atmospheric deposition of anthropogenically-mobilized mercury. There appears to be a geographic gradient in concentration of mercury and lead in walrus liver with concentrations lowest in the north (Foxe Basin) and increasing to the south (Hudson Bay) (Wagemann and Stewart 1994).

Mercury concentrations in the liver of beluga (from the periods 1981 to 1984 and 1993 to 1994 in the western and eastern Arctic), ringed seals (from the periods 1972 to 1973 and 1987 to 1993 in the western Arctic) and possibly narwhal (from the periods 1978 to 1979 and 1992 to 1994 at Pond Inlet) increased with time (Wagemann *et al.* 1996). The recently-collected belugas accumulated mercury in the liver approximately twice as fast as the earlier ones. Recently collected ringed seals accumulated mercury three times as fast as the earlier seals (Wagemann *et al.* 1996). Recent studies have shown that increasing concentrations of mercury correspond to increasing concentrations of selenium (Hild

Figure 3
Mercury profiles in four cores from Hudson Bay.

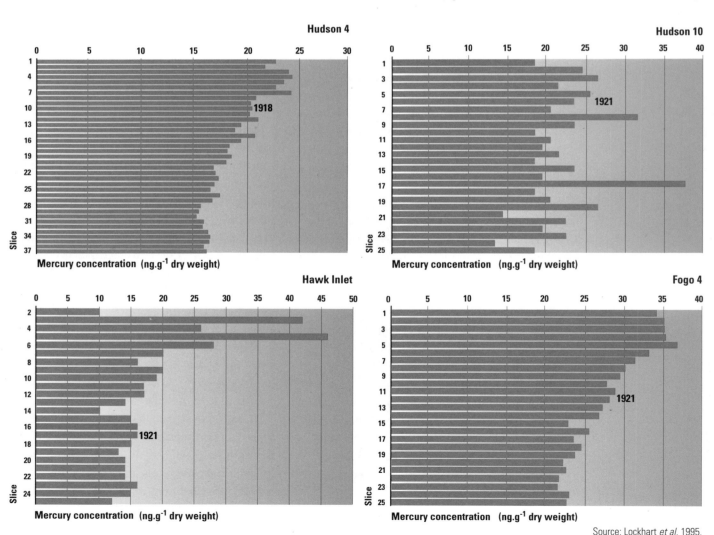

Source: Lockhart *et al.* 1995.

1995). The possible effects of selenium on mercury uptake and toxicity have not yet been adequately studied.

Vertical profiles of cadmium concentration measured in the Arctic Ocean usually show a range from about 30 to 70 ng.L^{-1} (Barrie *et al.* 1997). These concentrations seem to reflect natural background levels for these waters. Higher concentrations were associated with anthropogenic activity. For example, cadmium concentrations in seawater under ice of Strathcona Sound near the Nanisivik lead-zinc mine were 5 700 ng.L^{-1} at the 2 m water depth and decreased to 200 to 300 ng.L^{-1} at 10 to 20 m depth

(Wagemann *et al.* 1983). Strathcona Sound and Beaufort Sea near-shore sediments increased substantially in concentrations of lead, cadmium and zinc after the Nanisivik mine began operation (1979) compared with before operation (1975) (Fallis 1982; Thomas *et al.* 1982, 1983, 1984). Cadmium in sediment cores from Banks Island shelves, and the Chukchi and East Siberian seas often showed maxima (3 μg.g^{-1} dry weight) that exceeded the Canadian Marine Sediment Quality Guidelines of 0.7 μg.g^{-1}. These maxima were interpreted as natural due to cadmium precipitation with sulphide at certain depths in the core (Barrie *et al.* 1997). Cadmium profiles in Arctic marine sediments cannot be used to

Table 1   Mean concentrations of mercury and cadmium concentrations in muscle of Arctic marine fish.

Species	Location	Number	Mercury (μg.g^{-1} wet weight)	Cadmium (μg.g^{-1} wet weight)	Reference
Arctic char	Eastern Hudson Bay	9	0.06	0.003	[3]
(anadromous)	Cambridge Bay	5	0.05	-	[3]
	Chesterfield Inlet	6	0.05	-	[3]
	Rankin Inlet	4	0.11	-	[3]
	Frobisher Bay	5	0.08	-	[3]
	Dease Strait	10	0.03	-	[3]
Arctic cod	Arctic Bay	8	0.02	<0.07	Muir *et al.* 1986
	Kugmallit Bay	6	0.02	<0.05	Muir *et al.* 1986
	Cambridge Bay	1	0.02	<0.05	Muir *et al.* 1986
	Resolute Bay	2	0.04	<0.05	Muir *et al.* 1986
	Pangnirtung	6	0.03	<0.06	Muir *et al.* 1986
	Strathcona Sound	7	-	<0.01[1]	Bohn and McElroy 1976
	Strathcona Sound	50	-	<0.01[1, 2]	McDonald and Sprague 1988
	Inuvialuit Settlement Region	-	0.02-0.04	-	Muir *et al.* 1986
Pacific herring	Tuktoyaktuk	2	0.02	<0.05	Muir *et al.* 1986
Arctic flounder	Tuktoyaktuk	2	0.03	<0.05	Muir *et al..* 1986
Fourhorn sculpin	Tuktoyaktuk	2	0.18	<0.05	Muir *et al.* 1986
	Resolute bay	1	0.05	-	
Shorthorn sculpin	Strathcona Sound	3	-	0.28[1]	Bohn and Fallis 1978
Greenland cod	Cambridge Bay	7	0.04	<0.05	Muir *et al.* 1986
Fish doctor	Resolute Bay	2	0.08	<0.05	Muir *et al.* 1986

[1] - Converted to wet weight basis assuming a moisture content of 80%.

[2] - Whole body concentration.

[3] - Department of Fisheries and Oceans (DFO). 1994, Fish Inspection Service, Winnipeg, Manitoba (unpublished data).

calculate anthropogenic cadmium fluxes or trends as was done above for mercury (Barrie *et al.* 1997).

Although about half of the cadmium entering the Arctic Ocean via the atmosphere, rivers and ocean currents could be anthropogenic, these inputs are very small compared with the large natural pool of cadmium in the ocean (Barrie *et al.* 1997). Barrie *et al.* (1997) conclude that long-range anthropogenic loading of cadmium to the Arctic marine ecosystem is not presently a contaminant concern, and that problems arising

from human inputs of cadmium are likely to be from local sources.

Cadmium increased with trophic level in the food chain from phytoplankton (4.7 $\mu g.g^{-1}$) to carnivorous amphipods (6.3 $\mu g.g^{-1}$) in the northern Baffin Island, Lancaster Sound, and Jones Sound areas, but was unexpectedly low in Arctic cod for their trophic position (0.4 $\mu g.g^{-1}$ whole body) (Macdonald 1986; Macdonald and Sprague 1988). The few data on cadmium in Arctic marine fish muscle indicate that the concentrations are fairly low (Table 1). In marine

Table 2 **Concentrations of mercury, cadmium, and selenium in marine mammals from eastern Hudson Bay[1], and Pond Inlet[2].**

Species	Stage	Tissue	Number	Cadmium	Mercury	Selenium
				Metal Concentration Range or Mean ($\mu g.g^{-1}$ wet weight)		
Ringed seal[1] (Great Whale)		Muscle	3	0.009-0.20	0.058-0.093	0.54-0.69
		Liver	3	0.87-5.8	0.44-2.7	2.0-2.9
Ringed Seal[1] (Belcher)		Muscle	26	-	0.054-0.35	-
		Liver	27	-	0.13-26	-
Harp seal[b,2]	Mother	Liver	20	6.6	10	5
	Pup		20	<0.006	0.32[a]	0.53[a]
	Mother	Kidney	20	28	0.83	1.6
	Pup		20	<0.005	29	1.6
	Mother	Muscle	20	0.04	0.38	0.4
	Pup		20	<0.006	0.14	0.44
Walrus[1]		Muscle	4	0.006-0.058	.05-0.09	0.7-1.2
		Liver	4	1.1-5.5	35-3.7	1.2-2.2
Narwhal[c,2]						
(1979)		Muscle	58	0.19	0.85	0.44
(1979)		Liver	38	34	6.1	4.1
(1979)		Kidney	55	64	1.7	3.2
(1979)		Blubber	45	0.08	0.03	0.07
(1992)		Liver	9	47	-	-
(1992)		Kidney	9	76	-	-
Beluga[1]		Muscle	5	0.015-0.19	1.1-2.6	0.38-0.72
		Liver	4	1.8-19	15-99	4.3-25
		Muktuk	5	<0.001-0.047	0.16-1.5	5.2-10

a= < values not included in mean.
b= all harp seal values mean concentrations.
c= all narwhal values mean concentrations.

Source: Kingsley 1994; Wagemann *et al.* 1988; Wagemann *et al.* 1983; Muir *et al.* 1996.

mammals, the cadmium in liver is much higher relative to muscle than in fish. Highest concentrations were recorded in liver and kidney of narwhal from Pond Inlet. These are so high that there is concern that the cadmium may be chronically toxic to the narwhal themselves (Wagemann *et al.* 1983). Elevated cadmium concentrations in narwhal were confirmed in 1992 (Table 2).

Ringed seal and beluga from the eastern Arctic have higher concentrations of cadmium in their livers than do animals from the western Arctic (Figure 4; Wagemann 1994; Wagemann *et al.*

1996). This is opposite to the geographic gradient that was observed for mercury and is also attributed to variation in geological formations. There was no trend with time in the concentrations of cadmium in liver or kidney of belugas sampled in the period 1993 to 1994 compared with animals sampled in the period 1981 to 1984 (Wagemann *et al.* 1996). No time trend data for cadmium are available for ringed seal. Cadmium concentrations in the kidney of narwhal were among the highest in marine mammals in the Canadian Arctic (Table 2; Wagemann *et al.* 1996). No increase in cadmium in the tissues of narwhal were recorded over

Figure 4
Mean mercury and cadmium concentrations in ringed seal and beluga liver.

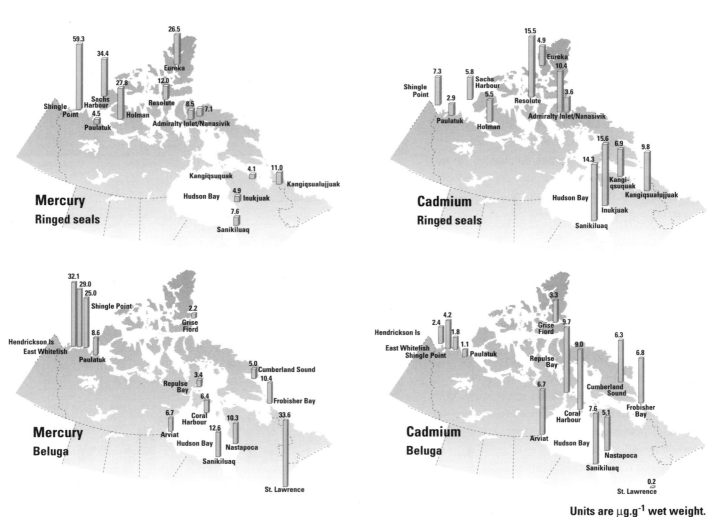

Units are $\mu g.g^{-1}$ wet weight.

Source: Muir *et al.* 1996.

the 14-year time span. The data were first taken during the period 1978 to 1979 and later from 1992 to 1994. Accumulation of cadmium in polar bear liver from the eastern Arctic was three times higher than for bears from the western Arctic (Braune *et al.* 1991). Cadmium concentrations increased with age of the bears. Cadmium concentrations in walrus liver showed a reverse trend to mercury, decreasing in concentration from northern to southern latitudes.

Harp seal mothers and pups showed different relationships for cadmium than for mercury and selenium (Table 2; Jones *et al.* 1976). Cadmium in the mother is bound intracellularly in the liver and kidney, and the blood reaching the placenta is low in cadmium. Thus, pups had low cadmium concentrations in liver and kidney even when this metal was in high concentrations in these organs in the mother. Mercury concentrations in the kidney of mother and pup were more similar to each other. Pups receive most of their mercury in the form of methylmercury, which is considered to be the most toxic form of mercury.

Primarily in response to controls on lead in gasoline in North America in the 1970s, lead in arctic air and glaciers reflects the decreasing trends in anthropogenic emissions of lead to the atmosphere (Barrie *et al.* 1997).

ORGANIC CHEMICALS
ORGANOCHLORINE COMPOUNDS

Organochlorine compounds appear in nearly all components of the Arctic marine ecosystem, although sometimes at very low concentrations. Although many organochlorine compounds have been detected in arctic fauna, the database is small and,

prior to 1990, was restricted almost entirely to tissues from mammals such as seals and polar bears (Muir *et al.* 1996).

HCH was the dominant organochlorine compound measured in Arctic air (Figure 5; Fellin *et al.* 1996), and its concentrations have

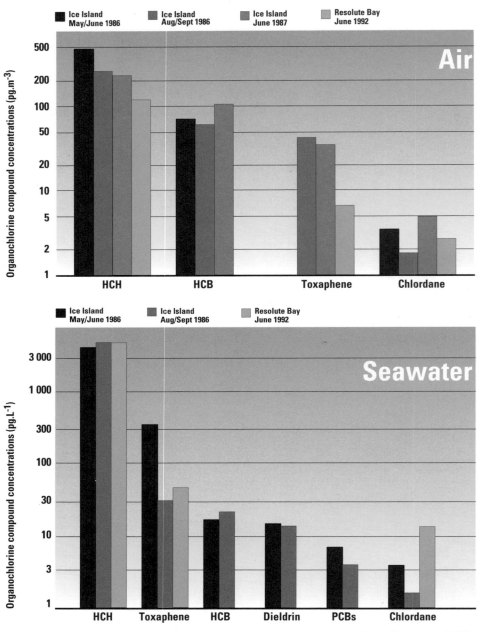

Figure 5

Various organochlorine compound concentrations in Arctic air and seawater from Ice Island and Resolute Bay.

Source: Hargrave *et al.* 1988;
Hargrave *et al.* 1989; Bidleman *et al.* 1995a.

declined with time (Bidleman *et al.* 1995b). Concentrations of HCH (total HCH as isomers α and γ-HCH) at Resolute Bay decreased from 880 pg.m^{-3} in 1979 to approximately 100 pg.m^{-3} in 1992 (Bidleman 1994a) bringing the concentrations to within the range of annual mean concentration of total HCH observed in air in southern Ontario (210 pg.m^{-3}; Hoff *et al.* 1992). HCH isomers are also the most prevalent organochlorine pesticide compounds in precipitation of the northern hemisphere, in Arctic snow (McNeely and Gummer 1984; Gregor and Gummer 1989; Welch *et al.* 1991; Swyripa and Strachan 1994), and in Arctic seawater (Figure 5; Hargrave *et al.* 1988, 1989; Bidleman *et al.* 1989, 1995b). From October to June the source of contaminants in the seawater is mainly snow and dryfall. Total HCH concentration in the seawater column at the Canada Basin decreased with increasing depths (Macdonald and McLaughlin 1993) but the total inventory in the water column has remained high.

Snow from ice caps and sea-ice surfaces in the Arctic contain several organochlorine compounds including α-HCH, β-HCH, γ-HCH, heptachlor epoxide, β-endosulfan, chlordane and dieldrin (Gregor and Gummer 1989). Decreasing amounts of organochlorine residues were observed in annual snow layers from 1970 to 1986 in the Agassiz Ice Cap on Ellesmere Island (Gregor and Gummer 1989).

HCB was second in concentration to total HCH in Arctic air (Figure 5; Hargrave *et al.* 1988; Fellin *et al.* 1996). Concentrations of HCB in Arctic air were within the same range as those observed in air in southern Ontario (greater than 54 pg.m^{-3}; Hoff *et al.* 1992).

Toxaphene tends to be less concentrated in air than total HCH and HCB but more concentrated in seawater than HCB (Figure 5; Bidleman *et al.* 1989, 1993; Barrie *et al.* 1993; Hargrave *et al.* 1988). Concentrations of toxaphene in Arctic air were similar to the annual mean concentration in air in southern Ontario (26 pg.m^{-3}; Hoff *et al.*

1992). Toxaphene was detected in Arctic snow (Welch *et al.* 1991; See Brown Snow Sidebar). Chlordane is ubiquitous in air of the Northern Hemisphere (Barrie *et al.* 1992). Concentrations of total chlordane in the Arctic (1.9 to 5.1 pg.m^{-3}) were much lower than in southern Ontario (40 pg.m^{-3}; Hoff *et al.* 1992).

There appears to be no evidence of a consistent long-term trend of PCB deposition to the ice cap or to the entire Arctic. In 1993, snow samples spanning 30 years from the Agassiz Ice Cap showed total PCB concentrations ranging from 1.2 to 6.7 ng.L^{-1} (Gregor *et al.* 1995). High inter-annual variability in PCB concentrations masked any time trend (Gregor *et al.* 1995). The total DDT in the "brown snow" falling on the District of Keewatin in the central Canadian Arctic was 2 to 10 times higher than in previous studies of Arctic snow (Welch *et al.* 1991).

Despite their prevalence in air, precipitation, snow and seawater, HCH isomers were the least concentrated residues found in biota, i.e., fish, around the Canadian Ice Island (Table 3; Hargrave 1994; Muir *et al.* 1996). The relatively low potential of these compounds for bioaccumulation is attributed to their low lipid solubility compared with PCBs and DDT-related compounds. This causes them to be proportionally less associated with particulates that are the main route of uptake of organochlorine compounds into the food chain.

HCB was among the least concentrated of the organochlorine pesticides in Arctic biota. HCB tended to be slightly more concentrated than total HCH in biota, particulate matter, all size classes and taxa of pelagic and benthic zooplankton, and amphipods and tissues from Arctic cod, char, and eelpout from around the Canadian Ice Island (Table 3; Hargrave 1994; Muir *et al.* 1996).

Toxaphene was the major organochlorine compound contaminant in marine fishes, followed by total PCBs, total DDT and total chlordane (Hargrave 1994; Muir *et al.* 1996). Toxaphene was the predominant contaminant measured in a variety of organisms (Table 3). In Arctic cod collected at three eastern Arctic locations in the mid-1980s, toxaphene was 5 to 10 fold more concentrated than DDT or PCB residues (Muir *et al.* 1996). Highest concentrations of organochlorine compounds in Arctic fishes were found in Greenland halibut (turbot). These predacious, bottom-feeding fish have a higher lipid content than char and sculpins (Muir *et al.* 1996). Turbot from the eastern Canadian Arctic and eastern Beaufort Sea had mean toxaphene concentrations of 380 ng.g^{-1} (wet weight). These concentrations are 3 to 5 times higher than in sea-run char muscle and 15 to 20 times higher than in Arctic cod (whole fish), reflecting the higher

trophic position of the turbot (Muir *et al.* 1996). The results for arctic char were consistent with the movement of chemicals in air masses from southern and central North America in a northeasterly direction across Quebec, Labrador, Baffin Island and Greenland (Muir *et al.* 1992).

PCBs (300 to 2 000 ng.g^{-1} wet weight), DDT (200 to 1 600 ng.g^{-1}), chlordane (150 to 1 600 ng.g^{-1}), toxaphene (150 to 660 ng.g^{-1}), and HCH (70 to 380 ng.g^{-1}) were the most commonly detected organochlorine compounds in blubber of ringed seals from a wide geographical range (Muir *et al.* 1996). Toxaphene (2 500 to 14 500 ng.g^{-1}) was usually the most prevalent in blubber of beluga whales, followed by DDT,

Table 3 Ranges of various organochlorine pesticide and PCB concentrations in lower trophic level marine biota.

	Region	Weight basis	Toxaphene	Total PCBs[3]	Total DDT[4] (ng.g^{-1})	Total Chlordane[5]	Total HCH[6]	HCB	Reference
Epontic particles	Ice Island (Axel Heiberg)	lipid	<100	40-360	20-70	10-40	10-280	6-30	Hargrave *et al.* 1992
	Barrow Strait	lipid	10-140	20-360	150-360	3-11	160-230	10-20	Hargrave 1994
Zooplankton[1] 25-125 µm	Ice Island (Axel Heiberg)	lipid	10-890	10-490	8-150	5-150	2-200	1-100	Hargrave *et al.* 1992
<509 µm	Iceland (Axel Heiberg)	lipid	20-1 360	10-110	10-60	10-50	10-280	10-100	Hargrave *et al.* 1992
<500 µm	Barrow Strait	lipid	200-1 400	4-20	2-20	8-107	90-180	5-130	Hargrave 1994
Amphipods[2] Pelagic	Ice Island (Axel Heiberg)	lipid	460	<440	<350	330	500	170	Hargrave *et al.* 1992
Pelagic	Barrow Strait	lipid	50-800	1-230	3-60	4-78	70-390	1-40	Hargrave 1994
Benthic (Eurythenes)	Arctic Ocean	lipid	3 000-35 000	5 700-34 000	2 200-27 000	750-4 500	6-420	60-260	Hargrave *et al.* 1992
Benthic	Barrow Strait	lipid	170-3 800	50-1 900	15-1 600	30-13 000	120-7 800	10-210	Hargrave 1994
Fishes Arctic cod	Lancaster Sound	lipid	340-640	66-95	66-120	55-110	39-49		Muir & Lockhart 1994
		wet	23-47	4-7	5-6	4-8	3-4		Muir & Lockhart 1994
	Barrow Strait	lipid	300-3 900	14-230	15-250	30-600	40-150	30-190	Hargrave 1994
Turbot	Cumberland Sound	lipid	1 800-3 500	910-1 600	630-1 000	510-960	59-120	160-280	Muir & Lockhart 1996
		wet	320-410	160-220	110-130	96-120	10-15	27-36	Muir & Lockhart 1996
	Beaufort Sea	lipid	1 700-3 200	880-1 600	660-1 300	660-1 250	78-150	140-270	Muir & Lockhart 1996
		wet	280-430	140-220	111-160	110-160	13-19	26-36	Muir & Lockhart 1996
Greenland cod	Wellington Bay	wet		0.79-20					Bright *et al.* 1995a, b
Fourhorn sculpin	Wellington Bay	lipid		170-520	93	450	100		Bright *et al.* 1995 a, b
		wet		2.4-7.3	1.3	6.3	1.4		Bright *et al.* 1995 a, b
Fourhorn sculpin	Cambridge Bay	lipid		2500	1 220	260	95		Bright *et al.* 1995 a, b
		wet		50	25	5.1	1.9		Bright *et al.* 1995 a, b
Fourhorn sculpin	Hall Beach	lipid		140	140	110			Bright *et al.* 1995 a, b
		wet		2.7	2.8	2.2			Bright *et al.* 1995 a, b
Bivales Clams	Sanikiluaq	lipid		62	33	56	<5.6		Cameron and Weis 1993
		wet		1.1	0.6	1	<0.1		Cameron and Weis 1993
Septentrion sp.	Manitounuk Sound	lipid	160	130	13	12	15	<0.5	Muir *et al.* 1995
		wet	3	2.5	0.25	0.24	0.3	<0.01	Muir *et al.* 1995

1   Zooplankton in the smaller size class were dominated by *Microcalanus* sp. and copepodites of *Calanus hyperboreus* with adults of the latter species in the larger size class.
2   Pelagic amphipods were *Parathemisto* sp. and *Gammarus wilkitzkii*. Benthic amphipods included *Onisimus* sp., *Tmetonyx cicada*, and *Anonyx nugax*.
3   As Aroclor 1254 equivalents for epontic particles, zooplankton and amphipods. All others as sum of 80 PCB congeners.
4   Sum of *p,p'*-DDT+ *p,p'*-DDE+ *o,p'*-DDE for epontic particles, zooplankton and amphipods. All others include *o,p'*-DDT.
5   Sum of cis-+ trans -chlordane +cis -nonachlor + oxychlordane + heptachlor + heptachlor epoxide.
6   Sum of α, β and γ -HCH.

Source: Muir *et al.* 1996.

PCBs and chlordane (Muir *et al.* 1996). Concentrations of organochlorine compounds in marine mammals varied with the geographical location of the animals, but no overall geographical trends were observed. Some geographical differences were attributed to feeding by marine mammals at higher trophic levels at some locations than at others.

Concentrations of PCBs and DDT in blubber of ringed seals at Arviat, Northwest Territories, were dependent on the sex and age of the seals (Figure 6; Muir *et al.* 1996). As male seals age, concentrations of these chemicals appeared to increase, while in females concentrations generally remained relatively stable. This increase in concentrations with age in male but not in female ringed seals is explained by the loss of these contaminants in females while lactating (Addison and Smith 1974). Male and female differences were not as obvious for total PCBs and DDT from 10 other Arctic locations during the period 1984 to 1989 (Figure 7).

Consistent temporal trends for organochlorine compounds in ringed seals have not been observed in the Arctic. This is because the available time series for existing data vary for different areas in the Arctic. Nevertheless, a large decrease in concentrations of total PCBs was observed between 1972 and 1981 and a small decrease between 1981 and 1991 in female ringed seals from around Holman Island, Northwest Territories (Figure 8). On the other hand, total

Figure 6
Total DDT and total PCB concentrations in ringed seal blubber from Arviat, Northwest Territories, versus age of seals.

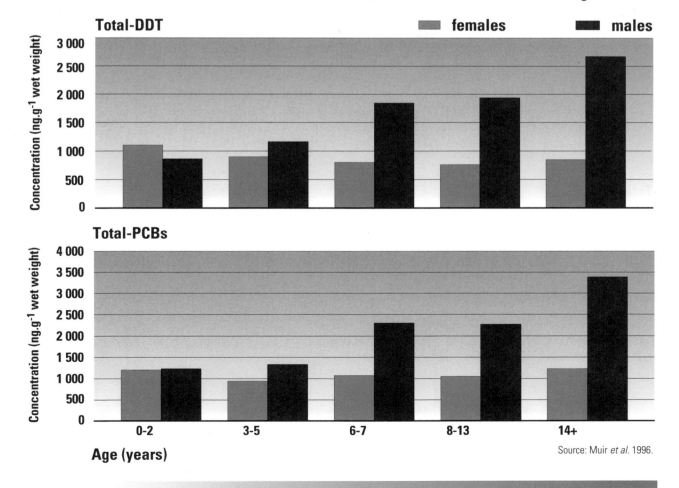

Source: Muir *et al.* 1996.

DDT declined slowly during both periods (Figure 8; Addison *et al.* 1986; R.F. Addison unpublished data). This downward trend was not apparent in concentrations of DDT, PCBs, chlordane and toxaphene in ringed seals from Cumberland Sound or Barrow Strait over 6 and 9 years, respectively (Muir *et al.* 1996). The rate of decline in total PCB concentrations observed in ringed seal from around Holman Island was similar to that seen in east coast harp seals from the Gulf of St. Lawrence (Addison *et al.* 1984), and almost certainly reflects the ban on PCB

manufacture and use during the early 1970s (Muir *et al.* 1996). The absence of decline for total DDT and total PCB compounds in ringed seals and narwhal in the Arctic during the 1980s and early 1990s may be explained by continuous transport of these chemicals to the Arctic, e.g., possibly from Asia where DDT is still used extensively (Muir *et al.* 1996). Shorter term studies of six and nine years reported by Muir *et al.* (1996) may have missed the steepest decline of DDT transport to the Arctic. Therefore, these studies may not have been able to distinguish differences in

Figure 7
Total PCB and total DDT concentrations in ringed seal blubber and total PCB and toxaphene concentrations in beluga and narwhal blubber.

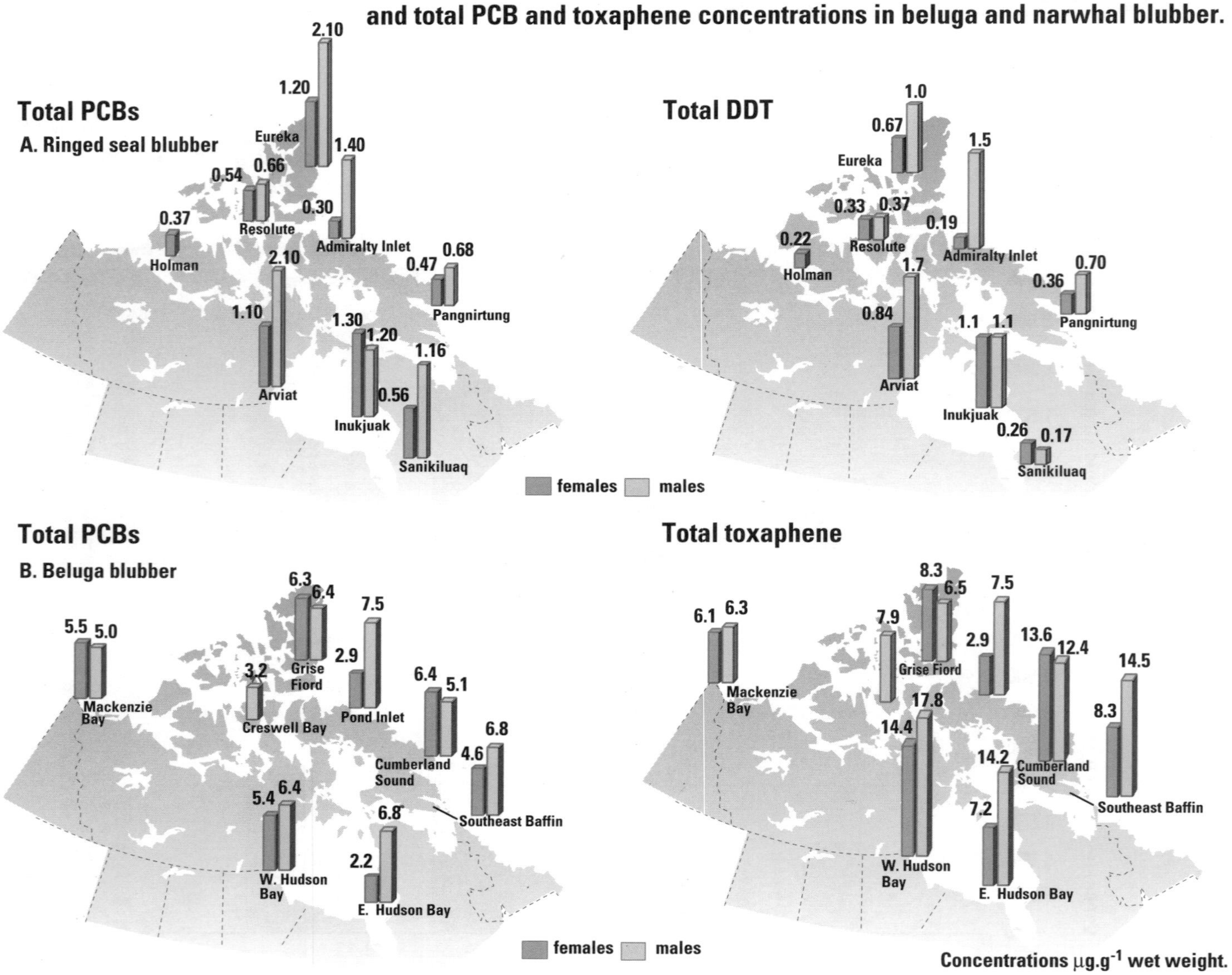

Total PCBs
A. Ringed seal blubber

Eureka 2.10, 1.20, 1.40, 0.54, 0.66, 0.30
Resolute — Holman 0.37, Admiralty Inlet, 0.47, 0.68, Pangnirtung
2.10, 1.10, Arviat 1.30, 1.20, 1.16, Inukjuak 0.56, Sanikiluaq

Total DDT

Eureka 1.0, 0.67, 1.5, 0.33, 0.37, 0.19, Resolute, 0.22 Holman, Admiralty Inlet, 0.36, 0.70 Pangnirtung
1.7, 0.84, Arviat 1.1, 1.1, Inukjuak 0.26, 0.17 Sanikiluaq

females ■ males

Total PCBs
B. Beluga blubber

Mackenzie Bay 5.5, 5.0
Grise Fiord 6.3, 6.4, 7.5, 2.9, 3.2 Creswell Bay
Pond Inlet 6.4, 5.1
Cumberland Sound 6.8, 4.6 Southeast Baffin
W. Hudson Bay 5.4, 6.4, 6.8, 2.2 E. Hudson Bay

Total toxaphene

Mackenzie Bay 6.1, 6.3
Grise Fiord 8.3, 6.5, 7.5, 2.9, 7.9
13.6, 12.4, 14.5
17.8, 14.4, 14.2 Cumberland Sound, 8.3 Southeast Baffin
W. Hudson Bay 7.2, E. Hudson Bay

females ■ males

Concentrations μg.g⁻¹ wet weight.

Source: Muir 1994, 1996; Cameron and Weis 1993; Cameron *et al.* 1997; Addison 1995.

contaminant concentrations in organisms as easily as studies done when larger decreases in contaminant transport occurred.

A long-term study of organochlorine compounds in male belugas from the Mackenzie delta region found no significant declines over a 10-year period and over a 20-year period for DDT (Muir *et al.* 1996). This may indicate that: belugas may be accumulating these contaminants at the same rate as 10 years previous; that they are exposed to the same amount of contaminant; that they are losing the contaminants too slowly to observe a change over the 10-year period; or that the period of measurement may have missed the time period when the most dramatic decreases in North America occurred.

Total PCBs and total DDT were significantly higher in male beluga from Cumberland Sound than in samples from Hudson Bay or the Beaufort Sea (Muir *et al.* 1996). Significant differences were not found among concentrations in females. Other differences were not detected between the five beluga stocks sampled between 1983 and 1989 (Figure 7; Muir *et al.* 1996). PCB concentrations averaged about 6 µg.g^{-1} wet weight in male belugas, and 4 µg.g^{-1} wet weight in females. These concentrations are about 12 times lower than in blubber of beluga from the St. Lawrence Estuary (Muir *et al.* 1996). Concentrations of total DDT (as well as mirex, toxaphene, and total chlordane) in beluga from the Arctic were also lower than those observed in

Figure 8
Trends in total DDT and total PCBs in blubber of female ringed seals from Holman Island in the western Canadian Arctic.

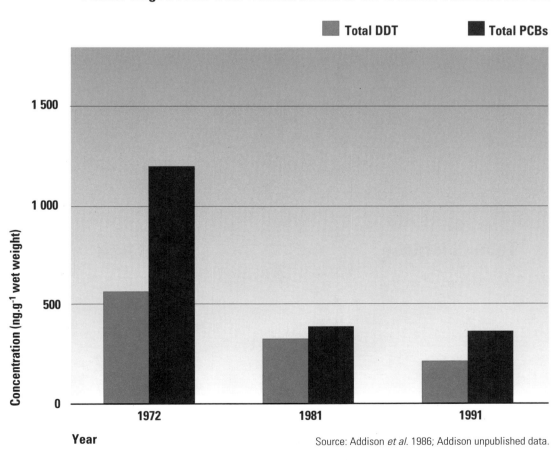

Source: Addison *et al.* 1986; Addison unpublished data.

beluga from the St. Lawrence Estuary (See the St. Lawrence Marine Ecosystem Chapter).

Biomagnification factors for some contaminants from particulate matter to polar bear lipid can be as high as 17 000 (Muir *et al.* 1996). PCBs and dichlorodiphenyldichloroethylene (*p,p'*-DDE) were first reported in polar bear tissues in the early 1970s (Bowes and Jonkel 1975); DDE-metabolites were later detected. A systematic survey in 1982 to 1984 of the geographical distribution of organochlorine compounds in polar bear fat and liver showed that concentrations tended to be highest in the southeast (Hudson Bay), and lowest in the northwest (Beaufort Sea) (Muir *et al.* 1996). Present in highest concentrations were PCBs and the pesticide

metabolite, oxychlordane, whereas chlordane-related compounds *p,p'*-DDE, dieldrin, HCB and *p,p'*-HCH were present in lower concentrations (Norstrom *et al.* 1988; Muir *et al.* 1988). A study from 1989 to 1993 in an area encompassing the western hemispheric Arctic, including the Eastern Siberian Sea to western Siberia in the east Arctic, showed that concentrations are relatively uniformly distributed over much of the study area. This indicates extensive transport and deposition to all areas of the Arctic and sub-Arctic (Muir *et al.* 1996). However, some geographic differences were apparent, with total chlordane tending to be highest in Hudson Bay (Figure 9). Although not shown in the figure, concentrations of DDE showed some geographic variability. Dieldrin (not shown in Figure 9) displayed a more pronounced

Figure 9
Spatial distribution of organochlorine compound concentrations in polar bear fat in northern Canada.

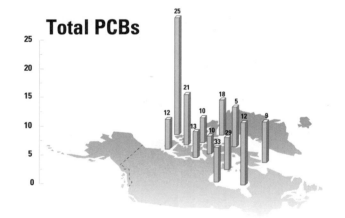

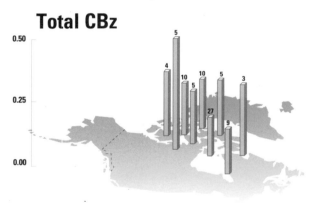

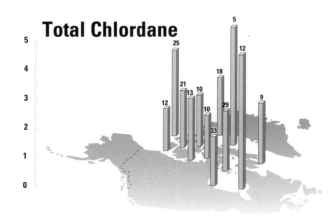

Numbers above bars are number of samples.
Geometric mean concentrations in μg.g⁻¹ lipid.

Source: Muir *et al.* 1996.

increase in concentrations from west to east than the other organochlorine compounds sampled. Highest dieldrin concentrations were detected in polar bears from the Hudson Bay at Belcher Island. These results may indicate atmospheric transport to this area after volatilization from soil surfaces, and perhaps the Great Lakes, in North America during the summer. Eastern Hudson Bay is the only area of those sampled that is in the direct path of average continental airflow in the summer (Barrie *et al.* 1992). Concentrations of total CBz and total HCH were similar in polar bears from all areas of the Arctic; however, unlike the other compounds measured, highest concentrations were found in the western regions (Figure 9).

Although the information on polychlorodibenzo-*p*-dioxins (PCDDs) and polychlorodibenzo-furans (PCDFs) in Arctic marine fish is very limited, the little there is indicates that concentrations are low, ranging from below detection limits to 2 pg.g^{-1} (Muir *et al.* 1996). Concentrations are also low in Arctic marine mammals (Norstrom *et al.* 1990). Concentrations of 2,3,7,8,-tetrachlorodibenzo-dioxin (2,3,7,8-TCDD), 2,3,7,8-tetrachlorodibenzo-furan (2,3,7,8-TCDF), and octachlorodibenzo-*p*-dioxin (OCDD) and 1,2,3,6,7,8-HxCDD were found at levels ranging from 2 to 41, 2 to 7, less than 15 to 43, and less than 4 to 9 pg.g^{-1}, respectively, in ringed seal blubber samples from various sites in the Arctic. In beluga, concentrations of these compounds were less than 2, less than 2 to 3, less than 8 to 17, and less than 4 to 5 pg.g^{-1}, respectively (Norstrom *et al.* 1990). Highest concentrations of 2,3,7,8-TCDD and OCDD in seals were found in the central Canadian Arctic Archipelago and the lowest were found in Hudson Bay, the reverse of PCB concentration distribution (Norstrom *et al.* 1990). The reason for higher concentrations of TCDD and OCDD in the Arctic than in the sub-Arctic may be transpolar movement of atmospheric particulate matter with combustion-related origins in Asia or Europe (Norstrom *et al.* 1990). PCDDs and PCDFs have also been detected in polar bear tissues (Norstrom *et al.* 1990).

OTHER AROMATIC COMPOUNDS

The particle-bound PAHs measured in air at Alert in 1992 showed clear seasonal differences with high values in winter and low values in summer. PAHs contribute to Arctic haze that consists of anthropogenically-generated compounds originating in temperate latitudes and transported to the Arctic region during the winter months (Fellin *et al.* 1996). Fluorene and phenanthrene were the PAHs present in highest concentrations at Alert in 1992 at mean concentrations of 17 and 28 pg.m^{-3} in summer and 590 and 210 pg.m^{-3} in winter, respectively. Other PAHs, such as pyrene, retene, chrysene and benzo[a]pyrene, (B[a]P), and perylene were present at 5 to 10-fold lower concentrations, but all showed strong seasonal variation (Fellin *et al.* 1996).

Recent historical trends in the deposition of PAHs to the Arctic were determined when the Agassiz Ice Cap on Ellesmere Island was sampled and total PAHs determined for the last 30 years (Figure 10; Peters *et al.* 1995). PAH concentrations ranged from 36 to 660 ng.L^{-1} most of which was napthalene. The flux of total PAHs to the Arctic has remained relatively constant since the 1970s at an annual total PAH loading of about 37 t per year assuming the area of Canada's Arctic and sub-Arctic is 3.4 x 10^6 km^2 (Barrie *et al.* 1992). Prior to 1972 to 1973, loading of total PAH was almost 7 times higher. The decline in PAH loading may be indicative of changes in global fossil fuel use with time.

Although atmospheric sources of PAH combustion products to the Arctic may be important in areas that are remote from the influence of large rivers, they are a minor contribution to the PAHs of the Mackenzie Shelf. The massive particulate flux provided by the Mackenzie River is the major source of PAHs to the Mackenzie Shelf (Yunker and Macdonald 1995). PAHs are present at higher concentrations than expected in the Mackenzie River delta and Beaufort Sea, yet are below concentrations associated with observable effects on bottom fish and other biota (Yunker and Macdonald 1995).

Hydrocarbons have been extensively measured in seawater at Cape Hatt, northern Baffin Island (Cretney *et al.* 1987a), and in the southern Beaufort Sea (Wong *et al.* 1976). Baseline concentrations of aliphatic and aromatic hydrocarbons have also been determined for the southern Beaufort Sea sediments (Wainright and Humphrey 1988). Recently, total PAHs (excluding perylene and retene) were determined to be 150 and 180 ng.g^{-1} in two marine sediment samples taken from the Canadian Ice Island (Lockhart 1994). Total PAHs in "brown snow" were determined to be 13 ng.L^{-1}, with naphthalene and phenanthrene the major compounds at 2.5 and 2.6 ng.L^{-1}, respectively.

Relatively few data on PAHs in Arctic marine biota have been published (Wong *et al.* 1976; Muir *et al.* 1992; Cretney *et al.* 1987b; Thomas 1988). PAH data in Arctic marine biota were reviewed by Muir *et al.* (1992) (Table 4). Naphthalene, phenanthrene, anthracene, fluoranthene and pyrene were all detected in the muscle of Arctic cod from Resolute Bay and inconnu from Tuktoyaktuk (Muir *et al.* 1992). PAH concentrations in these fish ranged from 0.1 ng.g^{-1} for anthracene to 6.3 ng.g^{-1} for naphthalene and 8.1 ng.g^{-1} for phenanthrene. Compared with organochlorine compounds, PAHs are generally of less concern because they tend to be metabolized by organisms and are not subject to bioaccumulation. Nevertheless, the high natural concentrations of PAHs in the Mackenzie delta may make biota there more susceptible to chronic or accidental spills of PAHs from oil production (Yunker and Macdonald 1995).

ECOLOGICAL CONSEQUENCES

Although Arctic biota contain a range of organic and inorganic contaminants, there has been insufficient research to elucidate the biological effects of contaminants on Arctic marine fish and mammals in their natural environment (Norstrom and Muir 1994). In contrast to the situation in the Great Lakes, there is currently no evidence to support or refute arguments that chemical contaminants are present in sufficient quantities to cause adverse biological effects in Arctic marine fish and mammals (Muir *et al.* 1996).

Some responses at the molecular, biochemical, or subcellular level are used as biomarkers of contaminants effect or exposure. There has been so little work on biomarkers in Arctic animals that their normal ranges have not been well defined (Muir *et al.* 1996). Potential effects of contaminants on marine mammals and fish of the

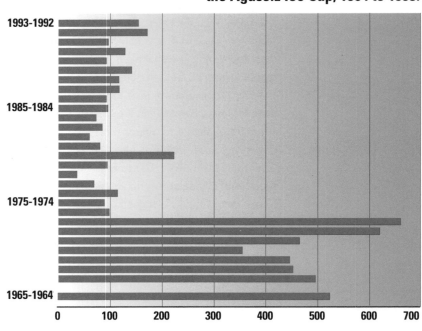

Figure 10
Trends in total PAH concentrations in ice from the Agassiz Ice Cap, 1964 to 1993.

Total PAH concentration (ng.L^{-1} snowmelt water)   Source: Peters *et al.* 1995.

Canadian Arctic could include a decrease in reproductive capacity, an increase in cancer/mutagenesis, and other pathological changes.

Cadmium at the high concentrations observed in marine mammals, i.e., in kidney of narwhal (Wagemann *et al.* 1996), is expected to be toxic but measurement of total metal concentration in a tissue alone is not a good measure of toxicity. The measurement of metallothionein as well as total cadmium gives a better indication of potential for cadmium toxicity (Wagemann 1994). The cadmium in the kidney of the narwhal was found to be mostly associated with metallothionein and thus believed to be relatively non-toxic (Wagemann *et al.* 1996). Liver tissue samples of narwhal from Repulse Bay and Iqaluit contained

Table 4   Aliphatic and aromatic hydrocarbon concentrations in Arctic marine biota.

Compound	Location	Year	Organism	Concentration range[a] (ng.g^{-1} wet weight)	N	Reference
Aliphatic hydrocarbons	Tuktoyaktuk Harbour/	1986-87	Flounder	7 600-710 000	24[b]	Thomas 1988
	Mason Bay	1986-87	Flounder	19 000-230 000	10[b]	Thomas 1988
Total PAHs	Tuktoyaktuk Harbour/	1986-87	Flounder	<10-420	18[b]	Thomas 1988
	Mason Bay	1986-87	Flounder	8-230	13[b]	Thomas 1988
Total PAHs	Beaufort Sea (Tuktoyaktuk)	1984	Inconnu	1.4-1.5	2	Muir *et al.* 1992
	Resolute Bay	1984	Arctic cod	4.5-15	5	Muir *et al.* 1992
Total PAHs	Beaufort Sea (Tuktoyaktuk)	1983	Least cisco	58	4	EnviroTest Laboratories1984
			Flounder	63	3	EnviroTest Laboratories1984
			Flounder	10	3	EnviroTest Laboratories1984
			Eelpout	6	3	EnviroTest Laboratories1984
			Whitefish	49	2	EnviroTest Laboratories1984
Total PAHs	Beaufort Sea (Liverpool Bay)	1983	Herring	24	10	EnviroTest Laboratories1984
Aromatic hydrocarbons	Cape Hatt, Baffin Island	1981	Bivalve	0.23-5.8	18	Cretney *et al.* 1987b
		1981	Bivalve	0.42-3.5	19	Cretney *et al.* 1987b
		1981	Bivalve	0.19-0.72	20	Cretney *et al.* 1987b
		1981	Bivalve	0.1-12	14	Cretney *et al.* 1987b
		1981	Sea urchin	9.0-46	20	Cretney *et al.* 1987b
Total PAHs	Cape Hatt, Baffin Island	1980	Bivalve	1-11	5	Cretney *et al.* 1987b
		1980	Bivalve	16	1	Cretney *et al.* 1987b
		1980	Starfish	39-410	4	Cretney *et al.* 1987b
Naphthalene	Tuktoyaktuk	1984-85	Inconnu	<1.0	Pooled[c]	Muir *et al.* 1992
	Resolute Bay	1984-85	Arctic cod	3.7-6.3	Pooled[c]	Muir *et al.* 1992
Phenanthrene	Tuktoyaktuk	1984-85	Inconnu	1.1-1.3	Pooled[c]	Muir *et al.* 1992
	Resolute Bay	1984-85	Arctic cod	0.6-8.1	Pooled[c]	Muir *et al.* 1992
Anthracene	Tuktoyaktuk	1984-85	Inconnu	<0.1	Pooled[c]	Muir *et al.* 1992
	Resolute Bay	1984-85	Arctic cod	<0.1-0.5	Pooled[c]	Muir *et al.* 1992
Fluoranthene	Tuktoyaktuk	1984-85	Inconnu	0.2-0.3	Pooled[c]	Muir *et al.* 1992
	Resolute Bay	1984-85	Arctic cod	<0.1-1.1	Pooled[c]	Muir *et al.* 1992

a = Whole organism basis unless otherwise specified.
b = Liver samples.
c = Muscle samples.

Source: Muir *et al.* 1992

similar, considered normal concentrations of metallothionein (420 and 390 $\mu g.g^{-1}$, respectively) while Pond Inlet narwhal had higher, more variable concentrations (average 1 000 $\mu g.g^{-1}$; maximum 2 300 $\mu g.g^{-1}$). The cadmium in the liver

Oil Spills in the Arctic

Because marine mammals come to the surface of seawater to breathe, they are likely to be exposed to any spilled oil. Seals (ringed seals, harp seals), seal lions, and walrus that depend upon blubber for insulation are more thermally tolerant of contact with oil than are fur seals that depend upon fur for maintaining body heat. Fur is compressed by oil and loses its insulating properties. Oil is also expected to cause significant thermoregulatory problems for free-ranging polar bears.

It is not clear whether whales avoid oil spills and slicks, but baleen whales and toothed whales are known to enter and feed in an oil slick. Heavy weathered oil may foul the baleen and lighter oils may physically damage it. Oil is not a serious threat to thermal regulation in whales because of their dependence on blubber (Richardson et al. 1989).

Releases of oil to the Arctic environment occur naturally, such as to the Mackenzie River at Norman Wells. Organisms have physiological means of responding to quantities of oil released naturally and possibly to cope with some anthropogenic loading. Ringed seals that were fed oil for 4 to 5 days showed rapid deposition of petroleum hydrocarbons in liver, kidney, muscle and blubber. The induction of MFO activity in liver and kidney led to rapid elimination of the hydrocarbons. Only traces of oil remained after one month (Richardson et al. 1989). Nevertheless, chronic or acute contamination by hydrocarbons as a result of oil spills and oil field development (in addition to natural oil seepage) is an on-going issue of concern in the Arctic.

in these narwhal was considered to be non-toxic because of its association with metallothionein (Muir *et al.* 1996).

A bioindicator that has been studied for Arctic marine organisms is mixed function oxydase (MFO) induction. There was a strong correlation between liver microsomal ethoxyresorufin *O*-deethylase (EROD) or aryl hydrocarbon hydroxylase (AHH) activities and blubber residues of several PCB congeners in a group of belugas from the western Canadian Arctic (Lockhart and Stewart 1992). While this is not proof of a causal relationship between PCBs and the bioindicators, it is probably the closest connection that can be made between contaminant residues and responses in animals that are not readily available for experimental study (Muir *et al.* 1996). The mean cytochrome P450 enzyme (CYP 450) content of polar bear liver, measured as EROD activity, was about 2 times higher than in beluga liver (White *et al.* 1994) and 10 times higher than in male hooded seals (Goksøyr *et al.* 1992). This may indicate that polar bears had elevated cytochrome P450 1A (CYP 1A) - mediated activity (Muir *et al.* 1996). The strong positive correlation between CYP 1A and some PCB congeners in polar bear livers suggests that these congeners may be responsible for the enzyme activity; some chlordane compounds such as oxychlordane and nonachlor and other PCB congeners may also have resulted in elevated cytochrome P450 1B (CYP 1B) activity in polar bears (Muir *et al.* 1996).

Payne (1993) hypothesized that the status of Vitamin A in marine mammals may be an important biological response in connection with exposure to contaminants. Judging from the relationship between chemically-mediated Vitamin A depletion and the deterioration of the health of belugas in the Gulf of St. Lawrence, Vitamin A depletion may be a biomarker of ecotoxicological importance for marine mammals in the Arctic.

Mathieu *et al.* (1995) detected 'uncharacterized' PAH-DNA adducts in brain, liver and kidney of Arctic beluga. This was the first evidence for PAH-related adducts in Arctic wildlife in general and in marine mammals in particular (Muir *et al.* 1996). Covalent bonding of many chemicals or their metabolites to DNA can result in the formation of DNA adducts. Such adducts are believed to be important in mutagenesis and carcinogenesis (Muir *et al.* 1996).

The relationships between the presence of chemical contaminants and overall ecosystem effects have not been fully studied. Although other stress factors may not be directly related to chemical contamination, they may nevertheless enhance the detrimental effects of contaminants. Such stressors include habitat destruction due to hydroelectric development, increased human settlement and activity, resource extraction, chemical spills, climate change, and over-harvesting (Welch 1995).

Conclusions

The coastline of the Arctic Ocean is twice as long as that of the Pacific and Atlantic coastlines combined. The Arctic coast includes more than one million km^2 of continental shelf, equivalent to the combined extent of the shelf in the Atlantic and Pacific waters within Canada's 200-mile economic zone. The vast coastline and area represented within the Arctic is a unique and fragile environment in which the living resources are limited.

Of all the ocean basins, the Arctic Ocean Basin is the least known from an ecological perspective. In the Canadian Arctic marine ecosystem, little is known about spatial and seasonal variability in primary production (Alexander 1995). In spite of the fact that species diversity is relatively low in the Arctic, the feeding relationships among species, including predator-prey relationships, can be complex and are not well known. This

information is required for sustainable management of important subsistence and commercial fish and mammal stocks (Beckmann 1995), and for understanding the Arctic marine pathways by which contaminants find their way into marine mammals and fisheries resources.

A dominant feature of the Arctic is the presence of multi-year and seasonal ice that influences the biological regimes of the region. The ice edge is an area where marine mammals and birds congregate to feed, and where the contaminants that are present can readily enter the food chain. Because of ice cover during half of the growing season and other factors, annual marine primary production is relatively low and it is highly seasonal. Fish and marine mammals have adapted to the cold temperatures and variable productivity by accumulating large lipid reserves. These reserves serve as energy stores and for thermoregulation in the mammals. Most of the persistent contaminants are lipophilic and accumulate in the lipid body reserves. Although Arctic food chains may be relatively simple, they are long, with up to three levels of carnivory. This leads to elevated concentrations of some contaminants as they biomagnify up the food chain. For example, some contaminants in polar bear lipids are 17 000 times the concentrations found in particulate matter in seawater. Compared with the concentrations of some organochlorine compounds dissolved in seawater, the bioconcentration factors in polar bears are even more dramatic, 7×10^4 for toxaphene, and 2×10^6 for PCBs (Muir and Norstrom 1994).

By far the largest source of chemical contaminants to the Arctic marine environment is LRTAP. Contamination of the Arctic has its origins in regions as far away as Asia and Europe where contaminants are produced, used, or emitted to the environment. Contaminants are also brought to the Arctic marine ecosystem by the major routes of ocean currents and inflow of rivers. Less important routes are direct run-off from the land, such as from local point sources, and direct disposal into the ocean. There are a few anthropogenic point sources in the Arctic marine ecosystem, such as mines, oil and gas exploration and production sites, and waste disposal sites.

Concentrations of mercury have been reported to be increasing in the abiotic environment (air and sediments) and in certain biota (beluga and ringed seal). The major source is considered to be LRTAP. Loading of anthropogenic mercury can be detected from sediment core profiles. The biological consequences of higher metal loading to sediments associated with anthropogenic activity are not known (Lockhart 1994). Nevertheless, there is a great deal of variability of natural background concentrations of mercury in biota. This variability is related to geological variability. The sedimentary geology of the western Arctic has higher concentrations of natural mercury than that in the central and eastern Arctic. This evidence that mercury is increasing in the Arctic is viewed with concern, but the amount of data available is too limited to associate the increases in the biota with the increased global mobilization of mercury by human activity with certainty (Barrie *et al.* 1997).

Cadmium concentrations are elevated in the organs of some marine mammals. Concentrations of cadmium found in the liver and kidney of narwhal from Pond Inlet were high enough to be potentially chronically toxic. Geological formations in the eastern Arctic are high in cadmium with resultant elevated concentrations in organisms. There is a contrasting east-west geographical variation for mercury. Mammals in the western Arctic have elevated concentrations of mercury compared with those in the central and eastern Arctic. Anthropogenically-mobilized cadmium is reaching the Arctic Ocean via rivers, ocean currents and the atmosphere, but is insignificant in amount compared with the large natural pool in the Arctic Ocean. Further information is needed on the concentrations of mercury and cadmium in the liver and kidney of marine mammals and fish from a wide geographic range.

Although the source is believed to be natural rather than anthropogenic, cadmium was found in the liver and kidney of narwhal from Pond Inlet at high enough concentrations to be potentially chronically toxic were it not for the accompanying high concentrations of metallothionein. Geological formations in the eastern Arctic are high in cadmium with resultant elevated concentrations in organisms. There is a contrasting east-west geographical variation for mercury. Mammals in the western Arctic have elevated concentrations of mercury compared with those in the central and eastern Arctic. Anthropogenically-mobilized cadmium is reaching the Arctic Ocean via rivers, ocean currents and the atmosphere, but is insignificant in amount compared with the large natural pool in the Arctic Ocean. Further information is needed on the concentrations of mercury and cadmium in the liver and kidney of marine mammals and fish from a wide geographic range.

Numerous organic contaminants that are neither produced nor used in the Arctic continue to be delivered there by LRTAP. Toxaphene is the major organochlorine compound contaminant in marine fish, followed by PCBs, DDT and chlordane. Past and present widespread use of organochlorine compounds in Canada and elsewhere makes it difficult to accurately assign sources of the contaminants, or to predict loadings to the Arctic environment.

Organochlorine compounds appear in nearly all components of the Arctic marine ecosystem, although some are at very low concentrations. HCH isomers and HCB are the most prevalent in air, seawater, snow and precipitation. Toxaphene, DDT and PCBs are the most prevalent compounds measured in marine fish, ringed seals and beluga. These differences in occurrence have to do with the relatively low lipid solubility of HCH compared with that of the organochlorine compounds that accumulate in biota.

Spatial distributions of organochlorine compounds in Arctic marine mammals are relatively well known. Nevertheless, there are very few data on concentrations of these substances and PAHs, in Arctic marine fish (Muir *et al.* 1996), and little knowledge of temporal trends. The mechanisms for transfer of organochlorine compounds between various compartments of the Arctic marine environment and food web are poorly understood (Hargrave 1994).

A major temporal trend in contaminants in the Arctic is the reported increase in mercury loading mentioned above. Dramatic decreases in loadings since the 1970s have been recorded for lead, HCH, and PAHs. PCBs have declined in marine mammals over that period but there is either a smaller or a lack of decline in DDT concentrations.

Although there are measurable responses of some bioindicators in marine mammals, they are less than those seen in animals from more contaminated areas of the world. The ecological significance of such responses is not clear. Consequently, there is a general lack of information by which to link contaminant concentrations and biochemical indicators of effects to biological effects on Arctic marine animals at the whole organism or population level. Nevertheless, there is concern that contaminants are approaching concentrations that may elicit adverse biological effects.

Today the Inuit and other northern residents combine aspects of traditional life styles with modern lifestyles. Traditionally, the consumption of country foods such as fish, land and marine mammals was a necessity. Despite increasing access to market or imported foods, aboriginal people still depend upon country foods. To the Inuit, country foods are directly associated with physical health and spiritual well-being. Nevertheless, the fish and marine mammals that are consumed are often large, long-lived individuals. These species would have

accumulated contaminants over long periods of time and could have the highest contaminant concentrations in the populations.

Developments in the North, including mineral and hydrocarbon exploration and production, tourism, ecotourism, hunting and guiding, and trophy fishing place added stress on fragile Arctic marine ecosystems. Increases in both permanent and transient populations and shifts in the population base may also present a cumulative risk to local ecosystems. These human activities could enhance the detrimental effects of contaminants or be negatively affected by the presence of contaminants.

Case Study:

CONTAMINANTS IN COUNTRY FOODS

Country foods include traditional native foods such as berries, game, marine mammals, birds and fish (Wein *et al.* 1991). Consumption of country foods can result in increased exposure to contaminants (Dewailly *et al.* 1989, 1994). Measurable concentrations of several organochlorine compounds and metals are present in biota, including country foods, throughout the Arctic and northern regions of Canada (Cameron and Weis 1993; Kinloch *et al.* 1992).

Consumption patterns of country foods differ considerably among northern communities (Mackey and Orr 1987) due to game availability, seasonal patterns, and food preferences. At Sanikiluaq on the Belcher Islands, Northwest Territories, beluga, ringed seal, arctic char, common eider and Canada goose were found to be important components of the native diet; 80% of all meals during a two-week study period in the summer consisted of country food with 27% of these meals consisting of sea-run arctic char (Cameron and Weis 1993).

Concentrations of organochlorine compound contaminants in the country foods of Sanikiluaq residents are found in Table 5. Of all country foods utilized, the fatty blubber tissues of ringed seal and beluga contained the highest concentrations of chemical contaminants, with total PCB and total DDT concentrations being the highest among the analyzed contaminants (Cameron and Weis 1993). Concentrations of organochlorine compounds in arctic char were much lower than those found in ringed seal and beluga blubber; total PCB and total DDT residues were also the major contaminants present in arctic char. Daily in-takes of total DDT and total PCBs of 0.23 and 0.15 μg.kg body weight[-1].day[-1] respectively, were estimated for 50 kg individuals at Sanikiluaq. This is below the acceptable daily intake level of 0.5 and 1.0 μg.kg body weight[-1].day[-1] (EPA 1984) for DDT and PCBs, respectively. In another dietary study, approximately 12% of the surveyed residents from Broughton Island, Northwest Territories, exceeded their acceptable daily intake for PCBs by consuming country foods; 90% of the total

PCB intake was obtained from four food sources, narwhal, seal, walrus and caribou (Kinloch *et al.* 1992). Fish contributed approximately 6.4% of the daily intake of PCBs.

Although there may be increased exposure to contaminants with the traditional Inuit diet, there are major benefits to consuming country foods. Traditional Inuit foods with the highest concentrations of contaminants are also the same species that contribute large amounts of meat and fat energy; they also provide vitamins (e.g. raw whole skin, known as muktuk, has as much vitamin C as a fresh orange) (St. Aubin and Geraci 1980), retinol, polyunsaturated fatty acids, iron, zinc, copper, and other essential nutrients (Kuhnlein and Kinloch 1988). Market foods, besides being expensive in the north, are also limited in freshness, variety and nutritional quality and cannot match the nutritional and cultural benefits provided by traditional Inuit foods (Kinloch *et al.* 1992).

Table 5 **Average organic contaminant concentrations in tissues of country foods collected at Sanikiluaq, Northwest Territories and estimated daily intake of total DDT and total PCBs.**

	Contaminant Concentration or (Intake of total DDT and total PCBs) (ng.g[-1] wet weight or μg.person[-1] day[-1])					
	Ringed seal fat	**Ringed seal meat**	**Beluga muktuk [1]**	**Beluga meat**	**Eider duck**	**Arctic char**
Total HCH	430	3.3	13	1.5	0.9	0.8
Total chlordane	210	13	27	1.2	2.1	0.8
Total DDT [2]	1 650	41	270	3.4	5.6	3.3
Total PCBs	1 300	43	145	6.8	5.6	8.1
Total Nonachlor	180	7.0	140	3.8	1.5	1.8
Dieldrin	187	5.0	35	4.6	2.0	1.4
Contaminant Intake [3] (μg.total DDT person[-1].day[-1])	—[4]	2.0	9	0.06	0.19	0.25
Contaminant Intake [3] (μg.total PCBs person[-1].day[-1])	—[4]	1.6	5	0.13	0.18	0.64
Lipid %	87	1.8	63	1.4	2.8	2

1 Muktuk is a term used to describe the fat and skin layer of the animal used for food.
2 Sum of *p,p'*-DDT, *p,p'*-DDD and *p,p'*-DDE.
3 Calculated from the amount (kg) consumed and the concentration of the contaminant in the food source for a person weighing 50 kg.
4 Consumption of ringed seal fat did not take place during the period of study.

Source: Cameron and Weis 1993.

References

AAGAARD, K., and E.C. Carmack. 1989. The role of sea ice and other fresh water in the Arctic circulation. J. Geophys. Res. 94C: 14485-14498.

ADDISON, R.F. 1995. Long-term trends in organochlorine residues in eastern and western Arctic seal blubber. *In* J.L. Murray, and R.G. Shearer [eds.] Synopsis of Research Conducted under the 1994/1995 Northern Contaminants Program. Environmental StudPies No. 73. Indian and Northern Affairs Canada, Ottawa.

ADDISON, R.F., and T.G. Smith. 1974. Organochlorine residue levels in Arctic ringed seals: Variation with age and sex. Oikos 25: 335-37.

ADDISON, R.F., P.F. Brodie, M.E. Zinck, and D.E. Sergeant. 1984. DDT has declined more than PCBs in eastern Canadian seals during the 1970s. Environ. Sci. Technol. 18: 935-37.

ADDISON, R.F., M.E. Zinck, and T.G. Smith. 1986. PCBs have declined more than DDT-group residues in Arctic ringed seals (*Phoca hispida*) between 1972 and 1981. Environ. Sci. Technol. 20: 253-55.

ALEXANDER, V. 1995. The influence of the structure and function of the marine food web on the dynamics of contaminants in Arctic Ocean ecosystems. Sci. Total Environ. 160, 161: 593-603.

BARRIE, L.A. 1986. Arctic air pollution: An overview of current knowledge. Atmosph. Environ. 20: 643-63.

BARRIE, L.A., D. Gregor, B. Hargrave, R. Lake, D.C.G. Muir, R. Shearer, B. Tracey, and T. Bidleman. 1992. Arctic contaminants: Sources, occurrence and pathways. Sci. Total Environ. 122: 1-74.

BARRIE, L.A., T. Bidleman, D. Dougherty, P. Fellin, N. Grift, D.C.G. Muir, R. Rosenberg, G. Stern, and D. Toom. 1993. Atmosph. Toxaphene High Arc. Chemosph. 27: 2037-46.

BARRIE, L., R. Macdonald, T. Bidleman, M. Diamond, D. Gregor, R. Semkin, W. Strachan, M. Alaee, S. Backus, M. Bewers, C. Gobeil, C. Halsall, J. Hoff, A. Li, L. Lockhart, D. Mackay, D. Muir, J. Pudykiewicz, K. Reimer, J. Smith, G. Stern, W. Schroeder, R. Wagemann, F. Wania, and M. Yunker. 1997. Chapter 2. Sources, Occurence and Pathways. pp 25-182. *In* J. Jensen, K. Adare, and R. Shearer, [eds.] Can. Arc. Contam. Assess. Rep. Indian and Northern Affairs Canada, Ottawa, Ont. 1997.

BECKMANN, L. 1995. Marine conservation: Keeping the Arctic Ocean on the agenda. North. Perspect. 23: 1-2.

BERKES, F. 1990. Native subsistence fisheries: a synthesis of harvest studies in Canada. Arctic 43: 35-42.

BIDLEMAN, T.F. 1994a. Modelling global-scale transport of hexachlorocyclohexanes: Review and preparation of supporting data, p. 31-38. *In* J. L. Murray, and R. G. Shearer [eds.] Synopsis of research conducted under the 1993/1994 Northern Contaminants Program. Environ. Stud. Rept., No. 72. Indian and Northern Affairs Canada, Ottawa, Ont.

BIDLEMAN, T.F. 1994b. Toxaphene in the Arctic: Atmospheric delivery and transformation in the lower food chain. p. 39-52. In J.L. Murray, and R.G. Shearer [eds.]. Synopsis of research conducted under the 1993/1994 Northern Contaminants Program. Environmental Studies Report, No. 72. Indian and Northern Affairs Canada, Ottawa.

BIDLEMAN, T.F., G.W. Paton, M.D. Walla, B.T. Hargrave, W.P. Vass, P. Erickson, B. Fowler, V. Scott, and D.J. Gregor. 1989. Toxaphene and other organochlorines in Arctic ocean fauna: Evidence for atmospheric delivery. Arc. 42: 307-13.

BIDLEMAN, T.F., M.D. Walla, D.C.G. Muir, and G.A. Stern. 1993. Selective accumulation of polychlorocamphenes in aquatic biota from the Canadian Arctic. Environ. Toxicol. Chem. 12: 701-9.

BIDLEMAN, T.F., R.L. Falconer, and M.D. Walla. 1995a. Toxaphene and other organochlorine compounds in air and water at Resolute Bay, N.W.T., Canada. Sci. Total Environ. 160/161: 55-63.

BIDLEMAN, T.F., L.M. Jantunen, R.L. Falconer, L.A. Barrie, and P. Fellin. 1995b. Decline of hexachlorocycohexanes in the Arctic atmosphere and reversal of air-sea gas exchange. Geophys. Res. Lett. 22: 219-222.

BOHN, A., and R.O. McElroy. 1976. Trace metals (As, Cd, Cu, Fe, and Zn) in Arctic cod (*Boreogadus saida*) and selected zooplankton from Strathcona Sound, Northern Baffin Island. J. Fish. Res. B. Can. 33: 2836-40.

BOHN, A., and B.W. Fallis. 1978. Metal concentrations (As, Cd, Cu, Pb, and Zn) in shorthorn sculpins (*Myoxocephalus scorpius* Linnaeus) and Arctic char (*Salvelinus alpinus* Linnaeus) from the vicinity of Strathcona Sound, Northwest Territories. Wat. Res. 12: 659-63.

BOWES, G.W., and C.J. Jonkel. 1975. Presence and distribution of polychlorinated biphenyls (PCBs) in Arctic and sub-Arctic marine food chains. J. Fish. Res. Bd. Can. 32: 2111-23.

BRAUNE, B.M., R.J. Norstrom, M.P. Wong, B.T. Collins, and J. Lee. 1991. Geographical distribution of metals in livers of polar bears from the Northwest Territories, Canada. Sci. Total. Environ. 100:283-99.

BRIGHT, D.A., W.T. Dushenko, S.L. Grundy, and K.J. Reimer. 1995a. Effects of local and distant contaminant sources: Polychlorinated biphenyls and other organochlorines in bottom-dwelling animals from an Arctic estuary. Sci. Tot. Environ. 160/161: 265-283.

BRIGHT, D.A., S.L. Grundy, and K.J. Reimer. 1995b. Differential bioaccumulation of non-ortho-substituted and other PCB congeners in coastal Arctic invertebrates and fish (*Myoxocephalus quadricornis, M. scorpius, Gadus ogac and Salvelinus alpinu*s). Environmental Science and Technology 29:2504-2512.

CAMERON, M., and I.M. Weis. 1993. Organochlorine contaminants in the country food diet of the Belcher Island Inuit, Northwest Territories, Canada. Arc. 46: 42-48.

CRETNEY, W.J., D.R. Green, B.R. Fowler, B. Humphrey, D.L. Fiest, and P.D. Boehm. 1987a. Hydrocarbon biogeochemical setting of the Baffin Island oil spill experimental sites. I. Sedim. Arc. 40 (Supp.1): 51-65.

CRETNEY, W.J., D.R. Green, B.R. Fowler, B. Humphrey, F.R. Engelhardt, R.J. Norstrom, M. Simon, D.L. Fiest, and P.D. Boehm. 1987b. Hydrocarbon biogeochemical setting of the Baffin Island oil spill experimental sites. III. Biota. Arctic 40 (Suppl. 1): 71-79.

DEPARTMENT of Fisheries and Oceans. 1995. Annual summary of fish and marine mammal harvest data for the Northwest Territories. Volume 6, 1993-1994. Winnipeg, Manitoba. xv + 86.

DEWAILLY, E., A. Nantel, J.P. Weber, and F. Meyer. 1989. High levels of PCBs in breast milk of Inuit women from Arctic Quebec. Bull. Environ. Contam. Toxicol. 43: 641-46.

DEWAILLY, E., P. Ayotte, C. Laliberté, D.C.G. Muir, and R. J. Norstrom. 1994. Human exposure to polychlorinated biphenyls through the aquatic food chain in the Arctic. Environ. Health Perspect. 101: 618-20.

DUNBAR, M. 1973. Stability and fragility in Arctic ecosystems. Arctic 26: 179-185.

EGENDE, I. 1995. Inuit food and Inuit health: Contaminants in perspective. Presentation to Inuit Circumpolar Conference, Seventh General Assembly, Nome, Alaska, 27 July 1995, cited in Chapter 4. Human Health 295-311. *In* J. Jenser, K. Adare, and R. Shearer, [eds.] Can. Arc. Contam. Assess. Rept. Indian and Northern Affairs Canada, Ottawa, Ont. 1997.

ENVIRO-Test Laboratories. 1984. Analysis of 29 fish specimens for a spectrum of hydrocarbons. Rept prep. Dept Fish. Oceans, Winnipeg. 114 p. + appendix.

ENVIRONMENTAL Protection Agency (EPA). 1984. Report from the Environment Effects Branch, Health and Environmental Review Division, Office of Toxic Substances. Washington, D.C.

FALLIS, B. 1982. Trace metals in sediments and biota from Strathcona Sound, N.W.T.: Nanisivik mar. monitor. prog., 1974-1979. Can. Tech. Rep. Fish. Aquat. Sci. 1082.

FELLIN, P., L.A. Barrie, D. Dougherty, D. Toom, D. Muir, N. Grift, L. Lockhart, and B. Billeck. 1996. Air monitoring in the Arctic: Results for selected persistent organic pollutants for 1992. Environmental Toxicology and Chemistry 15: 253-261.

GILMAN, A., E. Dewailly, M. Feeley, V. Jerome, H. Kuhnlein, B. Kwavnick, S. Neve, B. Tracy, P. Usher, J. Van Oostdam, J. Walker, and B. Wheatley. Chapter 4. Human Health. p. 295-377 *In* J. Jensen, K. Adare, and R. Shearer [eds.] Canadian Arctic Contaminants Assessment Report. Indian and Northern Affairs Canada, Ottawa 1997.

GOKSØYR, A., J. Beyer, H. Larsen, T. Anderson, and L. Förlin. 1992. Cytchrome P450 in seals: monooxygenase activities, immunochemical cross reactions and response to phenobarbital treatment. Mar. Environ. res. 34:113-116.

GREGOR, D.J., 1994. The historical record of persistent organic pollutants and trace metals in glacial snow/ice, p. 73-82. *In* J. L. Murray, and R.G. Shearer [eds.]. Synopsis of research conducted under the 1993/1994 Northern Contaminants Program. Environ. Stud. Rept, No. 72. Indian and Northern Affairs Canada, Ottawa, Ont.

GREGOR, D.J., and W. Gummer. 1989. Evidence of atmospheric transport and deposition of organochlorine pesticides and poly chlorinated biphenyls in Canadian Arctic snow. Environ. Sci. Technol. 23: 561-565.

GREGOR, D.J., A.J. Peters, C. Teixeira, N. Jones, and C. Spencer. 1995. The historical residue trend of PCBs in the Agassiz Ice Cap, Ellesmere Island, Canada. Sci. Total Environ. 160/161: 117-26.

HARGRAVE, B. 1994. Sources and sinks of organochlorines in the Arctic marine food web, p. 178-184. *In* J.L. Murray, and R. G. Shearer [eds.] Synopsis of research conducted under the 1993/1994 Northern Contaminants Program. Environ. Stud. Rept., No. 72. Indian and Northern Affairs Canada, Ottawa, Ont.

HARGRAVE, B.T., W. P. Vass, P. E. Erikson, and B.R. Fowler. 1988. Atmospheric transport of organochlorines to the Arctic ocean. Tellus 40B: 480-493.

HARGRAVE, B., W. Vass, P. Erickson, and B. Fowler. 1989. Distribution of chlorinated hydrocarbon pesticides and PCBs in the Arctic Ocean. Can. Tech. Rep. Fish. Aquat. Sci. 1644.

HARGRAVE, B.T., G.C. Harding, W.P. Vass, P.E. Erickson, R.Fowler, and V. Scott. 1992. Organochlorine pesticides and polychlorinated biphenyls in the Arctic Ocean food web. Arch. Environ. Contam. Toxicol. 22: 41-54.

HILD, C.M. 1995. The next step in assessing Arctic human health. Sci. Total Environ. 160/161: 559-69.

HOFF, R.M., D.C.G. Muir, and N.P. Grift. 1992. Annual cycle of polychlorinated biphenyls and organohalogen pesticides in air in southern Ontario: 1. Air Concen. data. Environ. Sci. Technol. 26: 266-75.

JONES, D., K. Ronald, D.M. Lavigne, R. Frank, M. Holdrinet, and J.F. Uthe. 1976. Organochlorine and mercury residues in the harp seal (*Pagophilus groenlandicus*). Sci. Tot. Environ. 5: 181-95.

KINGSLEY, M. 1994. Mercury and other inorganic contaminants in country foods in eastern Hudson Bay, p. 208-210. *In* J.L. Murray, and R.G. Shearer [eds.] Synopsis of research conducted under the 1993/1994 Northern Contaminants Program. Environ. Stud. Rept, No. 72. Indian and Northern Affairs Canada, Ottawa, Ont.

KINLOCH, D., H. Kuhnlein, and D.C.G. Muir. 1992. Inuit foods and diet: A preliminary assessment of benefits and risks. Sci. Total Env. 122: 247-78.

KUHNLEIN, H., and D. Kinloch. 1988. PCBs and nutrients in Baffin Island Inuit foods and diets. Arct. Med. Res. 47: 155-58.

LANNO, R.P., D.L. MacKenzie, C.D. Wren, and D. Gauthier. 1994. Assessment of the current knowledge and definition of the problems concerning contaminants in the Arctic environment. Sup. Serv. Can. 155.

LEWIS, E.L., D. Ponton, L. Legendre, and B. LeBlanc. 1996. Springtime sensible heat, nutrients and phytoplankton in the Northwater Polynya, Canadian Arctic. Contin. Shelf Res. 16: 1775-1792.

LOCKHART, W.L. 1994. Depositional trends: Lake and marine sediments, p. 138-151. *In* J.L. Murray, and R.G. Shearer [eds.] Synopsis of research conducted under the 1993/1994 Northern Contaminants Program. Environ. Stud. Rept No. 72. Indian and Northern Affairs Canada, Ottawa, Ont.

LOCKHART, W.L., and R.E.A. Stewart. 1992. Biochemical stress indicators in marine mammals. p. 158-164. *In* Synopsis of Research Conducted Under the 1991/92 Northern Contaminants Program. Environ. Stud. Rept No. 68. Indian and Northern Affairs Canada, Ottawa, Ont.

LOCKHART, W.L., R. Wagemann, B. Tracey, D. Sutherland, and D.J. Thomas. 1992. Presence and implications of chemical contaminants in the freshwaters of the Canadian Arctic. The Science of the Total Environment 122: 165-243.

LOCKHART, W.L., P. Wilkinson, B.N. Billeck, R.V. Hunt, R. Wagemann, and G.J. Brunskill. 1995. Current and historical inputs of mercury to high-latitude lakes in Canada and to the Hudson Bay. Wat. Air Soil Pollut. 80: 603-610.

MACDONALD, C.R. 1986. The influence of diet on the accumulation of cadmium in ringed seals (*Phoca hispida Shreber*) in the Canadian Arctic. Ph.D thesis, University of Guelph, Guelph ON.

MACDONALD, R.W., and J.M. Bewers. 1996. Contaminants in the arctic marine environment: priorities for protection. ICES J. Mar. Sci. 53: 537-563.

MACDONALD, R.W., and F.A. McLaughlin. 1993. Long-range transport of contaminants to the Canadian basin, p. 15-20. *In* J.L. Murray, and R.G. Shearer [eds.] Synopsis of Research Conducted under the 1992/1993 Northern Studies Program. Environ. Stud. Rept No. 70. Indian and Northern Affairs Canada, Ottawa, Ont.

MACDONALD, C.R., and J.B. Sprague. 1988. Cadmium in marine invertebrates and Arctic cod in the Canadian Arctic. Distribution and ecological implications. Mar. Ecol. - Progress Ser. 47: 17-30.

MACKAY, D., and F. Wania. 1995. Transport of contaminants to the Arctic: partitioning, processes and models. Sci. Tot. Environ. 160/161: 25-38.

MACKEY, M.G.A., and R.D. Orr. 1987. An evaluation of household country food use in Makkovik, Labrador, July 1980-June 1981. Arct. 40: 60-65.

MATHIEU, A., J.F. Payne, L.L. Fancey, and R. Santella. 1995. DNA adducts in beluga whales in Canadian waters. Presented at the first World Conference on Spontaneous Animal Tumors (Genoa, Italy, April, 28 1995-April, 30 1995).

MCNEELY, R., and W.D. Gummer. 1984. A reconnaissance survey of the environmental chemistry in east-central Ellesmere Island, N.W.T. Arct. 37: 210-223.

MITCHELL, M. 1956. Visual range in the polar regions with particular reference to the Alaskan Arctic. J. Atmos. Terrest. Phys. (Spec. Suppl.:) 195-211.

MORIN, R., and J.J. Dodson. 1986. The ecology of fishes in James Bay, Hudson Bay and Hudson Strait. *In* I.P. Martini [ed.] Canadian Inland Seas. Elsevier Oceanography Series 44: 293-325.

MUIR, D.C.G. 1994. Spatial and temporal trends of organochlorines in Arctic marine mammals. p 189-196. *In* J.L. Murray, and R.G. Shearer [eds.] Synopsis of Research Conducted under the 1993/1994 Northern Contaminants Program. Environmental Studies No. 72. Indian and Northern Affairs Canada, Ottawa, Ont.

MUIR, D.C.G. 1996. Spatial and temporal trends of organochlorines in Arctic marine mammals, p 143-146. *In* J.L. Murray, and R.G. Shearer [eds.] Synopsis of Research Conducted under the 1994/1995 Northern Contaminants Program. Environmental Studies No. 73. Indian and Northern Affairs Canada, Ottawa, Ont.

MUIR, D.C.G., and W.L. Lockhart. 1994. Contaminant trends in freshwater and marine fish, p. 264-271. *In* J.L. Murray, and R.G. Shearer [eds.] Synopsis of research conducted under the 1993/1994 Northern Contaminants Program. Environ. Stud. No. 72. Indian and Northern Affairs Canada, Ottawa, Ont.

MUIR, D.C.G., and W.L. Lockhart. 1996. Contaminant trends in freshwater and marine fish, p. 189-194. *In* J.L. Murray, and R.G. Shearer [eds.] Synopsis of research conducted under the 1994/1995 Northern Contaminants Program. Environ. Stud. No. 73. Indian and Northern Affairs Canada, Ottawa, Ont.

MUIR, D.C.G, and R.J. Norstrom. 1994. Persistent organic contaminants in Arctic marine and freshwater ecosystems. Arct. Res. U.S. 8: 136-146.

MUIR, D.C.G., R.J. Norstrom, and M. Simon. 1988. Organochlorine contaminants in Arctic marine food chains: accumulation of specific PCB congeners and chlordane-related compounds. Environ. Sci. Technol. 22:1071-1079.

MUIR, D.C.G., R. Wagemann, B.T. Hargrave, D.J. Thomas, D.B. Peakall, and R.J. Norstrom. 1992. Arctic marine ecosystem contamination. Sci. Total Environ. 122: 75-134.

MUIR, D.C.G., M.D. Segstro, K.A. Hobson, C.A. Ford, R.E.A. Stewart, and S. Olpinski. 1995. Can elevated levels of PCBs and organochlorine pesticides in Walrus blubber from eastern Hudson Bay (Canada) be explained by consumption of seals? Environ. Poll. 90:335-348.

MUIR, D., B. Braune, B. DeMarch, R. Norstrom, R. Wagemann, M. Gamberg, K. Poole, R. Addison, D. Bright, M. Dodd, W. Duschenko, J. Eamer, M. Evans, B. Eklin, S. Grundy, B. Hargrave, C. Hebert, R. Johnstone, K. Kidd, B. Koenig, L. Lockhart, J. Payne, J. Peddle, and K. Reimer. 1996. Chapter 3. Ecosystem Uptake and Effects. 183-294. *In* J. Jensen, K. Adare, and R. Shearer [eds.] Can. Arct. Contam. Assess. Rept. Indian and Northern Affairs Canada, Ottawa, Ont. 1997.

NORSTROM, R.J., and D.C.G. Muir. 1994. Chlorinated hydrocarbon contaminants in arctic marine mammals. Sci. Total Environ. 154: 107-128.

NORSTROM, R.J., M. Simon, D.C.G. Muir, and R. Schweinsburg. 1988. Organochlorine contaminants in Arctic marine food chains: Identification, geographical distribution and temporal trends in polar bears. Environmental Science and Technology 22: 1063-1071.

NORSTROM, R.J., M. Simon, and D.C.G. Muir. 1990. Polychlorinated dibenzo-p-dioxins and dibenzofurans in marine mammals in the Canadian north. Environ. Pollut. 66: 1-19.

OLSSON, M. 1995. Ecological effects of airborne contaminants in Arctic aquatic ecosystems; A discussion on methodological approaches. Sci. Total Environ. 160/161: 619-30.

OTTAR, B., Y. Gotaas, Ø. Hov, T. Iversen, E. Joranger, M. Oehme, J.M. Pacyna, A. Semb, W. Thomas, and V. Vitols. 1986. Air Pollutants in the Arctic. NILU OR Rept. 4/84, Norweg. Inst. Air Res., Lillestrøm, Norway.

PACYNA, J.M. 1995. The origin of Arctic air pollutants: lessons learned and future research. The Science of the Total Environment 160/161: 39-53.

PACYNA, J.M., and G.J. Keeler. 1995. Sourc. mercury in the Arct. Wat. Air Soil Pollut. 80: 621-632.

PATTON, G.W., D.A. Hinckley, M.D. Walla, and T.F. Bidleman. 1989. Airborne organochlorines in the Canadian high Arctic. Tellus 41B: 243-55.

PAYNE, J. 1993. Potential for effects on reproduction, carcinogenesis, mutagenesis and teratogenesis in Arctic mammals: Status of biomarkers in Arctic seals and whales, p. 136-143. *In* J.L. Murray, and R.G. Shearer [eds.] Synopsis of research conducted under the 1992/1993 Northern Contaminants Program. Environ. Stud. Rept, No. 70. Indian and Northern Affairs Canada, Ottawa, Ont.

PETERS, A.J., D.J. Gregor, C.F. Teixeira, N.P. Jones, and C. Spencer. 1995. The recent depositional trend of Polycyclic Aromatic Hydrocarbons and elemental Carbon to the Agassiz Ice Cap, Ellesmere Island, Canada. Sci. Total Environ. 160/161: 167-79.

POOLE, K.G., B.T. Elkin, and R.W. Bethke. 1995. Environmental contaminants in wild mink in the Northwest Territories, Canada. Sci. Total Environ. 160,161: 473-86.

RICHARDSON, W.J., J.P. Hickie, R. A. Davis, and D.H. Thomson. 1989. Effects of offshore petroleum operations on cold water marine mammals: A literature review. Prepared for the American Petroleum Institute, 1220 L St., N.W., Washington D.C. 20005. 328 p.

SCHLOSSER, P., J.H. Swift, D. Lewis, and S.L. Pfirman. 1995. The role of the large-scale Arctic Ocean circulation in the transport of contaminants. Deep-Sea Res. II 42: 1341-1367.

ST. AUBIN, D.J., and J.R. Geraci. 1980. Tissue levels of ascorbic acid in marine mammals. Comp. Biochem. Physiol. 66: 605-609.

STERN, G.A., D.C.G. Muir, C.A. Ford, N.P. Grift, E. DeWailly, T. F. Bidleman, and M.D. Walla. 1992. Isolation and identification of two major recalcitrant toxaphene congeners. Environ. Sci. Technol. 26: 1838-40.

SWYRIPA, M., and W.M.J Strachan. 1994. Current contaminant deposition measurements in Arctic precipitation (snow), p. 83-90. *In* J.L. Murray, and R.G. Shearer [eds.] Synopsis of research conducted under the 1993/1994 Northern Contaminants Program. Environ. Stud. Rept, No. 72. Indian and Northern Affairs Canada, Ottawa, Ont.

THOMAS, D.J. 1988. The Tuktoyaktuk Harbour benthic biological monitoring programme. Report prepared for Conservation and Protection. Environ. Can., Yellowknife, N.W.T. Sidney, B.C. SeaKem Oceanography Ltd.

THOMAS, D.J., R.W Macdonald, and A.B. Cornford. 1982. Arctic data compilation and appraisal. Vol. 2 Beaufort Sea: Chemical oceanography Can. Data Rep. Hydrogr. Ocean Sci. 5. 243.

THOMAS, D.J., P.F. Wainwright, B.D. Arner, and W.H. Coedy. 1983. Beaufort Sea sediment reconnaissance survey: A data report on 1982 geochemical and biological sampling. Arct. Lab. Ltd. Sidney, B.C. 459 p.

THOMAS, D.J., P.E. Erickson, and J.D. Popham. 1984. Chemical and biological studies of Strathcona Sound, N.W.T., 1982. Rept prep. Indian and Northern Affairs Cananada, Yellowknife, N.W.T. Sidney, B.C. SeaKem Oceanography Ltd.

THOMAS, D.J., R.W. Macdonald, and A.B. Cornford. 1986. Geochemical mass-balance calculations for the coastal Beaufort Sea. N.W.T. Rapports et Procès-Verbaux des Réun. Conseil Int. pour l'Explor. de la Mer. 186: 165-184.

THOMAS, D.J., B. Tracey, H. Marshall, and R.J. Norstrom. 1992. Arctic terrestrial ecosystem contamination. Sci. Total Environ. 122: 135-64.

WAGEMANN, R. 1994. Mercury, methyl mercury, and other heavy metals in muktuk, muscle, and some organs of belugas (*Delphinapterus leucas*) from the western Arctic. Final Summ. Rept Fish. Joint Mgnt Comm. Inuvialuit Settlem. Reg. ii + 38 p.

WAGEMANN, R., and R.E.A. Stewart. 1994. Concentrations of heavy metals and selenium in tissues and some foods of walrus (*Odobenus rosmarus rosmarus*) from the eastern Canadian Arctic and sub-Arctic, and associations between metals, age, and Can. J. of Fish. Aquat. Sci. 51: 426-36.

WAGEMANN, R., N.B. Snow, A. Lutz, and D.P. Scott. 1983. Heavy metals in tissues of the narwhal (*Monodon monoceros*). Can. J. Fish. Aquat. Sci. 40 (Suppl 2): 206-14.

WAGEMANN, R., R.E.A. Stewart, W.L. Lockhart, B.E. Stewart, and M. Povoledo. 1988. Trace metals and methyl mercury: Associations and transfer in harp seal (*Phoca groenlandica*) mothers and their pups. Mar. Mamm. Sci. 4: 339-355.

WAGEMANN, R., W.L. Lockhart, H. Welch, and S. Innes. 1995. Arctic marine mammals as integrators and indicators of mercury in the Arctic. Wat., Air, Soil Poll. 80: 683-93.

WAGEMANN, R., S. Innes, and P.R. Richard. 1996. Overview and regional and temporal differences of heavy metals in arctic whales and ringed seals in the Canadian Arctic. Sci. Total Environ. 186: 41-66.

WAINRIGHT, P. F., and B. Humphrey. 1988. Analysis of sediment data from the Beaufort Sea shorebase monitoring program 1982 to 1984. Environ. Stud. Res. Funds, Rept No. 090. Ottawa, Ont. 147 p.

WANIA, F., and D. Mackay. 1993. Global fractionation and cold condensation of low volatility organochlorine compounds in polar regions. Ambio 22: 10-18.

WEIN, E.E., J. Henderson Sabry, and F.T. Evers. 1991. Food consumption patterns and use of country foods by native Canadians near Wood Buffalo National Park, Canada. Arct. 44: 196-205.

WELCH, H.E. 1995. Marine conservation in the Canadian Arctic: A regional overview. North. Perspect. 23: 5-17.

WELCH, H.E., D.C.G. Muir, B.N. Billeck, W.L. Lockhart, G.J. Brunskill, H.J. Kling, M.P. Olson, and R.M. Lemoine. 1991. Brown snow: A long-range transport event in the Canadian Arctic. Environ. Sci. Technol. 25: 280-286.

WELCH, H.E., M.A. Bergmann, T.D. Siford, K.A. Martin, M.F. Curtis, R.E. Crawford, R.J. Conover, and H. Hop. 1992. Energy flows through the marine ecosystem of the Lancaster Sound region, Arctic Canada. Arctic 45:343-357.

WELCH, H.E., T.D. Siferd, and P. Bruecker. 1996. Population densities, growth, and respiration of the chaetognath, *Parasagitta elegans* in the Canadian high Arctic. Canadian Journal of Fisheries and Aquatic Sciences 53: 520-527.

WHITE, R.D., M.E. Hahn, and W.L. Lockhart. 1994. Catalytic and immunochemical characterization of hepatic microsomal cytochrome P450 in beluga whale (*Delphinapterus leucas*). Toxicol. Appl. Pharmacol. 126: 45-57.

WILSON, R.D., P.H. Monaghan, A. Osanik, L.C. Price, and M.A. Rogers. 1973. Estimate of annual input of petroleum to the marine environment from natural seepage. Trans. Gulf Coast Assoc. Geol. Soc. 23: 182-193.

WONG, M.P. 1985. Chemical residues in fish and wildlife species harvested in northern Canada. North. Aff. Prog., Environ. Stud. 46. North. Environ. Direct., Indian and Northern Affairs Canada, Ottawa, Ont.

WONG, C.S., W.J. Cretney, R.W. Macdonald, and Christensen. 1976. Hydrocarbon levels in the marine environment of the southern Beaufort Sea: Beaufort Sea Project. Tech. Rept No. 38. Dep. Fish. Oceans. Sidney, B.C. 113 p.

YUNKER, M.B., and R.W. Macdonald. 1995. Composition and origins of polycyclic aromatic hydrocarbons in the Mackenzie River and on the Beaufort Sea Shelf. Arct. 48: 118-129.

Additional reading

CAMERON, M.E., T.L. Metcalfe, C.D. Metcalfe, and C.R. Macdonald. 1997. Persistent organochlorine compounds in the blubber of ringed seals (*Phoca hispida*) from the Belcher Islands, Northwest Territories, Canada. Marine Environmental Research 43: 99-116.

CRETNEY, W.J., D.R. Green, B.R. Fowler, B. Humphrey, D.L. Fiest and P.D. Boehm. 1987. Hydrocarbon biogeochemical setting of the Baffin Island oil spill experimental sites. II. Water. Arctic 40 (Suppl. 1): 66-70.

ENVIRONMENTAL Protection Service (EPS). 1987. Summary of environmental criteria for chlorinated biphenyls (PCBs). Report EPS 4/HA/1. Revised October 1987. Environment Canada, Ottawa.

MUIR, D.C.G. 1993. Co-planar PCBs in Arctic marine mammals and fish, p. 285. *In* J.L. Murray, and R.G. Shearer [eds.] Synopsis of research conducted under the 1992/93 Northern Contaminants Program. Environ. Stud. Rept No. 70. Indian and Northern Affairs Canada, Ottawa, Ont.

MUIR, D.C.G., R. Wagemann, W.L. Lockhart, N.P. Grift, B. Billeck, and D. Metner. 1986. Heavy metal and organic contaminants in Arctic marine fishes. Min. Ind. Aff. North. Develop. Environ. Stud. Rept, No. 42. North. Aff. Prog. Sup. Serv. Can. 64 p.

MUIR, D.C.G., C.A. Ford, R.E.A. Stewart, T.G. Smith, R.F. Addison, M.T. Zinck, and P. Béland. 1990. Organochlorine contaminants in beluga (*Delphinaterus leucas*) from Canadian waters. Can. Bull. Fish. Aquat. Sci. 224: 165-90.

MUIR, D.C.G., C.A. Ford, N.P. Grift, R.E.A Stewart, and T.F. Bidleman. 1992. Organochlorine contaminants in Narwhal (*Monodon monoceros*) from the Canadian Arctic. Environ. Pollut. 75: 307-16.

YUNKER, M.B.,1988. Long range transport of organochlorines in the Arctic and sub-Arctic: evidence from analysis of marine mammals and fish, p. 107-128. *In* N. W. Schmidtke [ed.] Chronic effects of toxic contaminants in large lakes. World Conference on Large Lakes. Lewis Publishers Inc., Michigan.

YUNKER, M.B., and R.W. Macdonald. 1993. Sources and sinks of organochlorines in the Arctic marine food web, p. 81-86. *In* J.L. Murray, and R.G. Shearer [eds.] Synopsis of research conducted under the 1992/1993 Northern Contaminants Program. Environ. Stud. Rept, No. 70. Indian and Northern Affairs Canada, Ottawa, Ont.

YUNKER, M.B., and R.W. Macdonald. 1994. Toxaphene in the Arctic: Atmospheric delivery and transformation in the lower food chain, p. 39-52. *In* J.L. Murray, and R.G. Shearer [eds.] Synopsis of research conducted under the 1993/1994 Northern Contaminants Program. Environ. Stud. Rept, No. 72. Indian and Northern Affairs Canada, Ottawa, Ont.

Personal Communications

ADDISON, R.F., Institute of Ocean Sciences, Sidney, B.C., n.d.

WELCH, H.E., Department of Fisheries and Oceans, Central and Arctic, Winnipeg, Manitoba, n.d.

The PACIFIC
Marine and Freshwater
Ecosystems

Highlights

- The topography of today's B.C. coast is dominated by the Vancouver Island Range and the Coastal Range. The coastline is mountainous, with peaks rising to almost 4 000 m, and penetrated by deep inlets and fjords. Offshore, a complex system of islands is separated by deep channels. Rivers carry glacial till and large amounts of particulate matter to the sea.

- Oceanographic conditions on the outer B.C. coast are controlled by large-scale oceanic processes. However, freshwater inputs from rivers, and tides affect these large scale oceanographic processes on a more localized scale. These conditions significantly influence contaminant transport and accumulation.

- Commercial, sport, or subsistence fisheries are conducted on British Columbia's major rivers and inshore along the entire coastline, including the Strait of Georgia. These fisheries had a landed value of over $1 billion in 1994 and employed about 25 000 people.

- Contamination by polychlorodibenzo-*p*-dioxins (PCDDs) and polychlorodibenzo-furans (PCDFs) discharged from pulp mills led to fishery closures starting in 1988. By 1995, about 1 200 km^2 of fishing grounds had been closed. Some of these fisheries have since been re-opened as contaminant concentrations have declined.

- PCDDs and PCDFs discharged from pulp mills accumulate in sediments. Following reductions in these discharges in the late 1980s, contaminated sediment at some sites was gradually buried under layers of clean sediment. As a result, availability of PCDDs and PCDFs to crabs, prawns, oysters and shrimp was reduced, thus allowing the

Highlights

gradual re-opening of fisheries. However, where sediments are remobilized by tidal waves or bioturbation, PCDD and PCDF concentrations in biota remain elevated; hence, some fisheries remain closed. Sublethal biological effects are also found in fish exposed to pulp mill effluents.

■ Tributyltin (TBT) continues to be used in anti-fouling paints, particularly on commercial and naval shipping, although it was banned from use on most recreational vessels in 1989. TBT concentrations in sediments from harbours used by small boats have declined since the late 1980s, but sediments from commercial ports such as Vancouver remain contaminated. One effect of TBT contamination is imposex (development of male characteristics in females) in some marine snails. The frequency of TBT-caused imposex is declining in recreational areas, but remains elevated in commercial ports.

■ Acid mine drainage leaches metals, particularly copper and zinc. Shore-based tailings sites near abandoned mines, and tailings discharged to fjords also release metals. Although rates of metal leaching or release are high, most of the metals are confined to sedimentary basins in fjords. There is no evidence of widespread fishery contamination at present.

■ As the point source discharges of contaminants become better controlled and as environmental concentrations decline, future concerns about contamination may focus on the impact of the growing population density in the lower mainland and southern Vancouver Island. Population growth may put increasing pressure on the environmental quality of the Strait of Georgia. Increasing discharges of nutrients and organic matter or sewage may lead to eutrophication.

■ Hitherto undetected synthetic chemicals may survive sewage treatment and cause unforeseen problems, such as disruption of normal endocrine processes in biota or in those who consume the fish. Increasing use and discharge of pharmaceuticals through sewage by an increasingly aging population may also have unforeseen effects.

■ The growth of aquaculture may lead to increased accumulation of sludge and to releases of nutrients, antibiotics, or pesticides, which may affect inshore ecosystems.

Table of Contents

Pacific Marine and Freshwater Ecosytems

List of Figures

List of Sidebars

Introduction

In this chapter, the British Columbia coastal ecosystem is considered to consist of three components: the Strait of Georgia; the British Columbia open coast and fjords, including some local "hot-spots" such as Kitimat Arm; and the Fraser, Columbia, Nass, and Skeena river systems.

MAJOR CONTAMINANT ISSUES

Environmental regulations in general have reduced some of the conventional contaminants entering the region. However, less familiar contaminants whose production is increasing will require attention. These include non-ionic surfactants, some of which are potential endocrine disrupters. Therefore, the current scientific emphasis on the study of the "classic" range of organochlorine compounds, polycyclic aromatic hydrocarbons (PAHs) and metals will almost certainly change.

Eleven of British Columbia's pulp mills are situated on the coast. These mills discharge to the coastal environment and their effluents have a direct impact on inshore ecosystems. Polychlorodibenzo-*p*-dioxins (PCDDs) and polychlorodibenzo-furans (PCDFs) were released in pulp mill effluents until the late 1980s, when the federal government issued harvesting regulations and began developing legislation to reduce the release of these chemicals. Although residues of PCDD sand PCDFs remain in sediments near several pulp mills (Macdonald *et al.* 1992; Hagen *et al.* 1995), current emissions of these chemicals are very low. As a result, pulp mills are now considered to be of declining importance as a source of these contaminants.

Direct comparisons can be made between the use of dichlorodiphenyltrichloroethane (DDT)-group insecticides and the polychlorinated biphenyls (PCBs) in B.C. and Eastern Canada.

Studies of similar environmental compartments in both regions show lower concentrations of these chemicals in B.C. than in Eastern Canada. However, PCDDs and PCDFs are found at higher concentrations and are more widely distributed in B.C.

PAHs are introduced to coastal B.C. from various sources (Waldichuk 1983). Non-point sources, defined as those which are not well quantified (West *et al.* 1994), include general urban and industrial run-off in the lower mainland. These are most significant for the Strait of Georgia. The better-defined point sources are associated with particular locations, such as the aluminum smelter at Kitimat and the coal-loading terminal at Prince Rupert.

With the closing of many of the region's coastal mines, concerns over metal contamination from mine tailings have decreased. In addition, those contaminants already in place are considered to be well contained. This is due to the fact that contaminants from these mines were often discharged directly into fjords, such as the Observatory Inlet system, where the water circulation and sediment transport patterns do not promote great movement, thereby reducing the risk of widespread contamination (Macdonald and O'Brien 1996). However, acid mine drainage continues to be a problem in Howe Sound, where perhaps as much as 600 kg.d^{-1} each of copper and of zinc enter the sea (McCandless 1995). Similar though smaller discharges may occur at other abandoned mine sites. Mercury discharges from a chlor-alkali plant at Squamish, which contaminated at least the upper part of Howe Sound, have now ceased, though mercury may still remain in the system. Concern about metal contamination in the region now focuses primarily on organometallic compounds. The best example of this is tributyltin (TBT), which originates from anti-fouling paints.

A number of important contaminant issues are expected to emerge in the next few years. The significant population growth that is predicted for the southern mainland and southeast Vancouver Island in the next two decades will put additional pressure on the Strait of Georgia and may reduce its carrying capacity for natural fish stocks. The lower Fraser Valley is also expected to experience a population boom, raising concerns about traffic density and ground level ozone concentration. At the same time, demographics reveal an aging population. This will lead to an increase in the demand for pharmaceuticals and a corresponding increase in discharge associated with pharmaceutical use. In addition, the growing importance of aquaculture raises questions about the use of chemicals, including the use of pesticides to control sea lice and antibiotics to control disease, and their impact on the environment. The potential effects of these and other developments on the biological productivity of the ecosystem are yet to be seen.

OCEANOGRAPHIC FEATURES OF IMPORTANCE TO CONTAMINANT DYNAMICS

A number of marine environmental conditions affect on the B.C. coast, including the supply of particulate material; water and tidal movements; stratification; and biological activity (Thomson 1981).

Figure 1
Major topographic features of the Pacific region.

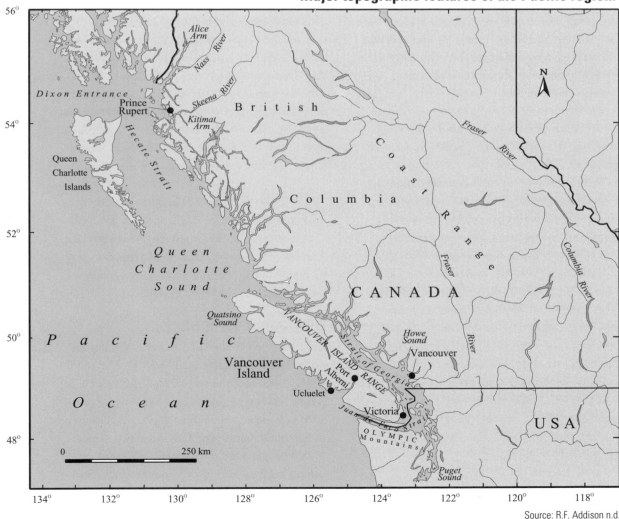

Source: R.F. Addison n.d.

The topography of today's B.C. coast originated about 150 million years ago when the Pacific and North American crustal plates collided. The consequent buckling and folding formed two mountain ranges, which over time have been further changed by upward movements, deposits, erosion and the movement of glaciers. These two mountain ranges are well identified today. The western mountain range is comprised of the Vancouver Island Range and its extensions to the Queen Charlotte Islands and the Olympic Range in the U.S. The eastern range includes the Coastal Range and its extension to the Cascades in the northwestern U.S. A trough, stretching from Alaska to the Gulf of California, separates the two ranges, and includes the Strait of Georgia and the Hecate Strait. The coastline is mountainous, with peaks rising to almost 4 000 m, and penetrated by deep inlets and fjords. Offshore, a complex system of islands is separated by deep channels. Rivers carry glacial till to the sea or, as in the case of the Fraser, Skeena and Nass rivers, large amounts of other particulate loads. Figure 1 shows the major topographical features of the region.

Oceanographic conditions on the outer B.C. coast are controlled by large-scale oceanic processes. The Aleutian Low atmospheric system in the northern North Pacific creates cyclonic (counter-clockwise) winds. During the winter, these winds blow predominantly from the southeast, creating an inshore current which sets to the northwest. The Fraser River via Juan de Fuca Strait, the Columbia River and local coastal run-off (Macdonald and Pedersen 1991) all provide important freshwater inputs to this surface flow. During the summer, as the Aleutian Low recedes and the more southerly North Pacific High pressure zone expands, anti-cyclonic winds blow from the northwest. These winds produce offshore currents that create coastal upwelling. This constant supply of deep, nutrient-rich cold water stimulates biological productivity in the region.

However, two localized conditions — freshwater inputs from rivers, and tides — affect these large scale oceanographic processes and create conditions that are very site specific. This makes it difficult to generalize about the movements of water masses. For example, the Fraser River flow is sufficiently large that the Strait of Georgia can be considered an estuarine system, with a net surface outflow of fresher water being replaced by deeper inflow of water with a higher saline content. Tidal oscillations are complicated by the convoluted coastline and abundance of islands and channels (Thomson 1981).

While there is little data from B.C. coastal areas, it is likely that wind is the predominant distribution mode of the more volatile organic contaminants. However, in the Strait of Georgia, which is surrounded by a large population with associated urban and industrial contaminant sources, the Fraser River provides the distribution mode through its suspended particulate material. The large natural supply of particulate matter from the Fraser River scavenges contaminants, depositing them predominantly within this coastal basin (Macdonald and Crecelius 1994). A similar process can be also seen on a more local scale at Howe Sound and Kitimat Arm.

Most information about chemical contaminants refers to the Strait of Georgia and to a few local "hot-spots" on the B.C. coast. In this chapter, the discussion of rivers focuses on two aspects: as a source of contaminants to marine ecosystems, and on effects in the rivers themselves.

BIOLOGICAL RESOURCES

The British Columbia commercial fishing and seafood industries generated about $1.2 billion in 1994. In addition, the recreational fishing and marine tourism industries generated over $200 million, accounting for about one-third of all tourism dollars spent in the province. About 25 000 people are directly employed in the fishing industry in B.C., which represents about

19 000 full-time jobs. About 20% of the commercial salmon fishery is conducted by Aboriginals. Of the commercial landings, six species of salmonids (sockeye, coho, chinook, chum, pink and steelhead) represent over 40% of the total landed value. Herring, groundfish (principally halibut and sablefish) and invertebrates (mainly geoduck and crab) each represent slightly under 20% of total landed value. Farmed salmon (of which Atlantic salmon accounts for about 68%) now represent about one-third the total landed value of salmonids (Anon. 1995).

The health of this economic base is, of course, affected by the health of the fishing grounds. In the late 1980s, chemical contaminants, particularly releases of PCDDs and PCDFs from pulp and paper mills, led to closures of the crab and shellfish fisheries. At the beginning of 1995, about 1200 km^2 of fishing grounds remained closed. By the end of 1995, about 40% of the grounds were reopened due to reductions in the concentrations of these contaminants.

Assessment of Chemical Contamination

THE STRAIT OF GEORGIA AND HOWE SOUND

The Strait of Georgia is about 200 km long and about 20 to 40 km wide, with a surface area of 6 800 km^2. It drains an area of about 290 000 km^2 (Thomson 1994). Flushing time in winter is 100 to 200 days; in summer, 50 to 70 days. While geographically connected to the Strait of Georgia, Howe Sound is a fjord which contains both a sedimentary basin and several local sources of contamination.

SOURCES

The largest centres of population surrounding the Strait of Georgia are in Vancouver (population 1.8 million) and Victoria (population 260 000). Several smaller communities stretch along the mainland coast north of Vancouver and on the southeast coast of Vancouver Island. Industrial waste enters the Strait not only from the cities, but from pulp mills and from isolated mines at points along the coast (Waldichuk 1983; Kay 1989; Macdonald and Crecelius 1994). The sediment carried down the Fraser River plays a significant role in controlling the distribution of particulate bound compounds. Local topographical features such as fjords may control the dispersal of contaminants originating from point sources.

DISTRIBUTION, TRENDS AND EFFECTS
METALS AND ORGANOMETALLIC COMPOUNDS

Metals enter the Strait of Georgia from several sources, including discharges from pulp mills and from metal working shops and shipyards in the urban areas of Vancouver, Victoria and Nanaimo, and from pulp mills. More general urban inputs include sewage and urban run-off. Tailings from an abandoned mine in Howe Sound had also been considered as a potential source of metal pollution, though this now appears unlikely (Drysdale and Pedersen 1992). Acid mine drainage in the abandoned workings maybe more important source of metals.

A general picture of metal distribution in the Strait of Georgia during the past century emerges from analyses of dated sediment cores from the Ballenas Basin, Thornbrough Channel and Howe Sound (Macdonald *et al.* 1991; Macdonald and Crecelius 1994). Since the Ballenas Basin has no immediate source of metal contamination, trends in metal distribution in that core represent contamination associated with increasing industrialization and population growth in the entire region. The other cores were taken at the mouth of Howe Sound. At the head of Howe Sound, a chlor-alkali plant near Squamish which had released significant amounts of mercury, and a mine at Britannia from which copper and zinc continue to be released, may have caused or continue to cause local contamination.

Lead concentrations in several sediment cores show a fairly general increase, particularly after about 1930, when lead-based anti-knock additives were first used in gasoline. These concentrations have decreased since the 1970s, following the elimination of lead additives from gasoline. Strait of Georgia sediments also show detectable copper contamination after about 1930; and in Howe Sound, where Britannia Mines had disposed of tailings since 1905, the sediments have contained much higher concentrations since about 1930 (Macdonald and Crecelius 1994).

The historical trend for zinc is similar to that of copper except that it is displaced about 10 to 20 years later, probably reflecting the later emphasis on zinc production at the Britannia Mine. Mercury concentrations in Ballenas Basin sediments have remained fairly constant over time, but in Howe Sound cores, mercury increased dramatically during the 1960s and then declined more slowly after the mid 1970s. This pattern reflects the release during the late 1960s of as much as 20 kg.d^{-1} mercury from the chlor-alkali plant at Squamish and the dramatic reduction in mercury discharges following controls imposed in the early 1970s (Thompson *et al.* 1980). Despite attempts to determine the relative contribution from the plant, much of the mercury released to Howe Sound cannot be accounted for in local sediments. Some of the mercury may have been either volatilized or moved out of Howe Sound. Establishing a time trend for cadmium from sediment core data is complicated by geochemical processes. Because cadmium tends to precipitate as an insoluble sulfide (Pedersen *et al.* 1989), elevated concentrations in sediments may reflect geochemical processes rather than contamination. For example, cadmium concentrations in surface sediment of up to 7 µg.g^{-1} dry weight in Ucluelet Harbour were interpreted as reflecting natural processes (Pedersen *et al.* 1989). In contrast, sediments from the Strait of Georgia have uniformly low cadmium concentrations (Macdonald *et al.* 1991). Near sites where particulate or elemental cadmium may have

entered the environment from boat repair, metalworking, or paint shops, increases in cadmium in surface sediments may reflect local contamination (Goyette and Boyd 1989). It is nevertheless difficult to interpret cadmium trends from sediment core data, due to the biogeochemical processes affecting the movement of cadmium, and its long residence time in the water.

Where data are available, metal distribution in biota generally follows the trends found by sediment analyses. During the 1970s, metal concentrations in a range of biota in Georgia Strait were generally quite low (Harbo *et al.* 1983). Thus, mercury was usually below 0.2 µg.g^{-1} wet weight in most species, except for dogfish. This is probably an example of a relatively high trophic level and long lived fish accumulating high concentrations from natural sources, as has been observed in swordfish and tuna on the Atlantic coast. Copper concentrations were usually under 5 µg.g^{-1} wet weight in molluscs, under 10 µg.g^{-1} wet weight in crustaceans, and below 2 µg.g^{-1} wet weight in fish. Zinc and cadmium concentrations in Pacific oysters were much higher than in other molluscs, but this species is well known for accumulating unusually high zinc concentrations. In other molluscs, zinc and cadmium concentrations were similar to those from biota in uncontaminated environments. Although zinc based whiteners had been used at the pulp and paper mill at Crofton until the 1970s, oysters sampled in the 1990s showed no elevation of zinc (J.A.J. Thompson unpublished data).

In contrast to biota from Georgia Strait proper, organisms sampled from Howe Sound during the early 1970s revealed evidence of chlor-alkali plant mercury discharges, and of tailings disposal from the Britannia Beach copper and zinc mine. Mercury concentrations in crabs and other biota sampled near the chlor-alkali plant at Squamish often exceeded 0.5 µg.g^{-1}, the recommended threshold for human consumption, forcing closure of the crab fishery. Samples taken further away from the plant showed slightly lower mercury concentrations (Harbo and Birtwell 1978, 1983). At Britannia Beach, disposal of tailings caused localized contamination of sediments, mainly within the inner Basin of Howe Sound (Thompson and Paton 1976; Drysdale and Pedersen 1992), which led to some localized contamination of resident aquatic biota in the mid 1970s (Drysdale and Pedersen 1992).

Currently the primary source of metals to Howe Sound is probably acid mine drainage from the Britannia Beach Copper mine (McCandless 1995). It has been estimated that up to 600 kg.d^{-1} each of copper and zinc enter Howe Sound, the largest input from such sources in North America (R. McCandless personal communication 1996).

Taken together, these results show that during the early 1970s, mercury discharges led to local contamination of fisheries, particularly of crabs. As a result, the crab fishery in the upper part of Howe Sound was closed from 1970 to 1979. Metals other than mercury do not appear to have presented any specific problems, except for the absence of colonization by fish of the area off Britannia Creek due to acid mine drainage (Levings and Riddel 1992). The decline in metal mining and discharges into the Strait of Georgia has resulted in lower metal concentrations in recent sediments. The earlier metal discharges, now buried under clean sediment, will become less biologically available as time passes. This is probably now reflected in lower metal concentrations in biota, though although no recent data are available to confirm this.

Dungeness Crab
(Cancer magister)

A highly prized commercial shellfish found along the west coast of North America, the Dungeness crab has broad fat legs. A blend of brown and tan in colour, Dungeness crabs live in bays, inlets, around estuaries, and on the continental shelf from shallow water to a depth of 180 m. Sometimes found on mud and gravel, they are most abundant on sand bottoms, frequently living in eel grass. Young crabs are initially pelagic (free swimming) for four months or longer, passing through five larval stages. Once settled on the bottom they grow through periodic moulting, maturing after 10 to 11 moults at 2 or 3 years of age. Males reach legal harvesting size (16.5 cm in shell width) after 12 to 13 moults or four years, while few females ever reach this size. Males grow to a maximum width of 23 cm and weigh about 2 kg. Dungeness crabs prey on clams, crustaceans and small fish. They find their food by probing their sensitive claws into the sand. Predators of the adult crab include halibut, dogfish,

sculpin and octopus. Coho salmon feed on crab larvae. In 1994 the total landed value for Dungeness crab in British Columbia was over $240 million.

Dungeness crab have been used extensively in British Columbia to monitor marine contamination. Since the crab lives on the sea bottom and feeds on bottom-dwelling organisms, it is exposed to contaminants which are associated with sediment and particulate material. As Dungeness crab are a commercially important species, analysis of their tissues provides a useful indicator for assessing the public health significance of marine contamination. Much of our current information about declining concentrations of PCDDs near some coastal pulp mills comes from monitoring the concentrations of these chemicals in Dungeness crab. Decisions to close and reopen fisheries are also based, in part, on analyses of these organisms.

Analyses of metals in Dungeness crabs sampled from the Fraser Estuary and Strait of Georgia in 1994 showed evidence of local contamination. Copper, cadmium, and silver all showed statistically significant elevations over the reference sites. Copper concentrations may reflect the relationship between the moulting cycle and control of the metal, rather than a contamination source. Silver indicates the presence of past and current sewage discharges. Figure 2 shows Iona high in such discharges. Elevated concentrations of silver in sediments off the Fraser delta clearly reflect the plume associated with the Iona outfall (K. Gordon personal communication). Silver concentrations at the impacted sites were around 8 µg.g^{-1} in crab hepatopancreas, about four times the concentration at reference sites. Cadmium (Figure 3), was above 10 µg.g^{-1} in hepatopancreas at impacted and industrial sites but below 5 µg.g^{-1} at reference sites (J.A.J. Thompson unpublished data).

TRIBUTYLTIN COMPOUNDS

TBT has been used for the last 20 years in marine anti-fouling paints. Because biological effects, such as shell thickening and chambering in oysters and masculinization of female whelks, were being found in marine organisms, the use of TBT-based anti-fouling paints was banned in Canada in 1989 on ships less than 25 m long (unless aluminum-hulled).

The Strait of Georgia is a popular venue for recreational sailors, and the Port of Vancouver is one of Canada's busiest commercial harbours. By comparing the distribution and effects of TBT at several recreational marinas, where TBT use and input should have declined after the 1989 ban, with Vancouver Harbour, which continues to be exposed to TBT from commercial shipping, it should be possible to measure the effectiveness of TBT regulation. TBT in sediment cores was analysed at several sites in the Strait of Georgia, and at sites on western Vancouver Island. Additionally, the frequency and extent of imposex in whelks were examined at appropriate sites. These data are described by Stewart and Thompson (1994) and by Tester *et al.* (1996).

In sediment cores from recreational harbours, inputs of TBT have declined since about 1989, apparently in response to the ban on TBT use. This is illustrated by the declines in TBT concentrations in surface sediments compared to older sediments (Figure 4a). However, in Vancouver Harbour, which continues to be exposed to TBT from commercial shipping, sediment cores show no decline in recent inputs (Figure 4b).

In whelks collected from around Vancouver Island, the extent of imposex has declined since the early 1990s, as indicated by percent "relative penis size" (Figure 5) and female penis length (Figure 6; Tester *et al.* 1996).

The rate of TBT degradation in deep sediments seems to be of the order of years to decades (Thompson and Stewart 1994). It is not clear whether the process is microbiological or abiotic.

Concentrations of TBT and its degradation products have been summarized from various unpublished reports by Thompson and Stewart (1994). Ranges tended to be wide, and in the absence of age/size data which may explain some of the variance of the results, it is difficult to infer trends. However, there seemed to be a general decline in concentrations between the late 1980s and the early 1990s. This may be partly explained by the ban on TBT use on small vessels introduced in 1989. Typical TBT concentrations in invertebrates from non-commercial harbours in 1992 and 1993 were less than 100 ng.g^{-1} dry weight (as tin). Concentrations in similar samples collected during the mid 1980s were as high as 1 000 ng.g^{-1} dry weight (as tin) (Thompson and Stewart 1994).

Figure 2
Silver concentrations in crab hepatopancreas from sites in the Fraser estuary.

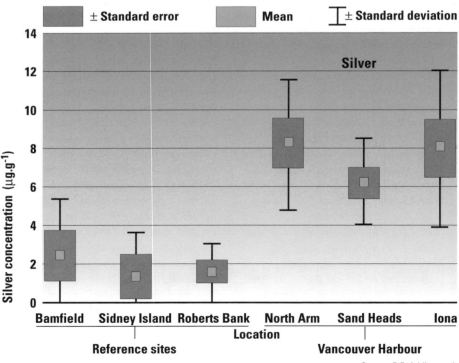

Source: R.F. Addison n.d.

In terms of spatial distribution, TBT in the Pacific region is usually associated with recreational or commercial shipping. At sites remote from marinas or commercial harbours, TBT concentrations are usually low and imposex in whelks is infrequent.

In terms of historical trends, it appears that the ban on the use of TBT on small boats has had a beneficial effect. Around recreational marinas, TBT concentrations are declining, as are incidences of imposex. Although data are limited, it seems that the environment of recreational marinas should recover from the effects of TBT over an interval of 20 to 50 years. However, although TBT degradation in sediments appears to occur over intervals of years to decades, concentrations of TBT in biota are declining within a matter of a few years, as is their biological impact, as shown by imposex frequency. This implies that although TBT may persist in sediments, it may not be biologically available to aquatic biota. It is speculated that

TBT behaves similarly to PCDDs and PCDFs in sediments. In oceanographically quiescent environments, TBT would be buried fairly rapidly and overlain by clean sediments, while in more active environments, where sediments are continuously agitated by tidal movements, TBT would continue to be biologically available.

ORGANIC CHEMICALS
ORGANOCHLORINE COMPOUNDS
PCDDs AND PCDFs IN PULP MILL EFFLUENTS

The manufacture of wood pulp and paper is a major industry in British Columbia, and eleven mills discharge wastes directly or indirectly into the Strait of Georgia. Four mills currently discharge into the Strait, two more discharge to Howe Sound, and another five are located on the Fraser and Thompson Rivers, which drain into the Strait of Georgia (Figure 7).

In the late 1980s, residues of PCDDs and PCDFs were reported in biota collected from the vicinity of some coastal pulp mills. Various studies (summarized in Yunker and Cretney 1995a, b) showed that these compounds were probably from several sources including: (a) PCDD and PCDF contaminants of chlorinated phenols which had been used to treat wood for sapstain control and from which chips were used as feedstock for pulping; (b) unchlorinated PCDDs and PCDFs in defoamers used at mills utilizing chlorine bleaching; (c) from the action of chlorine bleaching on natural compounds in the wood, or on chlorinated phenols, or on other chemicals used in the pulping process; and (d) from the chlorinated phenols which could have been converted to PCDDs and PCDFs during pulping. In addition, some PCDDs and PCDFs were emitted during general combustion and incineration processes.

Figure 3
Cadmium concentrations in crab hepatopancreas from sites in the Fraser estuary.

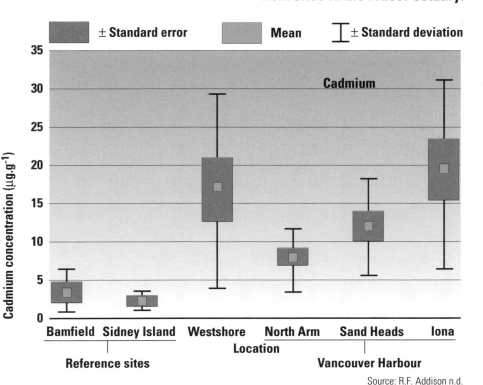

Source: R.F. Addison n.d.

The presence of potentially dangerous concentrations of PCDD and PCDF congeners with a 2,3,7,8-chlorine substitution pattern in the vicinity of pulp mills led to closure of some fisheries; the federal deregulation of the use of chlorophenates as anti-sapstain pesticides; federal prohibition on the use of contaminated chips and defoamers; and federal regulations prohibiting total PCDD and PCDF discharges over 15 and 50 pg.L^{-1}. These actions led to changes in the process chemistry used in the mills, primarily the replacement of chlorine as a bleaching agent with chlorine dioxide. These changes significantly reduced the discharge of these compounds (Figure 8).

These reductions in discharges are reflected in lower concentrations in local biota (Yunker and Cretney 1995a, b; Hagen *et al*. 1995). For example, in Dungeness crab from Howe Sound, concentrations of the PCDD and PCDF congeners attributable to chlorine bleaching had declined in 1993 to about 2% of their 1988 value.

In some cases, PCDD and PCDF concentrations have declined in sediments close to the sites of coastal pulp mills. Very generally, the mills can be divided into those which discharge into relatively quiescent oceanographic systems, and those which discharge into more energetic environments. In relatively quiet sedimentary environments, including basins such as upper Howe Sound, PCDDs and PCDFs combine with the particulate material, some of which is discharged by the mills (Hagen *et al*. 1995).

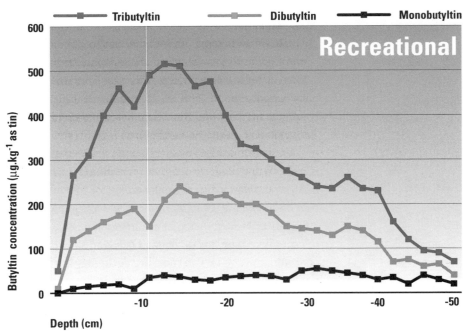

Figure 4a
Organotin concentrations in sediment cores from a recreational harbour.

Source: R.F. Addison n.d.

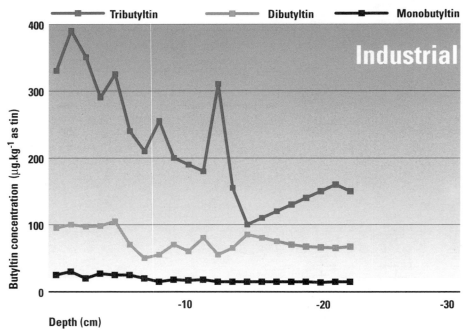

Figure 4b
Organotin concentrations in sediment cores from an industrial harbour.

Source: R.F. Addison n.d.

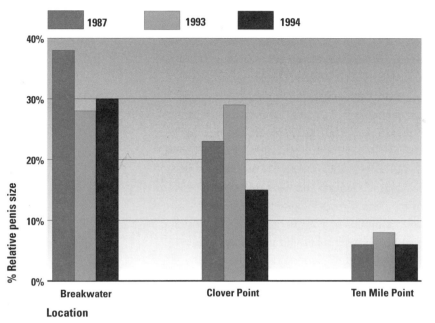

Figure 5

Imposex distribution in nucellid whelks at Breakwater, Clover Point and Ten Mile Point, 1987, 1993 and 1994.

Legend: 1987 | 1993 | 1994

Y-axis: % Relative penis size (0% to 40%)

X-axis (Location): Breakwater, Clover Point, Ten Mile Point

Source: R.F. Addison n.d.

Following the reductions in inputs since the late 1980s, PCDDs and PCDFs have tended to become buried under layers of cleaner sediment. Organisms that feed on the sediment surface, such as crabs, will accumulate decreasing amounts of these compounds. However, in many locations sediments are mixed by currents or by animals foraging at or near the surface (Macdonald *et al.* 1992). This mixing or "bioturbation" of surface sediments results in a slower decline in contaminant concentrations near the surface of the sediments, because the deeper, more contaminated sediments are returned to the sediment surface. At Crofton, where contaminated sediments are continuously being disturbed, crabs continue to accumulate PCDDs and PCDFs, even though the inputs of these chemicals to the coastal ecosystem is declining.

In an environment like Howe Sound, there may be zones in which PCDDs and PCDFs are not chemically broken down. Compounds discharged before the late 1980s are still present, though buried under an accumulation of clean sediment. At Crofton, where original inputs of these chemicals are constantly being disturbed, there may be a better chance of their eventual degradation by microbiological or abiotic processes. However the fisheries at Crofton may remain closed whereas some of those at Howe Sound have already been re-opened.

The spatial distribution of PCDD and PCDF concentrations depends largely on patterns of local sediment transport. Much of the contamination is confined to within 5 to 10 km of the source. Most sampling for these compounds has been undertaken close to the mills. Relatively few samples from sites distant from the mills are available. On the other hand, as clearly shown by Macdonald *et al.* (1992), tetrachlorodibenzo-furans (TCDFs) can be transported to sediments over basin-wide distances of 30 km or more. It is likely that these compounds are transported with

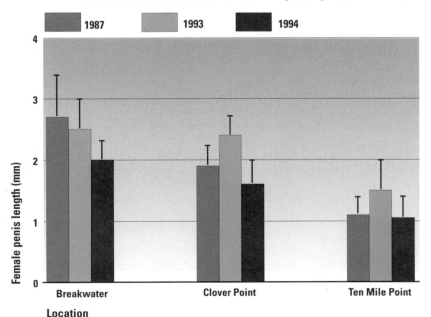

Figure 6

Imposex distribution in nucellid whelks at Breakwater, Clover Point and Ten Mile Point, 1987, 1993 and 1994.

Legend: 1987 | 1993 | 1994

Y-axis: Female penis length (mm) (0 to 4)

X-axis (Location): Breakwater, Clover Point, Ten Mile Point

Source: R.F. Addison n.d.

suspended sediments either in surface currents or as resuspended particles in bottom currents. Atmospheric trans-port may also contribute to contamination (Macdonald *et al.* 1992).

Pulp mills are not the only source of these contaminants in the region. Municipal incinerators may also contribute. Congener distribution and the historical trends (Macdonald *et al.* 1992) in incinerator-derived PCDDs and PCDFs are sufficiently different from those in pulp mill-derived material that broad differences in sources can be inferred. As pulp mill sources decline in importance, incinerator sources are expected to increase in importance. This is because their emissions enter the atmosphere directly, and are more susceptible to widespread distribution through atmospheric transport processes. This is well illustrated by comparing PCDD and PCDF concentrations in blubber from

two groups of harbour seals sampled in the early 1990s from the Strait of Georgia, which is affected by pulp mill effluents, and from Quatsino Sound on northwest Vancouver Island, where there are no local sources of PCDDs or PCDFs (R.F. Addison unpublished data). Concentrations of these chemicals in the Strait of Georgia samples showed a distinct pulp mill origin in the PCDD congeners, with relatively high concentrations of 1,2,3,6,7,8-hexachlorodibenzo-*p*-dioxin (HxCDD). Quatsino Sound samples contained much lower concentrations of PCDDs and PCDFs, usually about 5% of those found in Strait of Georgia and were more characteristic of incinerator input (Figure 9). High concentrations of 1,2,3,6,7,8-HxCDD have also been found in harbour porpoises sampled from the Strait of Georgia during the late 1980s (Jarman *et al.* 1996).

Figure 7
Location of pulp and paper mills in Georgia Strait.

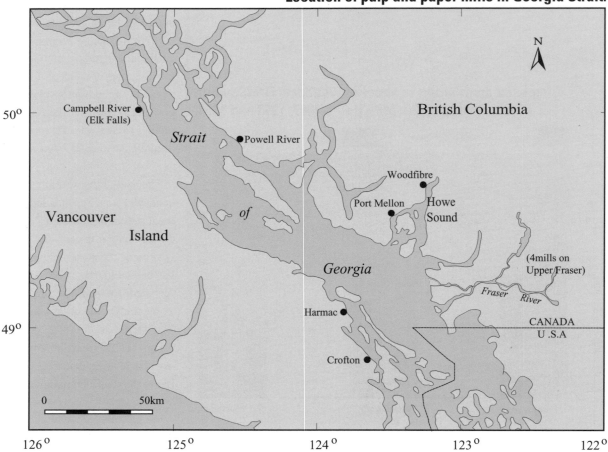

Source: R.F. Addison n.d.

PCDDs and PCDFs have been found in crab samples from commercial or industrial harbours, such as Vancouver and Victoria harbours, where their sources may reflect general historical use of pentachlorophenol (Yunker and Cretney 1995b).

PCDDs and PCDFs occur more frequently, and at higher concentrations, in west coast than in east coast marine biota. This point is illustrated by comparing analyses of PCDD and PCDF concentrations in benthic flatfish sampled from the British Columbia coast with ecologically comparable fish from several sites in eastern Canada. PCDDs and PCDFs are found in all west coast samples analysed, whereas they are present in only about half of the east coast samples. Concentrations are consistently lower in east coast samples (Figure 10a; (R.F. Addison unpublished data).

DDT-RELATED COMPOUNDS

Although data are limited, concentrations of DDT-group compounds were higher in the 1970s than at present. This is consistent with data from other locations in the northern hemisphere, and reflects the fact that DDT has been effectively banned from widespread use in the industrialized world since the mid 1970s (Dunlap 1981). It is possible that DDT-group compounds may enter the British Columbia environment by long-range atmospheric transport (LRTAP) from areas where DDT is still used, such as Asia; however, no data are available to estimate the importance of this route.

Relatively little information exists about the distribution of DDT-group compounds, and related chemicals, in the Strait of Georgia ecosystem. No data appear to exist describing DDT-group concentrations in marine sediments. In dated sediment cores from nearby Puget Sound, DDT concentrations reached a maximum in the 1970s of about 5 $ng.g^{-1}$ (Macdonald and Crecelius 1994) which is in the low range of those found in marine sediments (Addison 1980). DDT-group concentrations in Strait of Georgia sediments would be expected to be similar to those in Puget Sound.

Figure 8
Trends in PCDD and PCDF emissions from pulp mills, in British Columbia, 1989 to 1995.

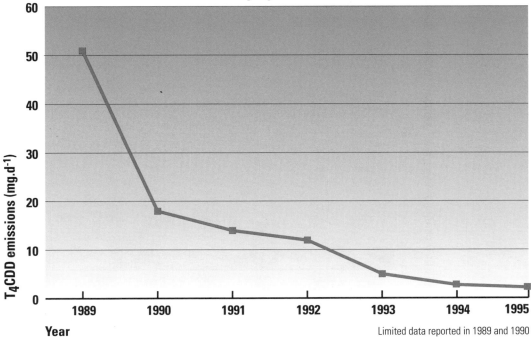

Limited data reported in 1989 and 1990

Source: R.F. Addison n.d.

In the early 1970s, concentrations of dichlorodiphenyldichloroethylene (*p,p'*-DDE) (the main metabolite of the insecticide *p,p'*-DDT) were around 40 ng.g^{-1} tissue wet weight in a wide range of species of different size, sex and condition (Albright *et al.* 1975). Nevertheless, a comparable survey of residues from a wide variety of fish analysed from across the U.S. at about the same time showed *p,p'*-DDE concentrations in these samples to be around 600 ng.g^{-1} . This value was more than 10 times higher than the Strait of Georgia samples.

It is more useful to compare similar species; mussels from the Fraser estuary had *p,p'*-DDE concentrations less than 4 ng.g^{-1} wet weight whole tissue, which is similar to the range

reported from this species in St. Margaret's Bay, Nova Scotia, an uncontaminated environment (Hargrave and Phillips 1976). Dungeness crab hepatopancreas from the Fraser Estuary contained *p,p'*-DDE in the range of 20 to 300 ng.g^{-1} wet weight; comparable figures for lobster hepatopancreas from St. Margaret's Bay were 400 to 940 ng.g^{-1} wet weight *p,p'*-DDE. More recently, liver from English sole collected in 1993 from Trincomali Channel, a relatively clean area in the middle of the Strait of Georgia, contained *p,p'*-DDT and *p,p'*-DDE at concentrations below 10 ng.g^{-1} tissue wet weight. Residue concentrations in three species of east coast flatfish collected at the same time ranged from about 20 to 90 ng.g^{-1} wet weight (Figure 10b; R.F. Addison unpublished data). Total DDT-group

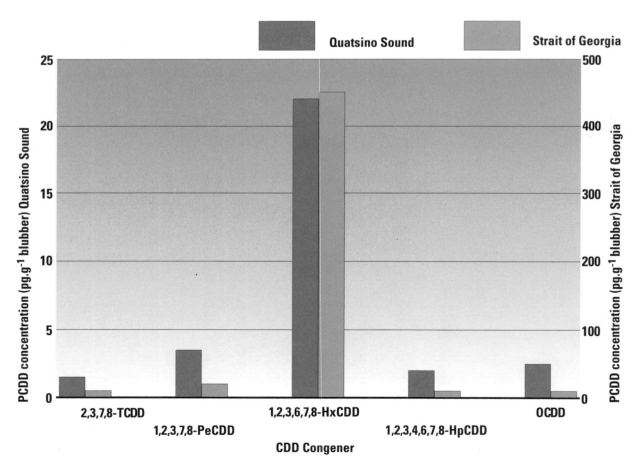

Figure 9
PCDD concentrations in harbour seals from the Strait of Georgia and Quatsino Sound.

Source: R.F. Addison n.d.

concentrations in liver from copper rockfish sampled in 1987 from the vicinity of fish farms in Sechelt Inlet were around 1 $\mu g.g^{-1}$ lipid (probably about 50 $ng.g^{-1}$ tissue wet weight, based on the lipid content calculated from other data in that report; Levings 1994). However, fish caught near fish farms may have accumulated contaminants from commercial fish food. The DDT-group concentrations recorded in copper rockfish from a reference site (about 260 $ng.g^{-1}$ lipid, or about 13 $ng.g^{-1}$ wet weight tissue) may be a better indication of background concentrations of contaminants. This figure is very close to that recorded in the Trincomali Channel flatfish sampled in 1993.

English Sole
(*Pleuronectes vetulus*)

Located in coastal British Columbia with small isolated populations at the heads of many inlets, the English sole has long been popular amongst local consumers who have learned to relish the iodine flavour that can be detected in some inshore catches. Distributed from Baja California to western Alaska and living between the surface and 550 m depth, English sole are known to make extensive migrations which may cover more than 7 km per day and more than 1 100 km in total. Despite this tendency to migrate, growth and tagging studies have provided good evidence to indicate the existence of several major segregated stocks and some isolated populations. Reaching a total length of 57 cm for females and 49 cm for males, both sexes attain sexual maturity at about half these lengths. Spawning in the waters off British Columbia occurs mainly from January to March with the young reaching a size of about 10 cm by August and 20 cm by the end of the following year. Young English sole are found in the intertidal zone and quite

shallow water. As they grow, they move into deeper water. Throughout their life they shift between shallow water in the spring and deeper water in winter. Growth patterns differ among localities. Food of this small-mouthed flounder consists mainly of clams and clam siphons, other small molluscs, marine worms, small crabs and shrimps, and brittle stars. English sole are first captured by the commercial fishery at lengths of 30.5 cm which corresponds to 3 years for females and 4 years for males.

Because English sole live near the sea bottom and feed primarily on bottom-dwelling organisms, analyses of their tissues can be used to monitor the distribution and effects of many sediment-borne contaminants. This is well illustrated by studies in Puget Sound, south of Vancouver Island in the State of Washington. These studies have shown that high concentrations of sediment-bound PAHs may cause biological effects, including liver tumours.

Harbour porpoises from the Strait of Georgia collected in the late 1980s contained about 8 $\mu g.g^{-1}$ total DDT; this was about half the concentration found in samples from California (Jarman *et al.* 1996).

In terms of spatial distribution, it is not surprising that concentrations of DDT-group compounds are lower in Strait of Georgia aquatic biota than those from the east coast. DDT was used extensively in eastern Canada to control the spruce budworm. At the height of the spraying in the mid 1950s, about 10% of the total annual U.S. production of DDT was sprayed on New Brunswick. Much of the DDT-group residues found in eastern Canadian biota are almost certainly derived from that source. Since the prevailing winds in eastern Canada are predominantly from the west, agricultural chemicals used in central Canada and the northern U.S. are carried into the region by atmospheric transport. Although DDT was used in British Columbia, the amounts applied were considerably smaller than in New Brunswick.

PCBs

In terms of both spatial distribution and temporal trends, the more highly chlorinated PCBs probably behave similarly to the DDT-group compounds. They have probably had roughly similar histories of manufacture and use in North America over the last 50 years. PCBs are dispersed mainly by atmospheric transport, and since British Columbia is upwind of industrialized North America, it is not surprising that even during the 1970s, PCB concentrations found in aquatic ecosystems were relatively low. PCB manufacture stopped during the early 1980s (de Voogt and Brinkman 1989). Although PCBs have continued to be used, the amounts released to the environment have probably been declining.

In sediments, Strait of Georgia data are available only for PCB #77; however, the trend in the distribution of this material follows closely that of the total PCBs measured in Puget Sound sediments (Macdonald *et al.* 1992; Macdonald and Crecelius 1994). The maximum value of total PCBs in Puget Sound sediments, around 30 µg.kg^{-1}, occurred around 1970. This concentration is at the low end of the range for PCBs in marine sediments (Addison 1980). Concentrations of PCB #77 reached a maximum around 1970 in cores from Ballenas Basin and Howe Sound: its distribution and flux were consistent with input from atmospheric transport (Macdonald *et al.* 1992). PCB manufacture also reached a maximum at about this time (Addison 1983).

In several species of estuarine fish collected from the lower Fraser River in 1987, PCBs (expressed as the commercial mixture Aroclor 1248) averaged around 60 ng.g^{-1} wet weight (Rogers *et al.* 1992). This appeared considerably lower than concentrations measured in other

Figure 10a
Geographic variation in HxCDD and TCDF concentrations at various locations throughout Canada.

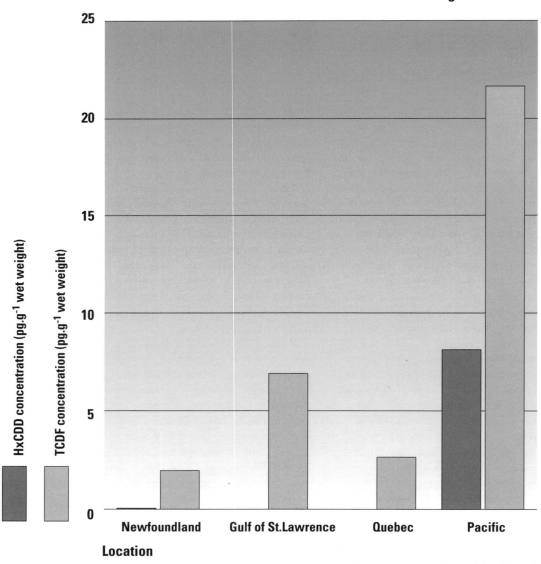

Source: R.F. Addison n.d.

species in the early 1970s (Albright *et al.* 1975) and expressed as Aroclor 1254 (mean value approximately 140 ng.g^{-1}). However, it is difficult to compare these data directly, since expressing a PCB congener mix as Aroclor 1248 usually yields a lower concentration than expressing it as Aroclor 1254. Nevertheless, a decline in total PCB concentrations from the early 1970s to the late 1980s is consistent with other data presented in this report.

In livers of copper rockfish sampled in 1987 around aquaculture operations, PCBs averaged about 130 ng.g^{-1} wet weight compared to about 30 ng.g^{-1} wet weight at a reference site (Levings 1994). The higher concentrations recorded at the fish farm sites may reflect the presence of PCBs in commercial fish food, so the lower value of about 30 ng.g^{-1} wet weight (equivalent to about 700 ng.g^{-1} lipid) is probably more representative of general background concentrations of

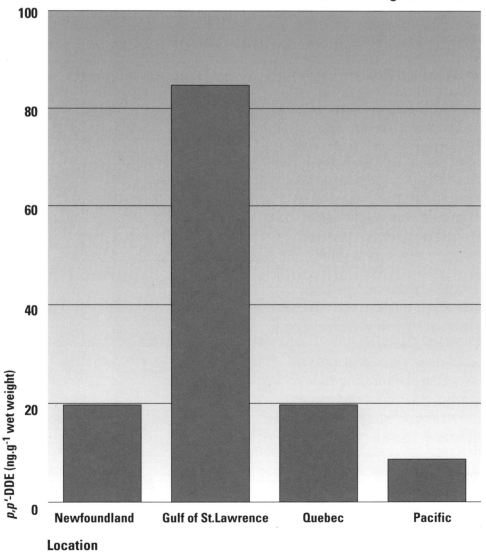

Figure 10b
Geographic variation in *p,p*'-DDE concentrations at various locations throughout Canada.

Source: R.F. Addison n.d.

contamination. In a more recent comparison, PCB concentrations in Trincomali Channel English sole sampled in 1993 averaged about 20 ng.g^{-1} tissue wet weight (expressed as Aroclor 1254); comparable concentrations in east coast flatfish ranged from about 40 to 70 ng.g^{-1} wet weight (Figure 10c; R. F. Addison unpublished data).

Harbour porpoises sampled from the Strait of Georgia in the late 1980s contained about 8 μg.g^{-1} blubber of PCBs (summed as individual congeners); this was slightly lower than concentrations found in California samples (Jarman *et al*. 1996).

OTHER AROMATIC COMPOUNDS
PAHs

Relatively few data describe PAH concentrations in sediments or biota from the Strait of Georgia. Macdonald and Crecelius (1994) reported total PAHs to be below 1 μg.kg^{-1} in sediment in cores from Ballenas Basin (in the Strait of Georgia) and Thornbrough Channel (at the mouth of Howe Sound), and slightly higher (around 2 μg.kg^{-1}) in sediment from Howe Sound. These values are about an order of magnitude lower than concentrations found in the more contaminated U.S. Puget Sound.

PAH concentrations have been measured in sediments and biota from Vancouver Harbour (Goyette and Boyd 1989). In surface sediments, concentrations ranged from 19 to 6 600 μg.g^{-1}; this wide range is to be expected in a harbour receiving various inputs from commercial and shipping activities. Concentrations of PAHs in tissue of fish and crab were usually below detection limits.

SUMMARY

Concentrations of PCDDs and PCDFs are higher in some fish and invertebrates from Georgia Strait than in other marine areas in Canada. This is primarily the result of discharges from pulp mills in the region which historically used chlorine bleaching or pentachlorophenol contaminated wood chips. However, following changes to mill processes in 1980s and 1990s, inputs of PCDDs and PCDFs have rapidly declined. In oceanographically quiescent sites, "old" PCDDs and PCDFs are being buried under

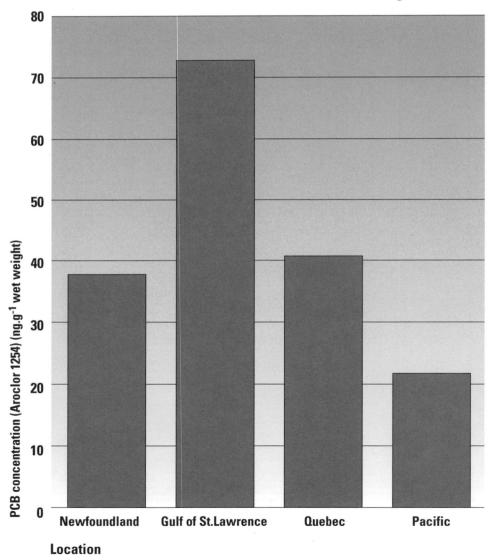

Figure 10c
Geographic variation in PCB concentrations at various locations throughout Canada.

Source: R.F. Addison n.d.

accumulations of clean sediment, and PCDD and PCDF concentrations in crabs which live and feed on the sediment surface are declining. However, where surface sediments are mixed, contaminated sediments are redistributed to the surface layers and PCDD and PCDF concentrations in crabs occupying this habitat decline more slowly.

Concentrations of the DDT-group of insecticides and of PCBs in the Strait of Georgia are generally lower than those in eastern Canada. From the few data available, concentrations appear to be declining at rates similar to those in eastern Canada.

Metal contamination has been a problem in the past, mainly as a result of mercury discharges from a chlor-alkali plant and from local mining activity. These sources have been reduced or controlled for several years, and metal contamination probably does not present a major concern in the Strait of Georgia, though the impact of acid mine drainage from the former Anaconda Mine to Britannia Bay remains to be evaluated.

TBT continues to be of concern in harbours such as Vancouver, where it is introduced from anti-fouling paints used on commercial shipping. The ban on the use of TBT based anti-fouling paints on small vessels seems to have resulted in a decline in TBT concentrations in recreational harbours, and in the frequency and extent of imposex in marine snails.

THE BRITISH COLUMBIA OPEN COAST AND FJORDS
THE OPEN COAST
SOURCES

Since the open coastline of British Columbia is relatively sparsely inhabited, contaminant issues tend to arise at the few sites where there are or have been mining or other industrial operations (Macdonald and Pedersen 1991). These sites include: Prince Rupert (pulp mill); Kitimat Arm (aluminum smelter, pulp mill); and Alice Arm (lead and molybdenum mines).

Because coastal fjords and basins effectively trap particulate matter, the outer coast tends to have fewer terrestrial inputs (Macdonald and Crecelius 1994; Macdonald and Pedersen 1991). As a result, particulate-bound contaminants entering from the land near urban or industrial areas tend to be removed before reaching the outer coast. However, the entire British Columbia coast is expected to receive contaminants from trans-Pacific atmospheric transport or advection from ocean currents, particularly those coming from the south (Macdonald and Pedersen 1991). At present, there are few data describing the scale of this process in British Columbia, although U.S.-based research on the Washington Shelf (reviewed in Carpenter and Pedersen 1989) provides a useful reference. For example, the Washington Continental Shelf has been shown to sequester excess radioactive fallout products delivered by on-shelf transport of water (Beasley *et al.* 1982).

DISTRIBUTION, TRENDS AND EFFECTS

The background contaminant concentrations in British Columbia coastal waters, some of which have been referred to in the discussion of the Strait of Georgia, tend to be very low, suggesting that long range transport has not been a significant route of contaminant supply to the region, and that local sources are more important. Metal concentrations in sediments and biota from various sites around the coast in the late 1980s (Harding and Goyette 1989) generally show no measurable contamination. Relatively few data describe organochlorine compound concentrations remote from point sources. Stout and Beezhold (1981) reported DDE at 18 ng.g^{-1} and PCBs at less than 8 ng.g^{-1} wet weight in a Pacific halibut caught in 1971; concentrations of the DDT-group in a Pacific cod caught in 1970 were not detectable. More recently, Jarman *et al.* (1996) reported high concentrations of total DDT and related organochlorine compounds in west coast marine mammals: in the blubber of mature male false killer whale, total DDT reached 1 900 µg.g^{-1}.

THE FJORDS
KITIMAT ARM

Kitimat Arm is the innermost part of the Kitimat Fjord system (Figure 11). Its general bathymetry and oceanography have been described by Macdonald *et al.* (1983). The region is the site of several industries, including an aluminum smelter established in 1954, and since expanded several times; a pulp mill; and a methanol manufacturing plant. The town of Kitimat developed around these industries and now houses about 20 000 people. Most of the industrial and residential concentration is at the head of Kitimat Arm, into which the Kitimat River discharges. The head of Kitimat Arm was once a separate basin but is now completely filled with glacial debris and sedimented material delivered by the Kitimat River (Bornhold 1983); two additional basins are located further out in Douglas Channel. Particle-bound contaminants discharged to the head of Kitimat Arm are likely to be sedimented out, either in association with natural particulate material from the Kitimat River, or with other discharged material, in the basins in the upper reaches of the fjord.

During 1978 and 1979, sediments from the upper Kitimat Fjord were sampled as part of a baseline assessment of Kitimat as a potential oil terminal (Cretney *et al.* 1983). PAHs had previously been identified as a byproduct of aluminum smelters. In 1952, the construction of a smelter in a pristine environment provided an opportunity to record spatial and temporal trends in PAH distribution and to identify some of the important processes controlling PAH movement. These surveys demonstrated that: surface sediment PAH concentration reached a maximum immediately south of the Alcan smelter (up to 15 $\mu g.g^{-1}$ sediment) and declined to less than 1 $\mu g.g^{-1}$ about 6 km further south. The source of the PAHs could not be attributed to the Kitimat River, as sediments immediately downstream from its

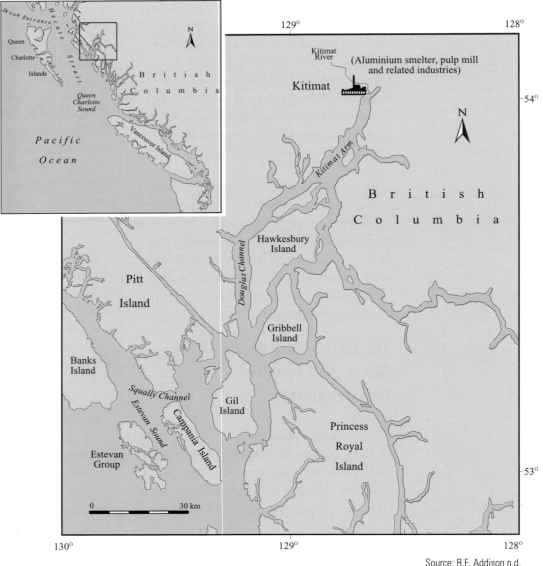

Figure 11
Location of Kitimat Arm, British Columbia.

Source: R.F. Addison n.d.

mouth had PAH concentrations less than 0.3 $\mu g.g^{-1}$; and dated cores showed a sharp increase in PAH concentrations from about the late 1940s to early 1950s, coincident with the start up of the smelter.

Concentrations of PAHs in surface sediments from the vicinity of the smelter are similar to, or lower than, those recorded in Sydney Harbour, Nova Scotia (See the Northwest Atlantic Marine Ecosystem Chapter), a bay heavily contaminated with PAHs (Addison *et al.* 1994). However, as sediment contamination may reflect the intensity of local point source discharges, it is more useful to compare concentrations more removed from obvious sources of contamination. The background concentrations found at Kitimat (less than 0.3 $\mu g.g^{-1}$) are similar to those recorded outside Sydney Harbour (less than 1 $\mu g.g^{-1}$) (Addison *et al.* 1994).

The profiles of sediment cores from Kitimat in 1993 show a maximum concentration of PAHs in the early 1970s and a decline of about 10 fold by 1993. During this period the aluminum smelter undertook process changes and improved emission controls to reduce environmental contamination by PAHs and other by-products of the smelting process (W. J. Cretney unpublished data).

PCDDs and PCDFs have also been investigated at Kitimat. These compounds were released from a local pulp mill; sawmill; and wood treatment facility some of which may have used materials contaminated with chlorophenols. Sediment cores from near the head of Kitimat Arm have PCDD and PCDF concentration profiles that are consistent with the history of industrial activity in the area. PCDD and PCDF concentrations in the cores increased almost 10 fold during the 1960s, coinciding with the development of the sawmill and the pulp mill. Concentrations reached a maximum in the early 1970s, then declined to pre-industrialization concentrations. This historical record is consistent with the operation of a sawmill and wood treatment facilities until

1981, and with the elimination of chlorophenol-contaminated woodchips as a feedstock (Yunker and Cretney 1995b).

Concentrations of PCDDs and PCDFs in Dungeness crab hepatopancreas have also fallen dramatically since the pulp mill ceased using chlorophenol contaminated feedstock and, in consequence, a fishery closure has been lifted. Unlike other coastal British Columbia pulp mills, the mill at Kitimat never used chlorine bleaching. The PCDD and PCDF congener pattern is therefore quite different from that seen in crab hepatopancreas influenced by bleached kraft mill effluent and is more like that seen in major British Columbia harbours. Although the pattern is consistent with that expected from the formation of PCDD and PCDF precursors in chlorophenol wood treatment formulations, the contaminants cannot be unequivocally attributed to the pulp mill.

OBSERVATORY-HASTINGS ARM FJORD

Alice Arm is part of the Observatory-Hastings Arm fjord system on the northwest coast of British Columbia (Figure 12). Mining and smelting operations were carried out first at Anyox (copper mining and smelting 1911 to 1935) and later at Kitsault (lead/molybdenum mining until 1981 to 1982) (Odhiambo *et al.* 1996). The Anyox operation produced copper rich tailings and slag, both of which were deposited in the intertidal zone and in Granby Bay. The Kitsault operation disposed of tailings first into Lime Creek and then later directly into Alice Arm via a 50 m discharge pipe (Pedersen *et al.* 1995). As a result of these operations, the sediments in Alice Arm and upper Observatory Inlet have elevated concentrations of metals including copper, zinc, lead and cadmium. During assessments of the later operation in Alice Arm (Burling *et al.* 1981), questions were raised as to whether the discharged tailings would spread significantly beyond Alice Arm to contaminate other water bodies. Based on sediment trap data (Macdonald and O'Brien 1996), the tailings discharge system eventually used by the mine was found to be effective in preventing tailings from escaping Alice Arm or from entering its surface waters. When the mine shut down operations in 1982 due to poor market prices, the movement of particulate matter into deep water returned to near normal within three months.

Odhiambo *et al.* (1996) used dated sediment cores from Alice Arm, Hastings Arm and Observatory Inlet to investigate the temporal trends and spatial distribution of metal contamination in the fjord system. They found no evidence that Kitsault tailings, which had high concentrations of lead and cadmium, were carried over the sill into upper Observatory Inlet to contaminate sediments outside Alice Arm, nor did they find evidence that copper in sediments from Anyox had reached Alice Arm. However, the copper-containing tailings disposed into Observatory Inlet during the earlier Anyox operations continue to migrate in Observatory Inlet and Hastings Arm. Contaminant-containing sediments at all sites are presently being diluted and buried by riverine inputs of sediment (Odhiambo *et al.* 1996), although at a much slower rate than for the Anyox operation.

PRINCE RUPERT

Environmental quality has been of some concern at Prince Rupert for the last thirty years. The concern developed in response to operation of a local pulp mill and also to proposed development of an oil terminal in the area (when it was planned to bring Beaufort Sea oil overland to Prince Rupert and then ship it by tanker). However, there seems to be no data describing PAH concentrations in the Prince Rupert area. The only contaminant data available refer to

PCDDs and PCDFs from a nearby pulp mill (Yunker and Cretney 1995a, b). Between 1989 and 1995, PCDD and PCDF concentrations in Dungeness crab hepatopancreas declined about 10 fold in response to changes in the mill's process chemistry which reduced PCDD and PCDF discharges (Hagen *et al.* 1995).

SUMMARY

Data describing contaminant distribution on the British Columbia open coast are sketchy; analyses have usually been undertaken only in response to specific concerns. Although there are instances of

Figure 12
Location of Hastings Arm, Alice Arm and Observatory Inlet, British Columbia.

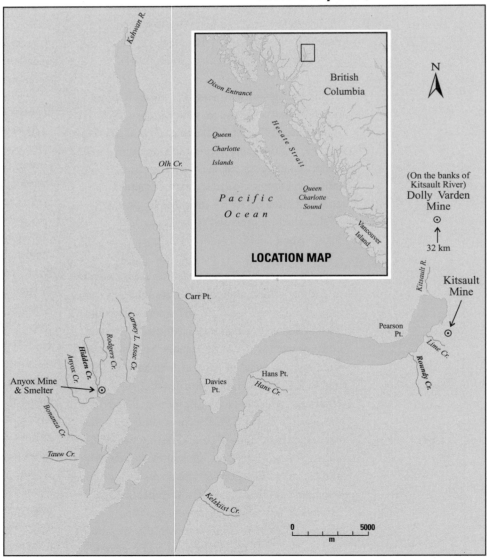

Source: R.F. Addison n.d.

local contamination, the available time trend data imply that contamination is declining in response to regulatory actions and improved industrial practices. Background concentrations of contaminants in the region are similar to those reported in other relatively uncontaminated areas.

THE RIVERS
FRASER RIVER

During the last thirty years, the main concern about contaminants in the Fraser River has arisen from pulp mill activities and their potential impact on various salmon and other species that spawn in the Fraser tributaries. Throughout the 1960s and 1970s, many studies showed that discharge of organic matter from the mills resulted in biological oxygen demand (BOD) and chemical oxygen demand (COD), often resulting in acutely toxic effects on migrating fish. However, steady reduction in such discharges has shifted the focus to the subtle effects of chemical contaminants. Recent concern has therefore concentrated on the impact of releases of PCDDs and PCDFs and related compounds. However, agricultural run-off has also been a longstanding problem causing degradation of smaller tributaries because of BOD and ammonia can reach toxic levels.

Juvenile chinook salmon that over-winter in the upper Fraser River are a useful monitoring species for recording changes in contaminant concentrations and detecting contaminant impacts. In early 1988, samples of these fish taken at uncontaminated sites above pulp mills on the Fraser River contained undetectable concentrations of PCDDs and PCDFs (Rogers *et al.* 1989). Samples taken downstream of the mills, however, contained measurable concentrations of 2,3,7,8-tetrachlorodibenzo-*p*-dioxin (2,3,7,8-TCDD) (up to 68 $pg.g^{-1}$ wet weight), 1,2,3,6,7,8-HxCDD (28 $pg.g^{-1}$ wet weight); and 2,3,7,8-TCDF (up to 370 $pg.g^{-1}$ wet weight), all of which are typical of some pulp mill effluents. Hepatic mono-oxygenase activity which was measured as ethoxyresorufin *O*-deethylase (EROD) induction

was also induced between about 10 and 50 fold over reference samples taken downstream from the pulp mills.

Following regulatory actions and changes to the process chemistry in the mills, PCDD and PCDF concentrations in juvenile chinook sampled downstream from the mills have declined steadily since 1988 (R. W. Gordon *et al.*, unpublished data). There has been a parallel reduction in the extent of EROD induction, with EROD activity (and cytochrome P450 1A concentrations) in downstream samples taken in 1995 the induction was measured at only about twice those in upstream samples (Wilson *et al.* 1995; Figure 13a,b).

Taken together, these results show that PCDD and PCDF plus other releases to the Fraser River, and their effects on juvenile chinook, have declined appreciably since the late 1980s. However, other subtle impacts remain. For example, in spite of the decline in and releases of other compounds, juvenile chinook sampled near pulp mills in the Upper Fraser River have increased interrenal nuclear diameters (D. W. Martens *et al.* unpublished data). This implies a potential effect on the endocrine system, since the interrenal tissue is the site of some steroid synthesis (Donaldson *et al.* 1984).

COLUMBIA RIVER

The Columbia River, in the southeastern corner of British Columbia, supports several species of fish, including rainbow trout, kokanee, mountain whitefish, white sturgeon, burbot and walleye. Although there is no commercial or subsistence fishery on the river, sport fishing is very popular. The main chemical contaminants in the lower Columbia River, from the Hugh Keenleyside Dam to the international border, include PCDDs and PCDFs associated with the discharge of effluent from the bleached kraft pulp mill at Castlegar and metals associated with the lead/zinc smelter and fertilizer complex at Trail. Effluents from these plants have been associated with contamination of fisheries that led to consumption advisories. The contamination had an impact on the river ecosystem and on fish health. A comprehensive study of the river through CRIEMP (the Columbia River Integrated Environment Management Program) has been undertaken.

Water column concentrations of metals are usually low and within criteria established by the British Columbia government for the protection of fish health. However, water quality criteria for cadmium, chromium, mercury, lead, zinc and copper were occasionally exceeded, usually downstream from Trail. Sediment metal concentrations were also usually elevated downstream of Trail. Most of the slag from the smelter is now diverted to land. Some PCDDs and PCDFs were also found at elevated concentrations in sediments downstream of Celgar.

In mountain whitefish collected in 1992, gross internal and external abnormalities, incidence of disease, and changes in histopathology were higher at a site downstream of the pulp mill at Castlegar and at a second site downstream of Trail than at a reference site on the Slocan River. PCDD, PCDF and PCB concentrations were also higher at these two sites than at the reference site, but disease incidence could not be correlated with organic contaminant concentrations. EROD was also increased at the downstream sites, presumably in response to the organic contaminants. Results of metal analyses were difficult to interpret. Copper, strontium and zinc concentrations were highest at the reference site, while mercury was highest below the pulp mill. This may reflect former use of mercury compounds at the pulp mill (Nener *et al.* 1995).

Mountain whitefish samples taken in 1994 were analysed in a fashion similar to the 1992 samples, with generally the same conclusions. PCDD and PCDF concentrations appeared to be declining slowly, but PCB concentrations remained constant (Antcliffe *et al.* 1997).

NASS AND SKEENA RIVERS

No data appear to have been published describing contaminants in these rivers, except in relation to Prince Rupert, as discussed above.

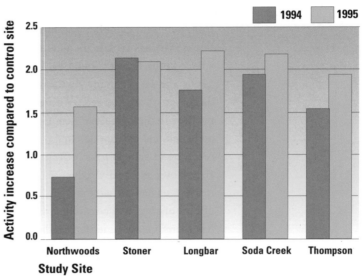

Figure 13a
Cytochrome P450 1A activity increase compared to control site.

Source: R.F. Addison n.d.

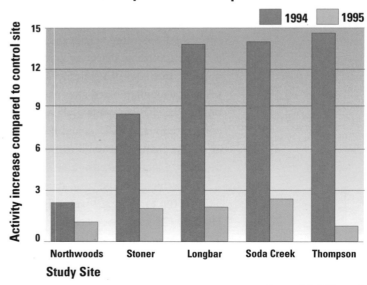

Figure 13b
EROD activity increase compared to control site.

Source: R.F. Addison n.d.

Conclusions

Contamination from various sources has affected fisheries and fish habitat throughout British Columbia, but mainly in the Fraser and Columbia Rivers, around the Strait of Georgia, and at local sites on the British Columbia coast. Contaminants have sometimes reached concentrations high enough to lead to the closure of fisheries, as in the case of mercury pollution in Howe Sound in the 1970s, and PCDD and PCDF contamination in the late 1980s. Consumption advisories for sport fishing have been issued, i.e., the Columbia River. These closures or consumption advisories have been imposed to protect the human consumer from the potential risk of exposure to these chemicals via contaminated food. Biological effects of contamination on biota have been demonstrated, as in the cases of imposex in whelks exposed to TBT, the oiling of some seabirds and perhaps mammals during the Nestucca spill of 1988, and biochemical changes in fish exposed to pulp mill effluents.

To the extent to which large scale east coast versus west coast comparisons are possible, concentrations of PCDDs and PCDFs in habitat and biota are higher in British Columbia than in eastern Canada, and those of the DDT-group and PCBs are probably lower. Metal and PAH concentrations are difficult to compare. Too few data exist to allow comparisons of TBT concentrations.

Following changes made by the pulp mills in response to the development of legislation introduced by the federal government, PCDD and PCDF discharges have declined since the late 1980s. This is reflected in declining PCDD and PCDF concentrations in some shrimp, prawn, oyster and crab stocks. This decline in turn has led to the lifting of closures imposed on some of these fisheries. However, in some sites, the mobilization of previously contaminated sediments results in crab being continuously exposed to PCDDs and PCDFs. Hence, these

fisheries remain closed. Similarly, following restrictions on the use of TBT in 1989, concentrations of TBT in recreational harbour sediments have declined, as has the frequency of its effects on whelks. However, in commercial harbours such as Vancouver, TBT concentrations remain elevated.

The successful regulation of some contaminants, at least at some sites, has meant that the future focus on contaminant issues may shift. Continuing contamination by PCDDs and PCDFs and by TBT at some sites will remain a concern, as well as continuing acid mine drainage and the release of metals at some sites containing improperly discharged tailings.

There are, however, new concerns. Population growth in the lower mainland and south-east Vancouver Island will lead to increasing discharges of domestic, municipal and industrial wastes. The organic matter and nutrients present in these discharges may lead to eutrophication in some parts of the Strait of Georgia.

Discharges of hitherto unexpected contaminants, such as the non-ionic alkylphenol ethoxylates used in detergents, which may be endocrine disrupters, are emerging as an issue in terms of their identification, analysis, and environmental significance.

The expected growth of the aquaculture industry will raise concerns about the impact of increasing use of antibiotics and pesticides, and of the accumulation of waste food and feces near aquaculture sites.

Although some components of pulp mill effluents, such as PCDDs and PCDFs, are being controlled, these and other industrial effluents continue to affect fish and fish habitat. The ecological significance of these effects is not known.

The significance of non-point sources of contamination, including LRTAP, needs to be assessed. Most current data and conclusions refer to specific issues or sites; relatively little information exists describing contaminant concentrations in sites remote from identifiable point sources, where contaminants are expected to be derived from atmospheric transport.

Although the distribution and trends of specific chemical contaminants can be defined using conventional analytical chemistry, these data do not provide general information about aquatic environmental quality. Methods and approaches designed to provide more integrative assessments of environmental quality, based on measurements of biological effects, need to be refined and applied, especially in the Strait of Georgia.

Case Study:

IMPROVEMENTS IN FISH HABITAT NEAR IONA ISLAND

Between 1963 and 1988 the Iona Island Sewage Treatment Plant discharged up to 400 000 m^3 per day of primary-treated sewage effluent into the shallow intertidal area off Iona Island in the Fraser River Estuary (Sturgeon Bank). Previous studies document a number of environmental effects, including toxic concentrations of chemical contaminants in the bottom sediments. Low concentrations of dissolved oxygen in the surrounding waters caused extensive fish kills. The shallow estuarine waters near the effluent discharge are important habitats for migrating juvenile salmon and many other fish species. In 1988, the effluent discharge to the mudflat area ceased after installation of a 5 km-long deep water diffuser. Environmental impact analyses conducted at the time predicted that discharge of the mainly domestic, primary-treated sewage effluent to offshore deep water would be superior to nearshore discharges.

To assess the degree of habitat recovery after installation of the deep water diffuser at Sturgeon Bank, several physical, chemical and biological variables were measured. Changes in the size of sediment particles, sediment chemical content, sediment toxicity, community structure of bottom dwelling invertebrate animals, the amount of oxygen in the water, and the abundance and diversity of resident fish communities were evaluated. Results of these projects were compared with results of similar studies completed from 1977 and 1982 when sewage effluent was actively being discharged to Sturgeon Bank.

Municipal Sewage

Municipal sewage or wastewater is a major source of contamination in both marine and freshwater ecosystems. Varying in content from municipality to municipality it may consist of effluents from residential, commercial and industrial sources containing suspended solids, nutrients, and toxic contaminants including heavy metals and bacteria, that may cause disease.

Suspended solids can be either organic (oxygen consuming) or inorganic. Most organic solids eventually decompose into elemental form and assimilate into plant growth. Inorganic solids can also be ingested by aquatic organisms under certain conditions but more frequently deposit onto the bottom of waterbodies including bays and harbours where they may interfere with fish habitat and plant growth. Suspended solids may also alter water quality by increasing turbidity, limiting light penetration, and reducing energy available to aquatic organisms.

Excessive discharges of nutrients such as phosphorus and nitrogen into calm, poorly mixed aquatic environments such as shallow estuaries can lead to excessive eutrophication

characterized by excessive algae and plant growth. Through a process of death and decay oxygen concentrations can be substantially lowered, leading to eventual fish kills.

Toxic contaminants including heavy metals, PCBs, PAHs, chlorinated phenols and chlorinated benzenes may persist and accumulate in sediments or biota regardless of where they are discharged. Many of these substances have the potential to bioaccumulate in aquatic organisms making them unfit for consumption by wildlife or humans. It is often difficult to determine the proportion of persistent substances in aquatic organisms that are attributable to sewage discharges, however, as most of these organisms accumulate these substances from a variety of sources.

Finally, microbiological contamination is a widespread concern. Contamination by bacteria and other microorganisms, especially in areas around sewage outfalls, continues to be a problem in waters used for recreation. Thus contamination has resulted in a number of shellfish closures on the Atlantic and Pacific coasts.

Studies in the 1990s led to the conclusion that there had been an overall improvement in fish habitat in this foreshore area of the Fraser River estuary. Concentrations of contaminants in sediments decreased by varying amounts (e.g., copper 12%, lead 41%, and mercury 56%) since the discharge ceased in the late 1980s. PAHs in sediments were within Canadian water quality guidelines for the protection of aquatic life set for these compounds. A similar pattern of environmental improvement was shown in sediment toxicity measured at identical sites in 1977 and 1994 using clams. Sampling of benthic invertebrate community structure showed an abundance of polychaete worms and amphipod crustaceans at stations where previously there had been none. Statistical analyses of the 1974 data found that a number of sites had no populations of aquatic life. However, a similar analysis of 1992 data found all sites had aquatic life. Post discharge monitoring of fish populations showed a significant increase in fish abundance compared to when the effluent was present. Relative to the 1980 data, more species of fish were caught at stations closest to the former outfall site (Nishimura *et al.* 1996).

Case Study:

THE EFFECTS OF LOW DISSOLVED OXYGEN CONCENTRATIONS ON SALMON IN ALBERNI INLET

Alberni Inlet is an arm of the Pacific measuring 45 km long on the west coast of Vancouver Island, British Columbia. The Somass River enters the head of the Inlet at Port Alberni, where a pulp mill is located. The inlet and river are used by sockeye salmon which migrate each summer to their spawning grounds in Sproat Lake and Great Central Lake.

During the unusually warm summer of 1990, about 100 000 salmon died in upper Alberni Inlet. Studies by DFO showed that the salmon preferred to occupy the cooler bottom water of the estuary rather than to migrate through the warmer and less saline surface water. Unfortunately, the bottom water of the Somass Estuary has consistently low oxygen concentrations; this is due partly to poor water exchange resulting from the natural topography of the area, and also to the oxygen demand of organic matter discharged from the pulp mill. Over the last forty years, oxygen concentrations in the bottom water have declined by about 60%. The prolonged exposure to the stressful conditions caused by low oxygen concentrations in upper Alberni Inlet may also have made the salmon more susceptible to parasite infections. Similar, but less dramatic losses, have occurred more recently.

Industry, provincial and federal authorities are working together to try to understand and remedy this situation. Water and oxygen "budgets" have been calculated by DFO for the Somass Estuary. The energy requirements of salmon during migration have been measured (also by DFO), so that the success of migration during different environmental conditions can be predicted. Mechanisms to introduce cooler oxygenated lake water to the Somass River have been devised by DFO in collaboration with the provincial government and with local organisations. Finally, the pulp mill operators have changed the mill processes to reduce effluent discharges and to provide more effluent treatment, and have begun a water quality monitoring programme in the summer time. All of these approaches should help to solve the problem in Alberni Inlet. However, one question that remains is whether climate change and global warming will continue exacerbate the existing problem, or work against the cooperative ventures of government and industry.

Case Study:

THE NESTUCCA OIL SPILL: FATE AND EFFECTS OF AN OIL SPILL

Perhaps the largest contaminant impact, at least in spatial terms, on the British Columbia open coast was that of the Nestucca spill (Duval *et al*. 1989). During the night of December 23, 1988, in bad weather, the fuel barge Nestucca collided with its tender tug and spilled about 875 000 L of Bunker "C" (No. 6) oil off Grays Harbour, Washington. By December 31, the oil had moved north and was stranded at Carmanah Point, on southwest Vancouver Island. Seventeen days later it had spread as far as Cape Scott, on the northwest tip of Vancouver Island; and on January 27, 1989 it was reported near Bella Bella. The spill raised concerns about contamination of habitat and fisheries and on the amenity value of the beaches in Pacific Rim National Park.

Because of circulation patterns, tides and winds, little oil penetrated the fjords and inlets and oiling was confined to exposed outer shorelines. The Broken Island group in Barkley Sound (part of Pacific Rim National Park) was contaminated by a fairly large slick, but a combination of physical clean-up measures and natural dispersion by wave action had removed much of the oil by the end of January 1989. Oil persisted longest in sheltered muddy beaches which are least exposed to wave action, while exposed rock shorelines cleaned up the most rapidly.

Most of the biological impact of the spill appeared to affect organisms in places where the oil accumulated. About 1 000 oiled and dead birds were collected from Vancouver Island beaches, though a subsequent report by Vermeer *et al*. (1992) estimated that the total kill in U.S. and Canadian waters was approximately 56 000. One sea otter was known to have died as a result of oil exposure. Other mammals, including seals and sea lions, did not appear to be severely affected. Some commercial fishery areas were closed to shellfish harvesting and crab fishing due to fouling of gear. There appeared to be no impact on the herring roe industry, nor was salmonid aquaculture affected. Overall, the impact of the spill, though widespread, was of low intensity. However, no follow up studies on possible long term effects have been conducted.

Case Study:

NATURALLY OCCURRING METALS IN MARINE SEDIMENTS: STUDIES IN UCLUELET INLET

Ucluelet Inlet, on the west coast of Vancouver Island (See Figure 1), has sediments containing cadmium reaching concentrations of up to 8 $\mu g.g^{-1}$. However, no anthropogenic source has been found in this area. A likely cause for the elevated concentrations of cadmium is natural geochemical processes. There is an appreciable supply of organically rich material which causes the sediment surface layers to become anoxic, and rich in sulfides. Dissolved cadmium diffuses from the overlying water into the sediments, and is precipitated, for example, as pyrites. These accumulate in the anoxic layer to become buried as the sedimentation process proceeds (Pedersen *et al*. 1989).

References

ADDISON, R.F. 1980. A review of organohalogens with respect to the Ocean Dumping Act. *In* J.J. Swiss, R.F. Addison, D.W. McLeese, and J.F. Payne [eds.] Regulated levels of scheduled substances in the Ocean Dumping Control Act - a review. Ocean Dump. Rept. No. 3. Dep. Fish. Oceans, Ottawa, Ont. 13-21p.

ADDISON, R.F. 1983. PCB replacements in dielectric fluids. Env. Sci. Technol. 17: 486A - 494A.

ADDISON, R.F., D.E. Willis, and M.E. Zinck. 1994. Liver microsomal mono-oxygenase induction in winter flounder (*Pseudopleuronectes americanus*) from a gradient of sediment PAH concentrations at Sydney Harbour, Nova Scotia. Mar. Env. Res. 37: 283-296.

ALBRIGHT, L.J., T.G. Northcote, P.C. Oloffs, and S.Y. Szeto. 1975. Chlorinated hydrocarbon residues in fish, crabs and shellfish of the Lower Fraser River, its estuary and selected locations in Georgia Strait, British Columbia - 1972-73. Pest. Monit. J. 9: 134-40.

ANON. 1995. The 1995 British Columbia Seafood Industry Year in Review. Gov. of B. C., Victoria, B.C. 8 p.

ANTCLIFFE, B.L., D. Keiser, J.A.J. Thompson, W.L. Lockhart, D.A. Metner, and R. Roome. 1997. Health assessment of mountain whitefish (*Prosopium williamsoni*) from the Columbia River in July 1994, p. 121-126 *In* Proceedings of the 23rd Annual Toxicity Workshop: October 7-9, 1996. Calgary, Alberta, Can. Tech. Rep. Fish. Aquat. Sci. 2144.

BEASLEY, T.M., R. Carpenter, and C.D. Jennings. 1982. Plutonium, 241Am, and 137Cs ratios, inventories and vertical profiles in Washington and Oregon continental shelf sediments. Geochim. et Cosmochim. Acta. 46: 1931-1946.

BORNHOLD, B.D. 1983. Sedimentation in Douglas Channel and Kitimat Arm, p. 89-114. *In* R.W. Macdonald [ed.] Proceedings of a Workshop on the Kitimat marine environment. Can. Tech. Rep. Hydrog. Ocean Sci. 18: 218 p.

BURLING, R.W., J.E. McInerney, and W.K. Oldham. 1981. A technical assessment of the Amax/Kitsault molybdenum mine tailings discharge to Alice Arm, B.C. Rep. Mining. Fisheries and Oceans, Government of Canada.

CARPENTER, R., and M.L. Pederson. 1989. Chemical cycling in Washingon's coastal zone, p. 367-509. *In* M.R. Landry, and B.M. Hickey [eds.]. Coastal oceanography of Washington and Oregon. Oceanographic Series 47. Elsevier, The Netherlands.

CRETNEY, W.J., C.S. Wong, R.W. Macdonald, P.E. Erickson, and B.R. Fowler. 1983. Polycyclic aromatic hydrocarbons in surface sediments and age-dated cores from Kitimat Arm, Douglas Channel and adjoining waterways, p. 162-195. *In* R.W. Macdonald [ed.] Proceedings of a Workshop on the Kitimat marine environment. Can. Tech. Rep. Hydrography Ocean Sci. 18.

DE VOOGT, P., and U.A.Th. Brinkman. 1989. Production, properties and usage of polychlorinated biphenyls, p. 3-45. *In* R.D. Kimbrough, and A.A. Jensen [eds.] Halogenated biphenyls, terphenyls, napthalenes, dibenzodioxins, and related products. 2nd Edition. Elsevier.

DONALDSON, E.M., U.H.M. Fagerlund, and J.R. McBride. 1984. Aspects of the endocrine stress response to pollutants in salmonids, p. 213-221. *In* V.W. Cairns, P.V. Hodson, and J.O. Nriagu [eds.] Contaminant effects on fisheries. Wiley, NY.

DRYSDALE, K., and T.F. Pedersen. 1992. Geochemical behaviour of a buried marine mine tailings deposit, Howe Sound, British Columbia. *In* C.D. Levings, R.B. Turner, and B. Ricketts [eds.] Proceedings of the Howe Sound Environmental Science Workshop. Can. Tech. Rep. Fish. Aquat. Sci. 1879: 270 p.

DUNLAP, T.R. 1981. DDT: scientists, citizens and public policy. Princeton Univ. Press, Princeton, NJ.

DUVAL, W., S. Hopkinson, R. Olmstead, and R. Kashino. 1989. The Nestucca oil spill: preliminary evaluation of impacts on the west coast of Vancouver Island. Report to Environment Canada and B.C. Ministry of Environment, ISBN 0-7726-0967-5.

GOYETTE, D., and J. Boyd. 1989. Distribution and environmental impact of selected benthic contaminants in Vancouver Harbour, British Columbia, 1985 - 1987. Regional Program Report 89-02, Environment Canada, Vancouver, B.C.

HAGEN, M.E., A.G. Colodey, W.D. Knapp, and S.C. Samis. 1995. Environmental response to decreased dioxin and furan loadings from British Columbia coastal pulp mills. *In* P. Adriaens, R. Clement, G. Hunt, D. Muir, and S. Ramamoorthy [eds.] Dioxin '95: 15th International Symposium on Chlorinated Dioxins and related Compounds. Dioxin '95 Secretariat, P.O. Box 46060, Petroleum Plaza PRO, 9945-108 Street, Edmonton, AB, Canada.

HARBO, R.M., and I.K. Birtwell. 1978. Mercury content data for crustaceans and fishes from Howe Sound, British Columbia, May 1977 to April 1978. Fish. Mar. Ser. Data Rept. 63.

HARBO, R.M., and I.K. Birtwell. 1983. Trace metal content data for crustaceans and fishes from Howe Sound, British Columbia. Can. Data Rept. Fish. Aquat. Sci. 379.

HARBO, R.M., I.K. Birtwell, and O.E. Langer. 1983. Trace metals in marine organisms from coastal waters of British Columbia (1971 to 1976). Can. Manusc. Rept. Fish. Aquat. Sci. 1691.

HARDING, L., and D. Goyette. 1989. Metals in Northeast Pacific coastal sediments and fish, shrimp and prawn tissues. Mar. Poll. Bull. 20: 187-189.

HARGRAVE, B.T., and G.A. Phillips. 1976. DDT residues in benthic invertebrates and demersal fish in St. Margaret's Bay, Nova Scotia. J. Fish. Res. Bd. Canada 33: 1692 - 1698.

JARMAN, W.M., R.J. Norstrom, D.C.G. Muir, B. Rosenberg, M. Simon, and R.W. Baird. 1996. Levels of organochlorine compounds, including PCDDs and PCDFs, in the blubber of cetaceans from the west coast of North America. Mar. Poll. Bull. 32: 426-436.

KAY, B.H. 1989. Pollutants in British Columbia's marine environment: A status report. State of the Environment Report No. 89-1. Environment Canada, Vancouver, B.C. Canada. 59 p.

LEVINGS, C.D. 1994. Some ecological concerns for net-pen culture of salmon on the coasts of the Northeast Pacific and Atlantic Oceans, with special reference to British Columbia. J. Appl. Aquacul. 4: 65-141.

LEVINGS, C.D., and B.E. Riddell, 1992. Salmonids and their habits in Howe Sound Basin - Status of Knowledge. *In* Proceedings of the Howe Sound Environmental Science Workshop. Can. Tech. Rep. Fish. Aquat. Sci. 1879. 65-81 p.

MACDONALD, R.W., and E.A. Crecelius. 1994. Marine sediments in the Strait of Georgia, Juan de Fuca Strait and Puget Sound: what can they tell us about contamination?, p. 101-134. *In* R.C.H. Wilson, R.J. Beamish, F. Aitkens, and J. Bell [eds.]. Review of the marine environment and biota of Strait of Georgia, Puget Sound and Juan de Fuca Strait. Can. Tech. Rept. Fish. Aquat. Sci. 1948.

MACDONALD, R.W., and M.C. O'Brien. 1996. Sediment trap evaluation of mine tailings transport. Mar. Geore. Geotech. 14: 97-109.

MACDONALD, R.W., and T.F. Pedersen. 1991. Geochem. Sedim. West. Can. Contin. Shelf. Cont. Shelf. Res. 11: 717-735.

MACDONALD, R.W., B.D. Bornhold, and I. Webster. 1983. The Kitimat Fjord system: an introduction, p. 2-13. *In* R.W. Macdonald [ed.] Proceedings of a Workshop on the Kitimat marine environment. Can. Tech. Rep. Hydrog. Ocean Sci. 18: 218 p.

MACDONALD, R.W., D.M. Macdonald, M.C. O'Brien, and C. Gobeil. 1991. Accumulation of heavy metals (lead, zinc, copper, cadmium), carbon and nitrogen in sediments from the Strait of Georgia. Mar. Chem. 34: 109-135.

MACDONALD, R.W., W.J. Cretney, N. Crewe, and D. Paton. 1992. A history of octachlorodibenzo-p-dioxin, 2,3,7,8-tetrachlorodibenzofuran and 3,3',4,4'-tetrachlorobiphenyl contamination in Howe Sound, B. C. Environ. Sci. Technol. 26: 1544-1550.

MCCANDLESS, R.G. 1995. The Britannia mine: historic landmark or environmental liability. The B.C. Prof. Eng., April 1995: 4-7.

NENER, J., D. Kieser, J.A.J. Thompson, W.L. Lockhart, D.A. Metner, and R. Roome. 1995. Monitoring of Mountain Whitefish (*Prosopium williamsoni*) from the Columbia River system near Castlegar, B. C.: health parameters and contaminants in 1992. Can. Manus. Rep. Fish. Aquat. Sci. 2036: 89 p.

NISHIMURA, D.J.H., G.E. Piercy, C.D. Levings, K. Yin, and E.R. McGreer, 1996. Changes in fish communities and water chemistry after cessation of municipal sewage discharge near the Iona Island foreshore, Fraser River estuary, B. C. Can. Tech. Rep. Fish. Aquat. Sci. 2096: 17 p.

ODHIAMBO, B.K., R.W. Macdonald, M.C. O'Brien, J.R. Harper, and M.B. Yunker. 1996. Transport and fate of mine tailings in a coastal fjord of B. C. as inferred from the sediment record. Sci. Tot. Env. 191: 77-94.

PEDERSEN, T.F., R.D. Waters, and R.W. Macdonald. 1989. On the natural enrichment of cadmium and molybdenum in the sediments of Ucluelet Inlet, B. C. Sci. Tot. Env. 79: 125-139.

PEDERSEN, T.F., D.V. Ellis, G.W. Poling, and C. Pelletier. 1995. Effects of changing environmental rules: Kitsault Molybdenum Mine, Canada. Mar. Geores. and Geotech. 13: 119-133.

ROGERS, I.H., C.D. Levings, W.L. Lockhart, and R.J. Norstrom. 1989. Observations on overwintering juvenile chinook salmon (*Oncorhynchus tshawytscha*) exposed to bleached Kraft mill effluent in the Upper Fraser River, B. C. Chemosphere 19: 1853-1868.

ROGERS, I.H., J.S. Macdonald, and M. Sadar. 1992. Uptake of selected organochlorine contaminants in fishes resident in the Fraser River Estuary, Vancouver, B. C. Wat. Poll. Res. J. Can. 27: 733-749.

STEWART, C., and J.A.J. Thompson. 1994. Extensive butyltin contamination in southwestern coastal B. C. Mar. Poll. Bull. 28: 601-606.

STOUT, V.F., and F.L. Beezhold. 1981. Chlorinated hydrocarbon levels in fishes and shellfishes of the Northeastern Pacific Ocean, including the Hawaiian Islands. Mar. Fish. Rev. 43:1-12.

TESTER, M., D.V. Ellis, and J.A.J. Thompson. 1996. Neogastropod imposex for monitoring recovery from marine TBT contamination. Env. Toxicol. Chem. 15:560-567.

THOMSON, R.E. 1981. Oceanography of the B. C. coast. Can. Spec. Publ. Fish. Aquat. Sci. 56: 291 p.

THOMSON, R.E. 1994. Physical oceanography of the Strait of Georgia-Puget Sound-Juan de Fuca Strait system, p. 36-98. *In* R.C.H. Wilson, R.J. Beamish, F. Aitkens, and J. Bell [eds.] Review of the marine environment and biota of Strait of Georgia, Puget Sound and Juan de Fuca Strait. Can. Tech. Rept. Fish. Aquat. Sci. 1948.

THOMPSON, J.A.J., and D.W. Paton. 1976. Further studies of mine tailings distribution in Howe Sound. Fish. Res. Bd. Can. Man. Rep. 1383: 15p.

THOMPSON, J.A.J., and C. Stewart. 1994. Organotin compounds in the coastal biota of B. C. - an overview. Can. Tech. Rept. Hydrogr. Ocean Sci. 155: 11 p

THOMPSON, J.A.J., R.W. Macdonald, and C.S. Wong. 1980. Mercury geochemistry in sediments of a contaminated fjord of coastal B. C. Geochem. J. 14: 71-82.

VERMEER, K., K.H. Morgan, and P.J. Ewins. 1992. Population trends of pelagic cormorants and glaucous-winged gulls nesting on the west coast of Vancouver Island, p. 60-64. *In* K. Vermeer, R.W. Butler, and K.H. Morgan [eds.] Proceedings of a Conference on the Ecology, Status and Conservation of Marine and Shoreline Birds on the West Coast of Vancouver Island. Can. Wildlife Serv., Ottawa, Ont. 75 p.

WALDICHUK, M. 1983. Pollution in the Strait of Georgia: a review. Can. J. Fish. Aquat. Sci. 40: 1142-1167.

WEST, P., T.M. Fyles, B. King, and D.C. Peeler. 1994. The effects of human activity on the marine environment of the Georgia Basin: present waste loadings and future trends. p. 9-33. *In* R.C.H. Wilson, R.J. Beamish, F. Aitkens, and J. Bell [eds.]. Review of the marine environment and biota of Strait of Georgia, Puget Sound and Juan de Fuca Strait. Can. Tech. Rept. Fish. Aquat. Sci. 1948.

WILSON, J., R. Addison, R. Gordon, D. Martens, and B. Glickman, 1995. MFO activity and contaminant analysis of overwintering juvenile chinook salmon in the Fraser River, B.C., p. 318. *In* Second SETAC World Congress. 16th Annual Meeting. Society of Environmental Toxicology and Chemistry. SETAC Press. 1010 North 12th Avenue, Pensacola, FL.

YUNKER, M.B., and W.J. Cretney. 1995a. Dioxins and furans in crab hepatopancreas: use of principal components analysis to classify congener patterns and determine linkages to contamination sources, p. 315-325. *In* M.R. Servos, K.R. Munkittrick, J.H. Carey, and G.J. Van Der Kraak [eds.] Fate and Effects of Pulp and Paper Mill Effluents. St. Lucie Press, Boca Raton, FL.

YUNKER, M.B., and W.J. Cretney. 1995b. Chlorinated dioxin trends between 1987 and 1993 for samples of crab hepatopancreas from pulp and paper mill and harbour sites in B. C. Can. Tech. Rep. Fish. Aquat. Sci.

Additional reading

ADDISON, R.F. 1996. The use of biological effects monitoring in studies of marine pollution. Environ. Rev. 4: 225-237.

MACDONALD, D.D., M.G. Ikonomou, A.L. Rantalainen, H. Rogers, D. Sutherland, and J. Van Oostdam. 1997. Contaminants in white sturgeon (*Acipenser transmontanus*) from the Upper Fraser River, British Columbia, Canada. Environ. Toxicol. Chem. 16: 479-490.

WILSON, R.C.H., R.J. Beamish, F. Aitkens, and J. Bell [eds.] 1994. Review of the marine environment biota of Strait of Georgia, Puget Sound and Juan de Fuca Strait. Can. Tech. Rep. Fish. Aquat. Sci. 1948. 388 p.

Personal Communications

ADDISON, R.F., Institute of Ocean Sciences, Sidney, B.C., n.d.

CRETNEY, W.J., Institute of Ocean Sciences, Sidney, B.C., n.d.

GORDON, K., n.d.

MARTENS, D.W. n.d.

McCANDLESS, R. M. Environment Canada, Vancouver, B.C., n.d.

THOMPSON, J.A.J., n.d.

Conclusions

Table of Contents

Conclusions

Introduction

Our understanding of how chemicals enter and move through the environment and how they affect fish, other aquatic life, and their habitat has dramatically increased in recent years. This chapter summarizes the major scientific conclusions on the fate and aquatic effects of the suite of contaminants that were reviewed in detail in the preceding chapters. It also discusses the significance of contaminants in aquatic ecosystems in relation to other factors that could affect the conservation and protection of fish and fish habitat and the sustainable utilization of fisheries resources. The chapter concludes with the identification of current and emerging chemical contaminant issues.

CANADA'S FISHERIES RESOURCES

Canada benefits from diverse fisheries resources in coastal, offshore and fresh waters. In 1993, fisheries production was valued at $2.9 billion of which 86% was exported to more than 100 countries around the world. Although Canada ranks 15th in the world in terms of tonnage of fish caught, it ranks fifth in tonnage in the export of fish and fish products, 90% of which is exported to the U.S., Japan and the European Community. In 1993, Canada's fishing industry employed approximately 76 000 fishermen, 1 500 aquaculturalists, 1 100 importers, 75 000 fish plant workers and supported 5 200 jobs in the aquaculture sector. In addition, more than 6 million recreational fishermen were licensed in Canada, including roughly 5 million Canadians and 1 million visitors. These activities generated $7 billion in economic activity and supported more than 150 000 full-time jobs (DFO 1995).

On a national basis, commercial fishing contributes about 0.5% to Canada's Gross National Product; however, it is a major contributor to the economies of coastal provinces and northern communities. For many coastal residents, fishing constitutes the only source of economic activity and represents the only viable means of earning an income. On the Atlantic coast, one in four residents lives in one of 1 300 fishing communities. Half of these communities depend primarily on fishing for their livelihood. In the North, fishing provides both a source of subsistence food and disposable income where other employment is limited or non-existent. Fishing is also an integral part of the cultural identity of many Canadians (DFO 1995).

DIVERSITY OF ECOSYSTEMS AND ENVIRONMENTAL VARIABILITY

Occupying 6.7% of the world's total land area, Canada is a large geographically and ecologically diverse country (World Coast Conference 1993). Freshwater lakes and rivers cover about 750 000 km^2 of Canada's surface area (Energy, Mines and Resources Canada 1989 [from Government of Canada 1996]). Lakes range in diversity from some of the largest freshwater lakes in the world to small, brown-water acidic lakes. For example, the Great Lakes ecosystem includes five of the 14 largest freshwater lakes in the world. Freshwater rivers range from massive systems such as the St. Lawrence, Fraser and Mackenzie Rivers to small streams. The 25 river basins of Canada drain an area of about 10 000 km^2 and discharge freshwater into the Atlantic, Pacific and Arctic oceans as well as into the Gulf of Mexico, James Bay and Hudson Bay. Sixty percent of this discharge flows into the Arctic Ocean and Hudson Bay (Pearse *et al.* 1985).

Canada has the longest coastline in the world, one that stretches nearly 240 000 km and borders on three oceans (Statistics Canada 1994). The Canadian coastline is characterized by a range of geological features, ranging from deep inlets and fjords to large coastal plans (such as the Grand Banks). Estuaries mix the fresh waters of rivers draining upland areas and the saline waters of the oceans. The oceanographic, limnological and biological processes occurring within and between these freshwater, coastal and marine systems are even more complex.

The variability and unpredictability of environmental conditions hampers our ability to distinguish between those changes that are due to human activities from those that are natural. As a result, it is often difficult to identify and predict long-term trends in concentrations of chemical contaminants and their spatial distribution. Determining contaminant effects on fish and fish habitat and on the sustainable utilization of fisheries resources is a complex problem. For these reasons, it is more useful to examine chemical contaminants on an ecosystem basis, which was done in preceding chapters of this report. This chapter summarizes some of the similarities across regions and ecosystems in an attempt to provide an integrated picture of chemical contamination and its effects on fish and fish habitat in Canada.

Most of Canada's population is found in the southern portion of the country. Residents here have the greatest water demands, however, northern Canada tends to have most of the fresh water supply. Therefore, the overall amount of freshwater that is readily accessible to Canadians is rather limited. This imbalance between supply and demand is most evident in the Great Lakes region where there are urbanized areas with dense population concentrations and industrial facilities, as well as agricultural and recreational lands. Most Canadians (two-thirds) live in the Lake Ontario basin and just six urban areas contain 75% of the Great Lakes Basin population.

In general, the highest concentrations of many of the chemical contaminants resulting from human activities are found in a small number of fresh water bodies located near urbanized areas while the vast majority of freshwater remains relatively pristine. Exceptions to this generalization include chemical contamination due to atmospheric input to more remote areas and the release of effluents from specific industrial point sources including, for example, pulp and paper mills, metal mining operations, smelters and the petrochemical plants.

Most of Canada's coastline is sparsely inhabited. As a result, concerns about chemical contamination have tended to focus on the more populated areas, especially those where there have been mining operations and other industrial activities. Since many coastal fjords, basins and estuaries retain much of the particulate material entering them, other outer coastal and offshore areas tend to have relatively lower concentrations of chemical contaminants. Exceptions include releases of chemical contaminants from offshore activities, accidents and spills and disposal of waste material.

CLASSES OF CHEMICAL CONTAMINANTS

This report focuses on a limited subset of the many organic and inorganic chemicals that are detected in Canadian freshwater and marine ecosystems. Emphasis is placed on those chemicals that are widespread, persistent, accumulate in aquatic biota and have the greatest likelihood of posing a risk to fish, fish habitat and the sustainable utilization of fisheries resources. Persistence of a chemical is related to its resistance to natural breakdown or metabolism. Chemicals that are not rapidly degraded in the environment are said to be persistent. In principle, because metals are part of the natural geological makeup of the earth's crust, they persist indefinitely in the environment. However, concentrations of specific metals can become elevated in aquatic ecosystems as a result of human activities.

In Canada, the major identifiable releases of chemical-containing effluents into the environment have been curtailed. In addition, the uses of some chemicals have been either banned or severely restricted. However, some of these chemicals are still being detected in fish and fish habitat, either as the parent chemical or as breakdown products. The classes of compounds described in this report include metals and organometallic compounds, organochlorine compounds used as pesticides, polychlorinated biphenyls (PCBs), polychlorodibenzo-*p*-dioxins (PCDDs), polychlorodibenzo-furans (PCDFs), and polycyclic aromatic hydrocarbons (PAHs).

Mercury, lead and cadmium represent metals that have been found to have elevated concentrations in fish and fish habitat due to both natural processes and anthropogenic activities. All three metals accumulate in aquatic biota. Information on other metals, such as nickel, zinc and copper, is included where appropriate.

Organometallic compounds in this report are represented by organotin compounds such as tributyltin (TBT) and alkyllead. TBT is used as a biocide in anti-fouling paints. Alkyllead use as an additive in gasoline was standard until 1989. Its use has been phased out since then.

Organochlorine pesticides have been detected in most biotic and abiotic elements of all Canadian ecosystems. Dichlorodiphenyltrichloroethane (DDT), dieldrin, toxaphene, hexachlorobenzene and mirex are examples of major organochlorine pesticides that are discussed in this report. DDT and its primary breakdown product, DDE, are routinely identified in all regions of Canada despite a general ban on their use in 1972. Organochlorine pesticides such as toxaphene and mirex are detected over a large geographic range, although their sources are either confined to a small region within Canada or are outside of the country. Other organochlorine pesticides include dieldrin, aldrin, endrin, chlordane, oxychlordane and hexachlorocyclohexane (lindane).

Organochlorine compounds such as the complex mixtures of PCBs, PCDDs and PCDFs are examples of industrial chemicals detected in the Canadian environment. Their presence originated either from the direct manufacture for specific applications (PCBs) or as the inadvertent byproducts of industrial processes such as pulp and paper production and chlorophenol manufacture (PCDDs and PCDFs).

PAHs are a group of organic compounds that can have both natural and anthropogenic sources. Because they are readily metabolized by vertebrates, PAHs are routinely detected usually only in bottom sediments and invertebrate species.

Sources

In general, elevated concentrations of chemical contaminants occur in Canada's heavily populated or industrialized regions. Some of these areas include the Great Lakes Basin, with its concentration of large cities and manufacturing facilities, the west coast, with its predominant forest products and pulp and paper industry, and the busy maritime shipping harbours of the Pacific and Atlantic. Chemical contaminants can also occur in sites remote from these locations. Atmospheric transport of materials to the Arctic or downstream transport of toxic substances from the Great Lakes via the St. Lawrence River system are some mechanisms which result in the distribution of contaminants over large areas of Canada.

Chemical contaminants enter aquatic ecosystems through releases from point and non-point sources. Most major point sources are anthropogenic in origin and include municipal and industrial effluents, solid waste discharges, leaching of materials from hazardous waste sites, and accidental releases or spills either directly into water or a watershed. For some metals, point sources may be natural in origin. Non-point sources may be natural or anthropogenic in origin. They include urban and agricultural run-off, mineral outcrops, and the long-range transport of chemical contaminants from distant sources. Riverine inputs, ocean currents, biotic transport and remobilization of chemicals from sediments have become increasingly significant routes of entry for the introduction of chemical contaminants into aquatic ecosystems.

Although anthropogenic point sources still contribute sizable quantities of chemical contaminants into aquatic ecosystems, many of these sources are being controlled or regulated. Accordingly, atmospheric and other non-point sources are now playing a greater role in the relative contribution of chemical contaminants to aquatic ecosystems. Unlike most point sources, which are relatively easy to identify, non-point sources tend to be more widely dispersed and may be located at substantial distances from a contaminant's destination.

The atmosphere is an important medium for the transport and deposition of chemicals into aquatic ecosystems, including mercury, lead, PCBs, and a number of organochlorine pesticides. DDT, for example, is still used in areas outside Canada and continues to enter the Canadian aquatic environment through long range transport of atmospheric pollutants (LRTAP). Many large lakes, with their huge surface areas, receive a significant proportion of contaminant loadings through atmospheric deposition. The Arctic has likewise become increasingly contaminated as a result of the continued transport and deposition of chemical contaminants via the atmosphere. When contaminants enter systems with long or slow turnover rates such as cold arctic systems or large volume lakes, the potential for long-term accumulation is increased.

Spatial Distribution and Temporal Trends

For the most part, the transport, dissemination, and consequent distribution of chemical contaminants in Canadian waters is primarily controlled through their association with suspended particulate matter or through air-water partitioning. In Canadian freshwater ecosystems, the dynamics of suspended particulate matter are influenced primarily by river flow rate and the topography of the various hydrographic basins. As a result, chemical contaminants tend to accumulate at the bottom of lakes and in those portions of rivers where the flow rate is less.

Sediment transport in Canadian marine ecosystems is influenced by a variety of factors, including topography, strength of ocean currents, winds, tides and freshwater run-off. The latter is particularly important because of its role in the transport of chemical contaminants downstream from freshwater to marine ecosystems. A significant proportion of contaminants that are released into fresh water in Canada is eventually exported to marine estuaries and basins via coastal rivers. For example, St. Lawrence River transport of contaminants from the Great Lakes downstream to the Estuary. In marine areas of Canada that are less influenced by coastal rivers, sediment transport and deposition, and the associated accumulation of sediment-bound chemical contaminants, are essentially controlled by ocean currents and bottom topography.

Although it is well recognized that individual aquatic ecosystems are not spatially homogeneous and that environmental conditions change with the seasons, it is generally less well understood that there are also marked variations over both short and long time periods. The diversity among and within Canadian ecosystems makes it difficult to compare contaminant concentrations over large geographic scales (spatial distribution) and over long periods of time (temporal trends) with any degree of certainty. Attributing specific causes to changes that are observed in ecosystems is even more difficult. However, keeping these challenges in mind, some general conclusions on the spatial distribution and temporal trends of chemical contaminants in aquatic ecosystems of Canada can be made.

For each of the classes of contaminants presented in this report, this chapter summarizes some of the specific problems associated with these chemicals in various regions of Canada. These summaries are not meant to be exhaustive, but they illustrate the variety of contaminant conditions that occur throughout a variety of regional ecosystems.

METALS

Elevated concentrations of metals, such as cadmium, mercury, lead, zinc, nickel and copper, found in freshwater ecosystems resulting from anthropogenic activities are often local or regional in nature. These metals are associated with activities such as non-ferrous mining, smelting and refining, and with industrialized urban centres. Increasing distance from the point source usually can be related to decreasing metal concentrations in sediment and biota. Background concentrations of metals vary with geology and often complicate or prevent an accurate differentiation between natural and anthropogenic sources.

Elevated concentrations of mercury in ecosystems are associated primarily with reservoir development, which results in the leaching of mercury from geological sources, and with certain industries, such as mercury cell chlor-alkali plants. Natural sources of mercury are also important. Generally, concentrations of mercury in the muscle of most freshwater fish are below the concentrations established for safe consumption. Nevertheless, mercury concentrations in some fish species remain high in several areas, necessitating restrictions on consumption.

Accumulation of lead and possibly other metals in the sediments of eastern Canadian lakes is related to the pattern of acidic deposition. Concentrations in sediment decrease in a northerly gradient from the Toronto-Montreal area to the James and Hudson Bay area, and do not reflect natural metal concentrations in stream and river sediments.

Elevated concentrations of metals in marine ecosystems are usually associated with particulate matter that has been carried by freshwater or may result when freshwater and salt water mix. Such mixing will lead to precipitation of insoluble metal complexes. Well-flushed coastal areas, such as in the northwest Atlantic, the Bay of Fundy, parts of the Gulf of St. Lawrence and some coastal areas in British Columbia, tend not to accumulate particulate material and precipitated metals. Consequently, metals generally occur in low concentrations in invertebrates, fish and marine mammals. However, the biomagnification of metals originating from natural sources can occur. For example, mercury accumulates in biota, particularly in long food chains. As a result, concentrations can exceed recommended consumption limits.

Elevated concentrations of metals in sediments and biota are generally the result of discharges into embayments or poorly-flushed areas. Point sources include industrial activities and sewage outfalls. For example, long-term discharges from a chlor-alkali plant in Howe Sound, British Columbia, and a copper and zinc mine at Britannia Beach, British Columbia, resulted in elevated metal concentrations in sediment and local biota. Elevated concentrations of mercury have been found in the Saguenay Fjord of the St. Lawrence marine ecosystem due to past industrial activities and natural weathering.

Metals can also accumulate in sediment deposition zones of large marine systems. For example, elevated concentrations of metals are found in the Laurentian Channel of the St. Lawrence River (lower estuary) marine ecosystem. Before 1971, concentrations in sediments from the Channel were 10 to 50 times higher than background values. Additionally, sediment cores collected from the Ballenas Basin in the Strait of Georgia had metal concentrations that corresponded to increasing industrialization and population growth.

East-west variations in geological formations in the Arctic result in higher background concentrations of mercury in biota in the western Arctic than in the central and eastern Arctic. High concentrations in the western Arctic are found particularly in the liver of ringed seals, beluga and polar bear. Cadmium concentration shows the reverse natural pattern.

Many of the discharges of metals directly into freshwater ecosystems from point sources, such as chlor-alkali plants and non-ferrous mines, smelters and refineries, are being eliminated or controlled. These actions are reflected in lower metal concentrations in surficial sediments and local biota. Inputs of metals such as cadmium, zinc and lead, have declined in freshwater and arctic lakes. However, despite such declines, concentrations of these metals in surficial freshwater sediments are still above background values.

Whereas inputs of metals to freshwater ecosystems from industrial point sources have declined in recent decades, discharges from industrialized urban centers and atmospheric deposition are becoming more important contributors. In contrast to metals such as lead, mercury has been increasing in some freshwater sediments and biota over the past few years. The atmosphere appears to be an increasingly important medium through which mercury originating from anthropogenic and natural processes is being delivered to freshwater ecosystems. For example, there has been a slow continuous increase in atmospheric deposition of mercury into arctic lakes. As well, the increasing number of reservoirs raises the probability that mercury concentrations will increase in freshwater sediments and biota as a result of the release of the metal from natural (geological) sources within the reservoir basin.

ORGANOMETALLIC COMPOUNDS

TBT is routinely found in both fresh and saltwater harbours in Canada as a result of its use in anti-fouling paints. Since its use on small boats (less than 25 m) was banned in 1989 there has been a decrease in concentrations in the sediments of some harbours along the British Columbia coast. A national survey of sediments in 265 recreational boating harbours in 1986 detected the

highest tributyltin concentrations in sites throughout the Great Lakes (Maguire *et al.* 1986; See the Great Lakes Ecosystem Chapter). Concentrations of TBT in sediment cores from recreational harbours have declined since 1989. However, TBT concentrations have not decreased in sediment cores from commercial ports such as Vancouver Harbour. Commercial ports are more likely than recreational harbours to accommodate TBT-treated boats. Declines in populations of marine oysters and whelks (snails) and incidences of increased male characteristics in female whelks have been linked to elevated concentrations of TBT in marine ecosystems. Surveys have shown increased concentrations of imposex in association with elevated concentrations of TBT. On the east coast, the concentrations of TBT tend to be lower than in Vancouver Harbour. However, whelks are missing from some habitats that are considered suitable for them. There is insufficient information on TBT in freshwater and in marine waters of the east coast to determine long-term trends.

ORGANOCHLORINE PESTICIDES

Concentrations of DDT in Atlantic offshore flatfish species, such as Greenland halibut (turbot) American plaice and yellowtail flounder, are lower in liver and muscle tissues than those reported for various flatfish species from Maine and the European Baltic Sea. Similarly, DDT concentrations in muscle tissue of cod from offshore Newfoundland were at the low end of the range reported from North Sea fish. Dieldrin, aldrin, endrin, chlordane, oxychlordane and mirex were detected at low concentrations in all inshore flounder samples analyzed. DDT concentrations in Bay of Fundy seabird eggs were comparable to those reported for Gulf of St. Lawrence populations but higher than concentrations found in birds off the coast of Newfoundland. In 1984, toxaphene concentrations in storm petrel eggs from Newfoundland were higher than concentrations in the Bay of Fundy. No recent measurements are available to describe the current pattern. DDT concentrations in marine mammals, including harp seals and grey seals from offshore sites in this region, are lower than concentrations reported in European populations and from communities offshore of Nova Scotia.

DDT concentrations in fish and eggs of fish-eating birds are higher in Lakes Ontario and Erie than in other Great Lakes. Mirex has been detected at significant concentrations in biota from Lake Ontario near the sites where it was originally manufactured. It has also been found in trace amounts in fish and sediments of upstream lakes (Erie, Huron and Superior). Mirex is also found in eels and beluga whales from the St. Lawrence River and Estuary, where the source is upstream in Lake Ontario. Toxaphene is detected in fish from all the Great Lakes, although the highest concentrations occur in the Lake Superior ecosystem. Atmospheric loadings to a large surface area body of water such as Lake Superior, combined with an extremely slow turnover time, account for higher concentrations in this lake.

Concentrations of DDT in sediments of Arctic freshwater lakes decrease with increasing northerly latitude. Organochlorine pesticide concentrations in fish are usually found to be higher at sites near more industrialized or developed areas compared to more remote regions (e.g., southern compared to northern Ontario lakes). In general, there is a decline in the concentrations of organochlorine pesticides in the livers of burbot with increasing northerly latitude from northwestern Ontario to the Northwest Territories. However, there is no decline in organochlorine concentration in lake trout tissues with increasing latitude. Interpretation of these spatial trends must await further investigation.

In the Arctic, concentrations of DDT in the blubber of male ringed seals from Hudson Bay are significantly higher than in males from more northerly locations. In general, higher concentrations of organochlorine pesticides were found in Hudson Bay samples than in samples from Arviat and Pangnirtung to the west. However, concentrations of organochlorine pesticides are about one order of magnitude lower

than in seals from the east coast of Canada. DDT concentrations in male beluga whales are significantly higher in samples from Cumberland Sound than in samples collected further south in Hudson Bay. However, concentrations of DDT, mirex, toxaphene and chlordane in beluga whale blubber are still lower than concentrations found in similar samples from the St. Lawrence Estuary.

A 1982 to 1984 survey of the geographic distribution of organochlorine pesticides in polar bear fat and liver showed that concentrations tended to be highest in the southeast range of sample collection (Hudson Bay), and lowest in the northwest (Beaufort Sea). A 1989 to 1993 study showed that organochlorine pesticide concentrations are uniformly distributed in the western Arctic. There were some geographic gradients with dieldrin concentrations in polar bear blubber, with concentrations being higher in the east than the west. Highest concentrations of lindane were found in polar bear samples from the western regions of the Arctic. Toxaphene concentrations in beluga whales in the Arctic show no geographical trend. Again, interpretation of spatial trends must await further investigation.

Organochlorine pesticides such as DDT are present at lower concentrations in the Pacific region than in eastern Canada. Specifically, concentrations of DDT in fish from the Strait of Georgia are lower than those found in similar fish species on the east coast of Canada. There are few recent data on organochlorine pesticide concentrations from the B.C. coastal environment to permit current comparisons to concentrations in other Canadian ecosystems.

DDT was generally banned from use in the western hemisphere in the early 1970s. Since that time, concentrations of DDT and its major breakdown products have declined in most fish, marine mammals and birds. This trend in declining concentrations has been found not only in Canada, but also in the U.S. and in Scandinavia, where long-term records have been kept. Concentrations of DDT in Great Lakes and

Atlantic coast aquatic biota have plateaued since the mid 1980s and early 1990s. However, DDT concentrations in Arctic cod appear to be increasing. Although scientists suspect that international sources are contributing to LRTAP to the Arctic, uncertainties exist with respect to sources, transport and ecological cycling.

Little information is available on temporal trends for toxaphene, except for the northern Atlantic where concentrations have increased in seabirds, and in liver from burbot from the Red River in Manitoba, where concentrations have decreased. Not enough data are available to determine the causes of these trends.

PCBs, PCDDs and PCDFs

In general, PCBs in biota and sediments are higher on the east coast than on the west coast. In the Great Lakes region, concentrations of PCBs are highest near industrial sources. Highest sediment PCB concentrations are found in the St. Lawrence River in southern Quebec and Baie des Anglais, Quebec. Among the Great Lakes, Lake Superior is least contaminated with PCBs, while Lake Erie and Lake Ontario are the most contaminated. In smaller lakes, concentrations in fish liver of more-highly chlorinated PCBs tend to decrease with increasing northerly latitude. This may be due to increasing distance from the sources of PCBs. Concentrations of PCBs in Arctic belugas are about one-tenth the concentration found in St. Lawrence belugas. PCDDs and PCDFs were not used in the Arctic, yet they are present in all biota.

In Canada, PCBs in biota have generally declined since the 1970s. In the Great Lakes, PCB concentrations in fish and birds have decreased substantially from maximum values detected in the 1970s. This is similar to trends observed in Atlantic Canada. PCBs in biota from some of the Great Lakes have recently fluctuated but are only a fraction of the high concentrations found in the 1970s. Concentrations of PCBs in water from Lake Superior have declined steadily between 1978 and 1992. Contaminant concentrations in

Ontario recreational fish generally showed a decline over time when the contaminant source was eliminated or curtailed.

Concentrations of PCDDs and PCDFs are highest near sources. For example, of the pulp and paper mills surveyed in the late 1980s, the highest concentrations of PCDDs and PCDFs in sediment were found near 10 of the mills; four in British Columbia, five in Ontario, and one in Quebec. None of the other mills surveyed in Quebec, the Prairies, Nova Scotia or New Brunswick had sediment concentrations as high as these mills. Higher concentrations of PCDDs and PCDFs have generally been found in biota and sediments on the west coast compared to the east coast. On a lakewide basis, concentrations in fish from Lake Ontario are higher than in fish from Lake Superior. Although information is limited, concentrations of PCDDs and PCDFs are low in marine shellfish, fish and mammals.

Concentrations of PCDDs and PCDFs in biota and sediments are declining near point sources. Declines in concentrations have also occurred in lake trout in Lakes Superior, Michigan and Huron, and for Lake Erie walleye collected from 1978 to 1992. However, there have been no significant declines in Lake Ontario lake trout over the same time period.

PAHs

PAHs are found in freshwater and marine ecosystems as a result of both natural process and anthropogenic activities. Their presence in high concentrations appears to be limited to areas close to identifiable point sources. High concentrations have been found in sediments in Halifax and Sydney Harbours, Nova Scotia; Baie des Anglais, Quebec; the inner basin of the Saguenay Fjord, Quebec; Hamilton Harbour and the Rainy River, Ontario; the Mackenzie River, the Northwest Territories and the Kitimat Fjord, British Columbia. In most of these areas, PAH concentrations decrease rapidly with distance from their source. However, low concentrations of PAHs have been found in areas further removed from identifiable point sources, such as in northern Canada, the Gulf of St. Lawrence, and offshore areas. Although few data are available, PAH concentrations in Arctic marine biota tend to be low. These low concentrations are probably the result of long range atmospheric and riverine transport, plus natural sources.

PAHs found in most fish throughout Canada are usually at low concentrations, sometimes close to the limit of detection, even in areas where high concentrations have been found in sediments. This is because PAHs are metabolized by fish and other vertebrates. However, PAH compounds, such as benzo[a]pyrene and its metabolites, have been detected in some aquatic biota, most notably invertebrates such as lobsters in industrial harbours and marine mammals such as belugas in the St. Lawrence Estuary.

In areas of Canada for which temporal trend information is available, concentrations of PAHs have decreased over the past few decades. For example, in the Saguenay and Kitimat Fjords, PAH concentrations in sediments peaked in the 1960s to 1980s and decreased afterwards when local aluminum smelters undertook process changes and improved emission controls. In the Arctic, PAH concentrations in ice samples decreased in the early 1970s but have remained relatively constant for the past 20 years. The lack of decrease in concentration may be due to LRTAP or natural sources such as petroleum seeps and natural combustion.

Biological Effects

The effects of chemical contaminants on aquatic biota can be manifested at various trophic levels within an aquatic ecosystem. These can range from molecular (biochemical), cellular (histological), organ (physiological), organism (death, behaviour change), population (reproductive or life-cycle) and community (composition) level effects. Biological effects that are observed both in the field and under controlled laboratory conditions range from an adaptive, transitory response, to an adverse,

irreversible change in an individual or a population. For the chemical contaminants discussed in this report, many of the observed changes are subtle and sublethal, and result from prolonged exposure which may ultimately lead to the demise of populations of aquatic organisms.

Some effects at the molecular and cellular, organism, population and community levels are briefly summarized and illustrative examples are given below. Short descriptions of the effects of these chemicals are given in the Glossary of Chemical Contaminants.

Effects may be observed initially at the molecular level, usually as a biochemical response. Specific biochemical indicators of exposure to individual chemicals or groups of chemicals have been used to determine geographical distribution and trends over time. For example, the activity of mixed function oxidases (MFO), particularly the hepatic aryl hydrocarbon hydroxylase (AHH) or ethoxyresorufin-*O*-deethylase (EROD) enzymes in fish, is an indicator of exposure to several chemical contaminants, including PCBs, PCDDs and PAHs. Surveys conducted in Ontario, British Columbia, Quebec and New Brunswick identified increased concentrations of EROD enzymes in the livers of fish collected downstream of pulp mills, particularly those that used molecular chlorine in the bleaching process. Elevated concentrations of another biochemical indicator, metallothionein, in kidney and liver tissues of fish is a direct response to exposure to metals in the environment. Surveys in freshwater ecosystems have demonstrated that elevated concentrations of hepatic metallothionein were in direct proportion to concentrations of metals in the water. The activity of δ-amino levulinic acid dehydratase (ALAD) enzyme system is inhibited upon exposure to lead. Within the Great Lakes ecosystem, surveys have shown that fish had depressed ALAD activity when exposed to both inorganic and organic forms of lead. Although the linkages between exposure to a chemical and biochemical changes are relatively well known, their relationship to eventual adverse effects in aquatic ecosystems is still subject to scientific debate. It is speculated, however, that continued exposure, coupled with persistent induction of some of these enzyme systems, may lead to adverse effects.

Depending on their position in the food web, fish accumulate chemical contaminants to different concentrations. Top predator fish species, which feed at the top of the food chain, accumulate the highest concentrations of available chemicals. The degree of accumulation varies with biological factors such as age, sex, diet and lipid content. Prey species, positioned in the middle of a typical food chain, usually have lower concentrations. The base of the food chain for most aquatic ecosystems includes the invertebrate community where the concentrations of most contaminants are the lowest. Excessive chemical accumulation at any trophic level within an ecosystem will initially result in adverse behaviourial or acutely toxic effects to individuals. The severity and extent of such an accumulation pattern has been observed at the population and community level for some fish and bird species in the Great Lakes and in the St. Lawrence Estuary.

Effects at the population level directly attributable to chemical contaminants can be illustrated by the thinning of the eggshells of birds exposed to DDT. The widespread use of DDT in the 1970s resulted in population declines in many bird species, including those in freshwater and marine ecosystems. Similarly, the increased incidence of shell thickening in oysters and the development of male characteristics in female marine snails exposed to TBT is another example of a population level effect of an identifiable chemical contaminant. This later condition, known as imposex, has been identified in populations of snails from harbours on both the west and east coasts of Canada.

Evidence of a direct causal link between exposure to a specific chemical contaminant and an identifiable effect on an organism, population or community is limited to a few cases. Instead, there are a number of biological responses that are occurring in aquatic ecosystem communities that cannot be attributed to any single compound or class of compounds. An example of one such response is the impairment of the reproductive potential of populations of some aquatic biota. Recent surveys of reproductive impairment in fish communities downstream of industrial sources, such as pulp and paper mills, have found decreased gonad size, reduced concentrations of sex steroids, decreased concentrations of serum testosterone in male fish, increased age to maturity, lower fecundity and an absence of secondary sex characteristics in males. It is not yet known whether these responses are transitory in nature or whether they will lead to adverse, long-term effects in exposed populations.

Aside from their biological and ecological effects in aquatic ecosystems, the presence of chemical contaminants in edible fish products at concentrations above those established for safe human consumption can have a major impact on the utilization of fisheries resources. While this report does not review scientific studies on chemical contaminants and associated human health effects, nor review the rationales for establishing concentrations for safe consumption, it does highlight incidents where elevated concentrations of chemical contaminants in fisheries products have resulted in closures or recommendations for restricted consumption. Examples include fishing closures, harvest restrictions and consumption advisories for freshwater and marine ecosystems. Most of the consumption advisories for recreational fish in the Great Lakes relate to the presence of high concentrations of PCBs in large top predator fish. In the late 1980s a number of crab and shrimp fisheries were closed on the west coast due to elevated concentrations of PCDDs and PCDFs.

Recent changes in pulp and paper manufacturing technology have allowed some formerly closed fisheries to be re-opened.

Domestic and foreign markets for Canadian fish products have been restricted in the past due to high concentrations of some chemical contaminants in edible portions of fish. For example, some European markets are currently closed to the import of commercial fish species due to contaminant concentrations which exceed foreign market guidelines (e.g., mirex in Great Lakes eels). Subsistence fisheries, especially in the Arctic, are also affected by elevated contaminant concentrations in aquatic biota. The diet of subsistence fishers usually consists of food items such as organ tissues and whole animals that have higher contaminant concentrations than seen in diets of other consumers.

Current and Emerging Issues

CHANGING SOURCES

With improvements in regulations and management practices, the input of chemical contaminants to Canadian waters has been significantly reduced over the past few decades. For example, changing technology in the pulp and paper industry has resulted in reductions of PCDDs and PCDFs in west coast fish species. The use of DDT was effectively banned throughout the western hemisphere in the early 1970s. As a result, concentrations of DDT and its major breakdown products have declined in fish, marine mammals and birds. This trend in declining concentrations has been found not only in Canada, but also in the U.S. and in Scandinavia, where long-term records have been kept. Similar trends exist for some industrial chemicals such as PCBs where concentrations have been declining since the mid-1970s in phytoplankton, shellfish, fish and marine mammals.

However, some chemicals that are banned or restricted in Canada continue to be manufactured and used on a large scale elsewhere in the world. Some of these chemicals are sufficiently volatile to be transported through the atmosphere. There is mounting evidence that concentrations are increasing in those aquatic ecosystems systems vulnerable to long range atmospheric transport (LRTAP) and deposition. Not only can LRTAP bring chemical contaminants into Canada, but it can also re-distribute them among aquatic ecosystems within Canada.

As industrial activities are better managed and controls on industrial point sources tighter, other sources, such as municipal effluents, will assume greater importance. Such sources still deliver significant amounts of organic waste material to aquatic ecosystems, including chemical contaminants. As populations increase and concentrate in certain areas of Canada, the input of known chemicals, as well as chemicals of untested ecological relevance, will increase. The increasing use and discharge of pharmaceuticals and endocrine-disrupting chemicals may have unforeseeable effects on aquatic biota.

As the inputs of chemical contaminants into Canadian waters are reduced and general environmental conditions improve, the remobilization of contaminants that have accumulated in freshwater and marine sediments is of growing significance. Remobilization of metals and organic contaminants can occur from sediments to the water column through natural processes, such as bioturbation and bioaccumulation by bottom-dwelling organisms, and by anthropogenic activity. Depending upon the physical-chemical properties of the chemical and the prevailing environmental conditions, a certain portion of chemicals will also enter the atmosphere. This process of remobilization from sediments and redistribution among water and air is exemplified in observations of the recent shift in PCB dynamics in Lake Superior where PCB output from the lake now exceeds input, essentially making the lake a source of PCBs (in Canada) rather than a sink.

NEW FISHERIES RESOURCES

As new markets are developed, fisheries resources will expand beyond traditional species and fish products. However, little or no information is currently available for new species to assess whether they are being exposed to accumulating chemical contaminants or whether there are any effects on these species or their predators.

New fisheries bring new harvesting methods and potential new disturbances to the natural environment, all of which may affect the mobilization, accumulation or remobilization of contaminants in aquatic ecosystems. Harvesting species that have never been harvested before may also alter community composition and interactions in the ecosystem, which may eventually affect movement of chemical contaminants in ecosystems in unknown ways.

RESOURCE-USE CONFLICTS

The amount of space with suitable environmental requirements for each potential new resource use is limited. Often, other users already occupy the site. Such competing demands on resources can have adverse effects on fish and fish habitat. For example, as industrial activities intensify in coastal areas, the potential for conflict between new and traditional uses of marine resources may increase. Activities carried out by one user group may release contaminants into the environment which are unacceptable to other users in the area. For example, the use of therapeutants and other control agents in aquaculture facilities could pose contaminant problems for other aquaculture species, or wild stocks which live in the same area. The development of aquaculture facilities may result in the introduction of feed into coastal bays and estuaries. This may alter local system ecology to favour retention of contaminants, and pose a problem for other resource users, or the aquaculture facilities themselves. High density shellfish aquaculture operations have the potential to alter the retention time and ultimate fate of particulate matter and their contaminant burden, resulting in unanticipated effects on aquatic life.

Physical disruption of the natural environment may also pose a threat to other activities within the same area. Activities associated with vessel traffic and harbours, including construction and maintenance of harbour and port facilities and associated dredged channels, often involve the movement or redistribution of sediments, thereby increasing the potential of re-introducing chemical contaminants into the aquatic ecosystem.

Chemical Contaminants in an Ecosystem Context

Although the concentrations of many organic and inorganic contaminants are decreasing in aquatic ecosystems, it does not necessarily follow that overall aquatic environmental quality is improving. As explained throughout this report, the presence of a contaminant in some component of an aquatic ecosystem does not necessarily imply that an adverse effect is occurring. Traditional monitoring of contaminants in aquatic biota has focused mainly on contaminant concentrations in edible tissue, such as muscle. While important for the protection of the health of consumers, these measures are less useful when attempting to determine the risk of chemicals to fish health or to consumers of non-traditional fishery products. Contaminants usually accumulate in liver, kidney, gill and lipids. It is these tissues and organs that should also be monitored to measure both the impacts on ecological health and the potential for human health effects.

Increasingly, studies are finding that prolonged exposure to low concentrations of chemicals affects fish health in subtle ways such as altering sex steroid production, delaying sexual maturity and depressing reproductive hormones. Such effects have been found in fish exposed to complex mixtures. The beluga whale community in the St. Lawrence River is exposed to a complex mixture of contaminants. Past studies have documented a range of contaminant related adverse health effects in this population of marine mammals.

The complexity of contaminant mixtures must be understood to assess their total impact on fish and fish habitat. Other factors affecting the action of contaminants include habitat quality and quantity, changes in the composition and complexity of the food web, and changes to the utilization patterns of the aquatic community affected by toxic chemicals. An understanding of contaminant loadings and the natural compensating ability of a system is needed to evaluate the total cumulative impact of chemical contaminants in aquatic ecosystems.

As knowledge is gained on the fate and effects of contaminants in aquatic ecosystems, our ability to more precisely assess their impact on ecosystems and fisheries resources will increase. The acquisition of knowledge applies to both existing chemicals and those new chemicals that will be produced or identified in the future.

References

DEPARTMENT of Fisheries and Oceans. 1995. Annual Report of the Department of Fisheries and Oceans for the year ending March 31, 1994. Min. Sup. Serv., Ottawa, Ont.

ENERGY, Mines and Resources Canada. 1989. Facts from Canadian maps, a geographical handbook. 2nd edition. *In* Government of Canada 1996. State Can. Environ. Min. Sup. Serv., Ottawa, Ont.

MAGUIRE, R.J., R.J. Tkacz, Y.K. Chau, G.A. Bengert, and P.T.S. Wong. 1986. Occurrence of organotin compounds in water and sediment in water and sediment in Canada. Chemosphere 15: 253-274.

PEARSE, P.H., F. Bertrand, and J.W. MacLaren. 1985. Currents of Change: Final Report Inquiry on Federal Water Policy. Environ. Can., Ottawa, Ont.

STATISTICS Canada. 1994. Canada Year Book 1994. Stat. Can., Ottawa, Ont.

WORLD Coast Conference. 1993. Canada country description for World Coast Conference 1993. International Conference on Coastal Zone Management, The Hague, The Netherlands, 1-5 November, 1993. 23 p.

GLOSSARY *of* terms

Glossary of Terms

Abundance A measure of the total number of individuals of a species within a defined group (population) or area (stock).

Acidic deposition (also known as acid rain or acid precipitation): Refers to deposition of a variety of acidic contaminants (acids or acid-forming substances) on biota or land or in waters of the Earth's surface. Deposition can be in either wet forms (e.g. rain, fog, snow, or dry forms (e.g., gas, dust particles).

Acute toxicity State of being toxic enough to cause severe biological harm or death within a short time, usually 96 hours or less. See also chronic toxicity.

Adsorption The attachment of one substance onto the surface of another.

Adduct A chemical compound that is formed from the chemical addition of two species.

Advection The transport of a property (e.g., heat) by fluid motions such as currents or winds.

Advection The process of transport of water solely by the mass motion of oceans or lakes, most typically by horizontal currents.

Age-class All the fish in a stock that are a particular age.

Alkalinity Alkalinity represents the acid-neutralizing capacity of an aqueous system.

Amphipod A member of the order Crustacea (*crustaceans*).

Anadromous Refers to species, such as salmon, that migrate from salt water to fresh water to breed (Upper Great Lakes Connecting Channels Study, Management Committee 1988).

Anoxia The absence of oxygen necessary to sustaining most life. In aquatic ecosystems, this refers to the absence of dissolved oxygen in water.

Anoxic bottom waters and sediments Waters and sediments whose oxygen content has been depleted by high levels of decomposition and biological activity.

Anthropogenic To do with humans — to be caused by humans.

Antifouling agents Various chemical substances added to paints and coatings to combat mildew and crustaceous formations, such as barnacles, on the hulls of ships.

Bathymetry The science of measuring ocean or lake depths in order to determine the sea floor topography.

Benthic Of or living on or in the bottom of a water body. See also benthos.

Benthic fish Fish such as flounder, cod and stingrays that spend much of their life at or near the sea bottom.

Benthos Plants and animals and bacteria that live on or in bottom sediments. Benthos can include benthic fish as well as true bottom dwellers such as shellfish and sea worms.

Bioaccumulation A general term describing a process by which chemical substances are ingested and retained by organisms, either from the environment directly or through consumption of food containing the chemicals. See also biomagnification.

Biochemical oxygen demand (BOD) The amount of dissolved oxygen required for the bacterial decomposition of organic waste in water.

Biodegradable Capable of being broken down by living organisms into inorganic compounds.

Biological diversity (biodiversity) The variety of different species, the genetic variability of each species, and the variety of different ecosystems that they form.

Biological production The production of organic matter by the organisms in a specific area.

Biomagnification A cumulative increase in the concentration of a persistent substance in successively higher levels of the food chain. See also bioaccumulation.

Biomarker measurement of biological changes in an organism in response to environmental stress.

Biomass The total mass of a species or group of species within a defined area. In fisheries, the area is often defined by the region inhabited by a particular fish stock, and the biomass measured in metric tonnes per stock.

Biota Collectively, the plants, bacteria, and animals of a certain area or region.

Bioturbation The movement and mixing of sediments by animals such as worms that burrow in and ingest sedimentary material.

Bloom (also known as algal bloom) A relatively high concentration of phytoplankton that is readily visible in a body of water as a result of proliferation during favourable growing conditions generated by nutrient or sunlight availability.

BOD See biochemical oxygen demand.

Breeding The act or ritual of sexual interaction leading to reproduction between adult animals. Fish, marine mammals and seabirds frequently breed in large groups in specific breeding or nesting areas.

Carcinogen A substance which induces cancer in a living organism.

Carnivory The eating of flesh or subsistence on nutrients obtained from animal protoplasm.

Carnivores Biological organisms that eat flesh.

Catadromous Fish such as American eels that live in fresh water and migrate to the sea to spawn.

Chlor-alkali plants Industrial plants that produce chlorine (for use in bleaching, chemical manufacturing [e.g., production of chlorinated organic compounds], water purification) and caustic soda (for use in making soap and paper), by electrolysis of brine. Worldwide, two processes are in common use: the diaphragm cell process and the mercury cell process.

Chronic toxicity Toxicity marked by a long duration that produces an adverse effect on organisms. The end result of chronic toxicity can be death, although the usual effects are sublethal (e.g., inhibition of reproduction or growth). These effects are reflected by changes in the productivity and population structure of the community. See also acute toxicity.

Congenital anomalies Birth defects.

Congener A chemical substance that is related to another substance, such as a derivative of a compound or an element belonging to the same family as another element in the periodic table.

Contaminant A substance foreign to a natural system or present at unnaturally high concentrations.

Crustal Dealing with the outermost solid layer of the earth.

Cytotoxic Having the ability to cause harmful effects to cells.

Demersal Living near or at the bottom of the sea.

Detection limit The smallest concentration or amount of a substance which can be reported as present with a specified degree of certainty by an analytical procedure.

Detritus Unconsolidated sediments composed of both inorganic and dead and decaying organic material.

Diadromous Fish such as Atlantic salmon that live in the sea and migrate to fresh water to spawn.

Diffuse source Source of contaminants that are widespread and do not come from a specific location.

Dioxin equivalent A relative measure of toxicity expressed as enzyme induction in rat liver tissue culture. Dioxin-like chemicals (e.g., PCBs, furans) are compared to the most toxic dioxin congener (2,3,7,8 – TCDD).

Dissolved organic carbon (DOC) Organic carbon that is dissolved in the water column.

Drainage basin The land mass drained by fresh-water, including wetlands, streams, rivers and lakes.

Echolocation The process of locating distant objects (e.g. prey) by means of sound waves reflected back to the emitter (e.g. predator).

Ecosystem An integrated and stable association of living and non-living resources functioning within a defined physical location. The term may be applied to a unit as large as the entire ecosphere. More often it is applied to a smaller division.

Eddies Water flows running counter to the direction of the main current. Eddies play an important role in the transport of dissolved substances and energy.

Effluent A liquid waste material that is a by-product of human activity (e.g., liquid industrial discharge or sewage), which may be discharged into the environment.

Embryotoxicity Adverse toxic effects of chemical agents on the developing embryo.

Endocrine Glands in animals that produce hormones.

Entrainment The process of drawing into and transporting by the flow of a fluid. For example, air bubbles are entrained into rapidly flowing and turbulent water.

Epidemiology The study of the prevalence and spread of disease in a community.

Epifauna Bottom dwelling animals (benthos), such as scallops, crabs, lobster and shrimp. See infauna.

Estuary A water passage where the tide meets a river current. An arm of the sea at the lower end of a river.

Eutrophication The process of over-fertilization of a body of water by nutrients that produce more organic matter than the self-purification processes of the water can overcome. Eutrophication is a natural process. It can be accelerated by an increase of nutrient loading to a water body by human activity.

Fauna The animal population of a specific environment.

Flora The plant population of a specific environment.

Fluvial Of or relating to, or living in a stream.

Food Chain A specific nutrient and energy pathway in ecosystems proceeding from producer to consumer. Along the pathway, organisms in higher trophic levels gain energy and nutrients by consuming organisms at lower trophic levels.

Food web The interrelationships between the different food chains found in a particular locality. Whereas food chains are simple linear relationships, food webs are more complex and encompass both predator/prey and predator/predator relationships.

Glycoprotein Any group of conjugated proteins containing both carbohydrate and protein units — also known as glycopeptide.

Grain size distribution The percentage (by weight or count) of mineral particles of different sizes in particular sediments. These particles and their sizes include clay (less than 2 μm), silt (2 to 64 μm), sand (64 μm to 2 mm), granules (2 to 4 mm), gravel (4 mm to 6 cm), cobbles (6 to 25 cm) and boulders (greater than 25 cm).

Groundfish (also known as bottom fish): Those species of fish that normally occur on or close to the seabed, such as cod and haddock.

Gyre A circular, horizontal movement of water formed by wind activity, upwelling or currents, especially when the latter pass obstructions or when two adjacent currents run counter to each other.

Habitat The environment in which a population or individual occurs. The concept of habitat includes not only the place where a species is found, but also the particular characteristics of that place, such as climate or the availability of suitable food and shelter, which make it especially well-suited to meet the life-cycle needs of that species.

Habitat degradation Physical or chemical harm done to natural areas that are important to an animal population for feeding, breeding or nesting. The source of harm can be local, such as siltation that occurs during construction of shoreline facilities, or more distant, such as the transport of toxic materials to the sea from terrestrial run-off.

Hardness A measure of dissolved salts in water related to concentrations of calcium and magnesium in water.

Heavy metal Metallic elements with relatively high atomic weights. These include lead, cadmium and mercury.

Hepatic To do with the liver.

Hydrocarbons Organic compounds containing only hydrogen and carbon. Crude oil consists, for the most part, of a complex mixture of hydrocarbons.

Histopathology Pathology that deals with tissue changes associated with disease.

Impoundment A body of water which has been backed up by a dam to create a reservoir for purposes such as hydroelectric power.

Indicator species An organism that may be used to indicate the presence or absence of a substance, such as heavy metals.

Infauna Bottom dwelling animals (benthos) that spend most of their lives buried in bottom sediments, such as sea worms, clams and oysters; also known as endofauna.

Inorganic Matter other than plant or animal, and not containing a combination of carbon / hydrogen / oxygen as in living things.

Insecticides Chemicals used to kill insect pests.

Interrenal Located between the kidneys.

Kraft mill A pulp and paper mill which utilizes chemicals in an alkaline medium to release cellulose fibres from logs by dissolving the lignin in a caustic solution of sodium hydroxide and sodium sulfide.

Leachate A solution formed by leaching.

Leaching Washing out of soluble substances by water passing down through soil, rock or ore. Leaching occurs when more water falls on the soil than is lost by evaporation from the surface. Rainwater running through the soil dissolves mineral nutrients and other substances and carries them via groundwater into water bodies.

Lesion An abnormal change in structure of an organ or part due to injury or disease.

Limit of detection The smallest concentration of a particular substance that can be determined by chemical analysis.

Lipid tissue Animal tissue made up primarily of organic substances such as fats, oils and waxes; commonly known as "fatty tissue".

Lipophilic compound A compound that has an affinity for fat or lipid tissues.

Loadings Total mass of contaminants delivered to a water body or to the land surface over a specified time (e.g., tonnes per year of phosphorus).

Long range transport of atmospheric pollutants (LRTAP) The movement of chemical substances to a remote location often over a number of days or weeks via the atmosphere.

Macrophyte A macroscopic plant usually found in an aquatic habitat.

Metabolic rate A measurement of an organism's chemical processes that create energy through decomposition of complex substances and the storage of complex substances and energy.

Nesting The construction and occupation by birds of temporary shelter in which to lay eggs and rear young. Among seabirds, nesting normally occurs annually, frequently within large colonies.

Neoplasm An aberrant new growth of abnormal cells or tissues; a tumor.

Non-point source Source in which pollutants are discharged over a widespread area or from a number of small inputs rather than from distinct, identifiable sources. See also diffuse source.

Nutrient A chemical that an organism must take in from its environment because it cannot produce it or cannot produce it as fast as it needs it.

Organic Describes compounds based on carbon, and also containing hydrogen, with or without oxygen, nitrogen, or other elements.

Organochlorine Chlorine-containing organic compound, in some cases also containing oxygen and other elements such as phosphorus.

Papilloma A growth of proliferating epithelial cells.

Particle-active compound Compound that is readily attracted to and binds with solid particles, either in the air or in water.

Particulates Fine solid particles which remain individually dispersed in gases (and stack emissions).

PCB congener Any of over 209 possible configurations of polychlorinated biphenyls (PCBs) differing only in the number or position of the chlorine atoms in their molecular structure.

Pelagic Pertaining to organisms that swim or drift in a sea or lake in mid-water or close to the surface as distinct from those that live on the bottom (see benthic). Includes plankton, many fish species, and oceanic birds. Pelagic fish such as herring are free swimming species that feed at mid water or surface depths. Migratory pelagic fish such as tuna and swordfish are species that travel regularly between feeding, breeding and spawning areas.

Persistent toxic substance Any toxic substance that is difficult to destroy or that degrades slowly; generally those with a half-life in water greater than eight weeks.

pH A numerical expression of the concentration of hydrogen ions in solution. pH 0-<7 is acidic, pH 7 is neutral, and pH>7-14 is basic or alkaline.

Photosynthesis The process by which chlorophyll containing organisms such as algae use the energy of sunlight to produce new organic material (carbohydrates) from inorganic carbon dioxide and water.

Phytoplankton Minute, microscopic aquatic vegetative life; plant portion of plankton; the plant community in marine and freshwater situations that floats free in the water and contains many species of algae and diatoms. See also zooplankton.

Piscivorous Feeding on fish.

Plankton Collective term for organisms that drift around in water because they are not capable of swimming against currents in the water. See also phytoplankton and zooplankton.

Point source A source of pollution that is distinct and identifiable.

Polychaete Describing an invertebrate in the phylum Annelida with paired lateral appendages on most segments.

Polynyas Recurring area of open water or reduced ice cover surrounded by frozen seas.

Population A closely associated group of individuals of the same species that may occupy a common area, such as cod on the Grand Banks.

Population biomass The total mass of a single species that occupies a common area or space.

Predator An animal that hunts and consumes other animals for food.

Primary production The amount of new organic matter produced from inorganic material by organisms using photosynthesis.

Primary wastewater treatment First step in sewage treatment to remove large solid objects by screens (filters) and sediment and organic matter in settling chambers. See also secondary wastewater treatment and tertiary wastewater treatment.

Productivity The rate at which a population, species or ecosystem produces organic material over a specific time period.

Recruitment The entry of maturing juvenile fish into the adult (sexually mature) portion of the population.

Remobilization The transfer of dissolved contaminants from sediments to the overlying water through chemical processes. Remobilization may be enhanced by processes such as physical mixing or bioturbation.

Reproductive success The proportion of reproductive attempts which produce healthy, independent young.

Resuspension The remixing of sediment particles back into the water by storms, currents, and biological and human activities.

Salinity The quantity of dissolved salts in water.

Secondary wastewater treatment After primary wastewater treatment, removal of biodegradable organic matter from sewage using bacteria and other microorganisms, inactivated sludge, or trickle filters. See also tertiary wastewater treatment.

Sediment Solid material that may contain inorganic and organic matter. Sediment may be suspended in water or deposited at the bottom of a water body.

Slimicide A chemical used to kill or arrest the growth of moulds.

Spawning biomass The sexually mature portion of a fish population that is ready to reproduce. In some species (e.g., haddock), spawning adults congregate into a distinct group referred to as a spawning aggregate.

Species A group of individuals that share certain identical physical characteristics and are capable of producing fertile offspring.

Stable nitrogen isotope ratios Used as a measurement of trophic relationships between different biota in freshwater and marine systems. All matter contains nitrogen. When an organism eats something the heavier isotope of nitrogen ^{15}N, increases from the prey to the predator due to the preferential excretion of the lighter isotope (^{14}N) during metabolic processes. Relative trophic positions of different biota can therefore be determined.

Stratification The separation of a body of water into vertical layers (strata) that have differing densities due to variations in temperature, salinity and (to a lesser degree) pressure. Stratified marine systems of this type are generally characterized by warmer, less dense waters on top and colder, denser waters on the bottom.

Sublethal Involving a stimulus below the level that causes death.

Sustainable development Development that ensures that the current use of resources and the environment does not damage prospects for their use by future generations.

Sustainable use Use of an organism or ecosystem at a rate within its capacity for renewal or regeneration.

Teratogen An agent that increases the incidence of congenital malformations (birth defects).

Tertiary wastewater treatment Removal of nitrates, phosphates, chlorinated compounds, salts, acids, metals, and toxic organic compounds after secondary wastewater treatment. See also primary wastewater treatment.

Toxic substance A substance that can cause death, disease, birth defects, behavioural abnormalities, genetic mutations or physiological or reproductive impairment in any organism or its offspring or that can become poisonous after concentrating in the food web or in combination with other substances.

Toxicity The inherent potential or capacity of a material to cause adverse effects in a living organism.

Toxicology The study of the adverse effects of chemical agents on biological systems.

Trophic Relating to processes of energy and nutrient transfer from one or more organisms to others in an ecosystem.

Trophic level Functional classification of organisms in a community according to feeding relationships; the first trophic level includes green plants, the second level includes herbivores, and so on.

Trophic status A measure of the biological productivity in a body of water. Aquatic ecosystems are characterized as oligotrophic (low productivity), mesotrophic (medium productivity), or eutrophic (high productivity).

Turbidity A decrease in water clarity, thus preventing penetration of light. Turbidity results from suspended matter such as clay, silt, small organic and inorganic debris, plankton and microscopic organisms.

Upwelling The rising to the sea surface of cold, dense subsurface water. Upwelling can occur where two currents diverge or where surface water is displaced by physical forcing. Upwelled water, in addition to being cooler, is also often rich in nutrients, so that regions of upwelling are generally also regions of rich fisheries. Tidal upwelling occurs when tide induced currents flow away from the coastline, displacing surface waters.

Vertical mixing The transport and distribution of properties such as temperature, salinity and chemical composition between surface and subsurface waters by forces such as waves, tides and currents. Vertical mixing is generally most intense in the surface layer of water and can be inhibited in deeper layers.

Volatile anthropogenic chemicals Chemical compounds made by humans that evaporate readily at ordinary temperatures or on exposure to air.

Volatilized The conversion of a chemical substance from a liquid to a gas.

Water column The volume of water between the surface and bottom of a water body.

Wet deposition A process of precipitation whereby acidic chemicals, including dilute sulfuric and nitric acids and sulphates, are deposited in the form of rain, snow, fog, etc. See also acidic deposition.

Zooplankton Microscopic animals that move passively in aquatic ecosystems. See also phytoplankton and plankton.

Information sources:

McGraw-Hill Dictionary of Scientific and Technical Terms, Third Edition. 1984.

The State of Canada's Environment 1996. Minister of Public Works and Government Services, Ottawa 1996.

GLOSSARY

of

chemical contaminants

Table of Contents

Glossary of Chemical Contaminants

Glossary of Chemical Contaminants

ORGANIC CONTAMINANTS

In this report, chemical contaminants are grouped into organic contaminants and heavy metals and organometallic contaminants. Organic contaminants include chlorine-containing compounds, such as the familiar pesticide DDT, and the group of industrial chemicals known as PCBs. These compounds are termed organochlorine contaminants in the text. Non-chlorine containing organic contaminants, such as polycyclic aromatic hydrocarbons or PAHs, are also included under organic contaminants.

Strictly speaking, heavy metals such as lead, mercury and cadmium, are not considered contaminants unless their concentrations exceed some natural level. For this reason, the term heavy metal is used instead of inorganic contaminant or metal contaminant.

Where possible, simple names for all chemicals are used. On occasion, it is necessary to use specific chemical names in order to adequately explain the complicated processes that affect the transport of these chemicals in the environment and their effects on aquatic life. Examples of this include dioxins and furans, PCBs and some pesticides, as are explained later in this chapter.

ORGANOCHLORINE COMPOUNDS

Organochlorine compounds describe a range of synthetic organic chlorine containing compounds that share, to varying degrees, properties which govern their environmental behaviour. They are insoluble in water (hydrophobic), but are soluble in fat (lipophilic). When they enter water they become adsorbed to particulate material, and accumulate in sediments. When they enter organisms, they accumulate in fatty tissues including blubber, organs, blood, and stored fats. Their chemical structure (derived from hydrocarbons) makes them stable and resistant to chemical or biological degradation. As a result they persist in the environment, often for decades. They are volatile enough that they can be transported through the atmosphere. As a result, they are found in regions where they have never been deliberately used, such as in the Canadian arctic.

CHLORDANE

Technical chlordane is a mixture of various chlorinated hydrocarbons including heptachlor, nonachlors, and chlordane (Brooks 1974; Cochrane and Greenhalgh 1976; Noble 1990). Until December, 1985 it was used in Canada to control a variety of insects and insect larvae on ornamental plants, vegetables, bulbs,

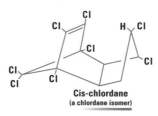

Cis-chlordane
(a chlordane isomer)

strawberries, lawns and turfs (CCREM 1991). It was also used to control wood-boring insects in houses and other buildings. In 1985, all uses of chlordane were suspended with the exception of its use in termite control. All uses of chlordane were later terminated at the end of 1995.

Although chlordane is volatile, it readily adsorbs to particulate matter and is found in aquatic sediments. It is also persistent and can bioaccumulate in aquatic organisms. Chlordane degrades to oxychlordane which is more persistent and

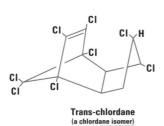

Trans-chlordane
(a chlordane isomer)

more toxic than most other organochlorine pesticides (Stickle *et al.* 1979; Noble 1990). Chlordane, and especially two of its breakdown products, oxychlordane and heptachlor epoxide, have been shown to be lethal to birds in laboratory tests and have been implicated in the deaths of a number of birds in the wild (Hoffman *et al.* 1995).

DDT

DDT (Dichlorodiphenyltrichloroethane) was introduced as an insecticide in the 1940s (Carson 1962; Noble 1990). The parent compound, *p,p'*-DDT, is usually found in the environment with two metabolites, *p,p'*-dichlorochlorophenylethane (DDD) (*p,p'*-DDD) and *p,p'*-dichlorodiphenyldichloroethylene (*p,p'*-DDE), and with the corresponding *o,p'*-isomers (present in technical-grade DDT); these are often grouped together as "total-DDT". Because DDE is more stable in the environment than DDT, the age of

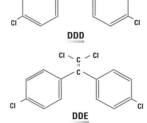

DDT residues can be estimated roughly by the ratio of DDE to DDT. The use of DDT in Canada (and most western countries) was restricted in the early 1970s, but DDT is still used widely as an insecticide in other countries, and continues to enter the Canadian environment

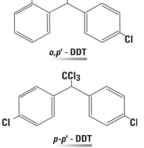

through long-range atmospheric transport (Environment Canada *et al.* 1991). All uses of DDT in Canada were terminated at the end of 1990.

DDE is bioaccumulated in biological tissues and is biomagnified as it moves through aquatic food chains (Noble 1990). DDE has been identified as the cause of eggshell thinning in many fish-eating birds and has since been found to cause sterility, alter hormonal balances in bird and wildlife embryos and to change the activity of liver enzymes in many organisms (IJC 1991, Environment Canada/US EPA 1995).

DIBENZO-DIOXINS

Polychlorodibenzo-dioxins (commonly referred to as dioxins or PCDDs) are formed as a by-product of the manufacture of chlorinated phenols (themselves used as disinfectants, wood preservatives) or during incineration or other high temperature processes involving chlorine and phenolic compounds. Seventy-five compounds (congeners) exist, and those which attract most attention have a 2,3,7,8 chlorine substitution; 2,3,7,8-tetrachlorodibenzo-dioxin (2,3,7,8-TCDD) is one of the most toxic substances known. The combined toxicity of PCDDs and some related compounds is often reported as "2,3,7,8-TCDD toxic equivalents" (TEQs) by summing the individual toxicities of PCDD congeners calculated from experimental studies (See Sidebar on Toxic Equivalents in the Inland Freshwater Ecosystems Chapter).

In Canada, pulp mills used to be a major source of PCDDs. Following bans on the use of pentachlorophenol (a preservative for wood chips), and after changes to the pulping process chemistry used by pulp mills, they have become a less important source of PCDDs. Municipal incinerators plus some old landfill sites near chemical manufacturing facilities now seem to be the major sources of these compounds. Some chlorophenols, or compounds derived from them such as the chlorophenoxyacetic acid herbicides 2,4-D and 2,4,5-T contain PCDDs and polychlorodibenzo-furans (PCDFs) as impurities.

PCDDs and PCDFs (discussed below) have similar toxic effects on the organisms that consume them. While their level of toxicity may greatly vary according to their chemical make-up, all forms demonstrate a degree of toxicity in aquatic organisms. Dioxins, especially 2,3,7,8-TCDD are highly toxic to many animals in low doses and are believed to be responsible for the fatal edema disease in Lake Ontario Herring Gulls in the 1970s (IJC 1991) Other impacts on aquatic organisms may include MFO induction, reproductive impairment, weight loss, suppression of the immune system, and hormonal alterations. While effects on humans are not well understood, PCDDs are nonetheless considered to be very hazardous chemicals (IJC 1991).

Dibenzo-*p*-dioxins
(Basic structure)

DIBENZO-FURANS

Like PCDDs, polychlorodibenzo-furans (commonly referred to as furans or PCDFs), are usually formed as a by-product of incineration or other high temperature processes involving chlorine and phenolic compounds. Their physical properties, and hence environmental behaviour, are similar to those of the PCDDs, and the two groups of compounds are usually analyzed together. One hundred and thirty five PCDF congeners exist. At present, the main source of PCDFs to the environment seems to be from municipal incinerators.

Dibenzo-furans
(Basic structure)

Like PCDDs, PCDFs can also act as endocrine disrupters and thus alter the hormonal balances and reproductive structures of wildlife (Environment Canada/US EPA 1995). Although less toxic than dioxins they are nevertheless considered to be very hazardous to living organisms (IJC 1991).

DIELDRIN

Dieldrin is one of a group of synthetic organochlorine insecticides which was widely used in the 1960s and 1970s to control crop pests found in soil and on lawns and golf courses. Most uses of dieldrin except for termite control, however, were phased out in the 1970s. All uses of dieldrin in Canada were terminated at the end of 1995.

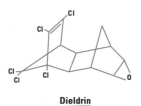

Dieldrin

In addition to being an insecticide itself, dieldrin is a metabolic product of the related compound, aldrin which was also applied to soil and plants to control insects (CCREM 1991). Dieldrin is persistent and bioaccumulates in aquatic organisms thus explaining its continued occurrence in many environmental samples. Both aldrin and dieldrin are toxic and have been implicated in numerous incidents of wildlife mortality (Hoffman *et al.* 1995). Dieldrin, in particular, has been linked to the death of adult bald eagles in the Great Lakes Basin (IJC 1991).

HEPTACHLOR

Heptachlor was widely used in Canada on agricultural crops and for protecting seeds and bulbs against insects and insect larvae. Like dieldrin it was often applied to soil (Noble 1990).

Heptachlor

The technical mixture of heptachlor contains mainly the parent compound but also several related compounds. Like other organochlorine compounds, it can move about the environment and be taken up by aquatic organisms. Once inside an organism it is rapidly metabolized. However, the metabolic product, heptachlor epoxide, is persistent and is as toxic as the original compound. Heptachlor has been demonstrated to lower reproductive success and to cause mortality in a variety of bird species, including Canada geese, as well as induce mortality in predators, such as foxes, that feed on these birds (Hoffman *et al*. 1995). With the exception of its use on narcissus bulbs, all uses of heptachlor were suspended at the end of 1976. All uses of heptachlor in Canada were terminated at the end of 1990.

HEXACHLOROBENZENE

Hexachlorobenzene (HCB) is a by-product of several industrial chlorination processes, often associated with chlor-alkali plants (CCREM, 1991). It is among the most volatile of organochlorine compounds (hence its wide distribution) but is also very lipid soluble and hence is readily bioaccumulated. Between 1948 and 1972 HCB was used as a seed dressing for several crops to prevent fungal diseases. Although

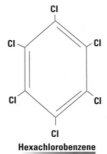

Hexachlorobenzene

HCB has not been used commercially in Canada since 1972 it is still released in small quantities into the environment as a by-product of the manufacture and use of chlorinated solvents and pesticides. All uses of HCB as a pesticide in Canada were terminated at the end of 1981.

Hexachlorobenzene can interfere with enzymes that control the production of hemoglobin, a component of blood, and can suppress immune system components. HCB can also affect the nervous system, liver, and reproductive system of various aquatic organisms and has been shown to produce cancer in laboratory animals (IJC 1991, Environment Canada/US EPA 1995).

HEXACHLOROCYCLOHEXANE

Hexachlorocyclohexane (HCH) describes a group of compounds with the same molecular formula ($C_6H_6Cl_6$), but which differ in the positions of the chlorine atoms in relation to the cyclohexane ring; these are "isomers" and the most common are known as the alpha (α), beta (β), and gamma (γ) forms.

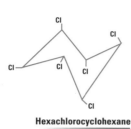

Hexachlorocyclohexane

Technical HCH (1,2,3,4.5,6-hexachlorocyclohexane) consists of 50 to 70% α-HCH, 5 to 14% β-HCH, 10 to 18% γ-HCH, δ-HCH and impurities. α-HCH seems to be formed from γ-HCH in the atmosphere by the action of sunlight. Lindane, a mixture of isomers which initially contains greater than 90% γ-HCH, is the only form of HCH used in North America, where it is used as a seed treatment, to control flea beetles, wereworms, lice and ticks on livestock. It was used as a household insecticide until 1970 (CCREM, 1991). Other HCH isomers have been banned for use in North America and most other circumpolar countries since the late 1970s. All the HCHs are volatile. This property leads to their wide distribution by atmospheric transport. HCHs are appreciably water soluble and are found in all the world's oceans. Technical HCH is still used extensively throughout other parts of the world. Although much less hazardous than some of the other organochlorine compound pesticides which it replaced, HCH nonetheless has been implicated in reduced egg production and hatchability, eggshell thinning, and increased mortality in some bird species (Hoffman *et al.* 1995), in liver and thyroid tumors in some mammals.

MIREX

Although once used in the U.S. to control fire ants, mirex has never been registered as a pesticide in Canada (Noble 1990). Other non-pesticide uses, under the name of Dechlorane, used in both Canada and the United States have included, its use as a flame retardant in plastics, as a smoke generator in

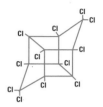

Mirex

pyrotechnics, in anti-fouling paints, rodenticides and antioxidants and various industrial products (Verschueren 1983; CCREM 1991). In 1978, the import, manufacture and processing of mirex was banned in Canada for any uses that would lead to its dispersion into the environment (Environment Canada 1985; CCREM 1991). The main routes by which mirex enters the Canadian environment are two facilities in the U.S. which discharge their effluent into rivers draining into Lake Ontario and from toxic chemical dumpsites within the Lake Ontario basin (CCREM 1991; Sergeant *et al.* 1993). By exposure to sunlight mirex is converted slowly to photomirex, which is also toxic (CCREM 1991).

Mirex is a very persistent chemical (Metcalf *et al.* 1973; Kaiser 1978; Neely 1980). It is insoluble in water and is not volatile (Smith *et al.* 1978; CCREM 1991). It readily adsorbs to organic materials in the aquatic environment and can bioaccumulate in aquatic organisms (CCREM 1991). Mirex has potential for chronic toxicity because it is only partly metabolized and is eliminated slowly (Hoffman *et al.* 1995). Mirex has been demonstrated to cause reproductive problems in birds and cancer in laboratory animals (IJC 1991; Environment Canada/US EPA 1995).

POLYCHLORINATED BIPHENYLS

Polychlorinated biphenyls (PCBs) were first synthesized in the late 1920s, and were used until the early 1970s throughout the world. They were used where chemically stable compounds available in a range of physical properties were needed such as in heat transfer mixtures and hydraulic fluids (Government of Canada 1991). They were also used as electrical insulators (CCREM 1991). Two hundred and nine individual PCBs (congeners) exist, each with a different pattern of chlorine substitution, though usually only about fifty are found in environmental samples. Modern analytical methods allow the measurement of each congener (numbered as CB 118, 153, 138) but in the older literature PCBs were usually reported as commercial mixtures (Aroclor 1254, Clophen A50). Therefore, it is sometimes difficult to compare modern PCB measurements with those in the older literature.

Manufacture, importation and most non-electrical uses of PCBs were banned in Canada in 1977 (Government of Canada 1991). Nevertheless, some PCBs still remain in use or in storage in Canada. Some PCBs have escaped into the environment over the years, primarily as a result of inadequate disposal, especially in landfill sites as well as the result of accidental spills or when PCB-impregnated products are incinerated (CCREM 1991, Government of Canada 1991).

The characteristics of PCBs that make them desirable for industrial applications also make them a hazard to the aquatic environment. Because they are chemically stable and not readily metabolized by aquatic organisms they are very persistent and tend to accumulate in the fatty tissues of organisms. Once absorbed into aquatic organisms, PCBs are often carcinogenic and may alter physiological functions including growth, molting and reproduction as well as the ability of organisms to eliminate foreign organic compounds and waste products. PCBs, in particular, have been linked to a number of health problems in wildlife including embryo mortality, hormonal alteration and physical deformities and are suspected of causing developmental problems in human infants (IJC 1991, Environment Canada/US EPA 1995).

Cl$_x$ — Cl$_x$

PCB
(Basic structure)

TOXAPHENE

Toxaphene (polychlorinated camphenes) is actually a complex mixture of many different compounds (CCREM 1991). There may be as many as 32,768 possible isomers of toxaphene. Currently, typical analytical methods can only detect about 150 to 200 of these individual isomers. Toxaphene is a broad spectrum pesticide that was used in Canada until 1980 to control insect pests and in experimental fish eradication programs (Eisler and Jacknow 1985; Noble 1990). Because of its volatile nature, it can be widely dispersed in the atmosphere and like the other chlorinated hydrocarbons, it is readily bioaccumulated.

Toxaphene

Toxaphene is acutely toxic to fish, but less toxic to mammalian species. It has, however, been identified as an animal carcinogen (US EPA 1980; Environment Canada/US EPA 1995) and is suspected of being responsible for a number large bird die-offs in areas where it has been used (Hoffman *et al.* 1995).

NON-CHLORINE CONTAINING ORGANIC COMPOUNDS

POLYCYCLIC AROMATIC HYDROCARBONS

Polycyclic aromatic hydrocarbons (PAHs) are a group of aromatic compounds that arise from natural and anthropogenic sources. Some are components of oil, and may be released from natural oil seeps or during spills. Others are by-products of incomplete combustion, including municipal and industrial incineration, car exhaust, forest and grass fires and smelter and refinery operations. Some may be directly introduced into the environment such as from creosote used to preserve marine pilings. Some PAHs are natural products of plants, bacteria and fungi (Hoffman *et al.* 1995).

PAHs may be divided into low-molecular weight PAHs and high molecular-weight PAHs depending upon their physical make-up (CCREM 1991). Vapour pressures and water solubilities of PAHs likewise vary in accordance with their molecular weights and physical structures.

PAHs are usually found concentrated near specific point sources. Most of the PAHs which enter the aquatic environment are localized in rivers, lakes, estuaries and coastal marine waters, and their concentrations tend to decrease rapidly with distance from their source (CCREM 1991). There is usually a direct relationship between PAH concentrations in the aquatic environment and the degree of industrialization or urbanization in a watershed (NRCC 1983).

PAHs are introduced to the environment usually bound to particles, most commonly soot (carbon particles). While water- and land-based discharges may represent significant and localized inputs, atmospheric deposition is another significant route of entry of PAHs into aquatic environments (NRCC 1983, CCREM 1991). In the aquatic environment, PAHs tend to adsorb to particulate material and settle out in aquatic sediments. Since microbial degradation of PAHs in cold sediments is slow, these compounds persist, especially in colder climates.

All PAHs are toxic to some extent and can cause a number of effects in aquatic organisms (Environment Canada 1991). Some PAHs, especially benzo[a]pyrene (B[a]P) are carcinogenic to fish, animals and humans (IJC 1991). Fish taken from PAH-contaminated waters, for example, often have an increased frequency of liver and other tumours (Environment Canada *et al.* 1991). PAHs also can disrupt the glycoprotein coating on the surface of gills and other epithelia of fish, taint the flesh of lobsters, suppress the immune systems of clams, and reduce the egg production and hatching success of both marine vertebrates and invertebrates (J. Kiceniuk personal communication).

In vertebrates, liver mono-oxygenase enzymes metabolize PAHs, usually to hydroxylated products some of which can bind to DNA and initiate carcinogenesis (Dunn 1991). In this situation, it is not clear whether mono-oxygenase activity represents detoxification (by metabolizing PAHs as a first step in their excretion) or an exacerbation of their toxicity by converting them to a more toxic product. In either case, the result is that PAHs tend not to be accumulated to any significant extent in vertebrates. Their break-down products or metabolites, known as adducts, however, may be more harmful than their parent compounds. Invertebrates lack a mono-oxygenase system, and although they can metabolize PAHs by other processes, they accumulate them more than vertebrates.

METALS AND ORGANOMETALLIC COMPOUNDS

In certain forms and in sufficient concentrations, heavy metals such as cadmium, lead, mercury, and tin can be toxic to living organisms (Government of Canada 1991). These metals are naturally present in the environment but are not considered to be biologically essential. While their natural concentrations are usually too low to pose any threat to aquatic life, human activities have greatly increased their concentrations in some aquatic ecosystems (Nriagu and Pacyna 1988; Nriagu 1990). Metals in aquatic biota can, however, have natural or anthropogenic sources. As a result, it is sometimes difficult to differentiate the importance of diverse sources.

In uncontaminated aquatic sediments, metals are usually in mineral lattices or bound tightly to organic molecules, and so are not readily biologically available (Environment Canada 1991). Dissolved metals from riverine or effluent sources tend to precipitate upon entering seawater. Although metals in these forms generally pose little direct threat to aquatic life, under certain conditions they can become transformed or methylated into biologically available forms which are toxic (Environment Canada 1991).

CADMIUM

Cadmium is often found at elevated concentrations in aquatic organisms (Noble 1990). High concentrations in aquatic life can result from both natural cycling or from human activities. The main natural sources of cadmium include the windblown transport of soil particles, forest fires and volcanic emissions. Industrial sources include base metal mines, smelters, fertilizer plants and thermal electric plants (Noble 1990). Cadmium is used in electroplating other metals or alloys for protection against corrosion, in the manufacture of pigments, nickel-cadmium storage batteries, solders, electronic equipment, lubricants, photography supplies, glass, ceramics and as a stabilizer in plastics (CCREM 1991). Some cadmium compounds are used in the production of polyvinyl chloride (PVC) and in tubes for television sets (CCREM 1991).

Metallic cadmium is insoluble in water; however, several cadmium compounds are soluble. In general, cadmium is found attached to organic and inorganic particulate matter. Its mobility in aquatic ecosystems is increased by low pH, low water hardness, low levels of particulate matter and low salinity (U.S. EPA 1979; CCREM 1991).

Cadmium is accumulated by aquatic organisms, including macrophytes, phytoplankton, zooplankton, invertebrates and fish (Florence 1982; CCREM 1991). Water hardness may influence the uptake of cadmium by aquatic organisms; tissue concentrations decrease with increasing water hardness (Kincade and Erdman 1975; CCREM 1991). A reduction in water pH such as that resulting from acid rain can lead to increased uptake of cadmium especially in bacteria and algae (CCREM 1991). Bioaccumulation of cadmium in aquatic organisms also tends to increase with increasing water temperature (Remacle *et al.* 1982; Rombough and Garside 1982; CCREM 1991).

While cadmium bioaccumulates in the tissues of aquatic organisms, however, it does not biomagnify in aquatic food chains.

The main clinical signs of cadmium toxicity in aquatic organisms are anemia, retarded gonad development, enlarged joints, scaly skin, liver and kidney damage, and reduced growth. While the toxicity of cadmium to fish and other aquatic organisms is greatly influenced by species type, it is also known that age and life cycle stages are important in regulating its impact on specific organisms. Fish embryos and newly hatched alevins, for example, are relatively tolerant to cadmium but young birds may be more susceptible than older birds. Despite these variations, however, because the excretion of biologically incorporated cadmium is relatively slow, in general most organisms continue to accumulate it over their lifetime.

LEAD

Lead is mobilized in the environment by natural processes such as the weathering of rocks, volcanic activity and radioactive decay of uranium and thorium. Until leaded gasoline was phased out in 1990, the combustion of motor fuels containing tetraethyl lead was the largest anthropogenic contributor of lead to the environment (Government of Canada 1991). Other anthropogenic activities resulting in important contributions to environmental lead include mining and smelting of lead-bearing ores and metals, use of lead plumbing and solder, thermal power plants, sewage sludge, paints and ceramic glazes, and the disposal of lead-zinc batteries in landfill sites. Lead also enters the environment through the use of lead shot and lead sinkers for fishing (Noble 1990).

Beyond increasing the total amount lead released into the environment, anthropogenic uses have substantially altered the natural environmental distribution of lead both locally and globally (Hoffman *et al.* 1995). High lead emissions from urban centres, for example, have increased urban and near urban concentrations to several thousand times greater than those found in natural concentrations. Lead reaches the aquatic environment through precipitation, fallout of lead dust, street run-off and industrial and municipal wastewater discharges (U.S. EPA 1976; Jaques 1985; CCREM 1991).

The chemical form of lead in aquatic ecosystems varies and depends on factors such as pH, dissolved oxygen and the presence of coexisting inorganic and organic compounds. In general, lead compounds that are soluble are removed from solution by adsorption to sediments and particulate matter suspended in water (Leland *et al*. 1974; CCREM 1991).

Lead is a non-essential, highly toxic metal, and all its known effects on biological systems are deleterious (Hoffman *et al*. 1995). Its bioaccumulation by aquatic organisms, however, depends on a wide range of factors including the amount of lead present in the immediate environment, its physical and chemical form, the physical and chemical properties of the water body into which it has been deposited, and the biology of the exposed organisms. Organic lead compounds which accumulate in lipids, for example, tend to be taken up and accumulated more readily than inorganic lead compounds (Hoffman *et al*. 1995). Similarly, organic lead compounds are generally more toxic to aquatic organisms than inorganic compounds, and toxicity tends to increase with the degree of alkylation.

Although biomagnification does not appear to occur with lead, there is significant evidence that it does bioaccumulate in a number of aquatic organisms, including benthic bacteria, freshwater plants, invertebrates and fish (CCREM 1991). One method now used to indicate lead exposure in fish is through the measurement of erythrocyte d-amino levulinic acid dehydratase (ALAD)

activity (Hodson *et al*. 1983). Although the effects of lead may vary according to a number of factors such as those described above, low concentrations of lead have been associated with anemia, reduced growth rates, kidney and liver damage, the depression of immune systems, and possibly some impairment of nervous systems especially in younger organisms, including humans (Scheuhammer 1987; Noble 1990). Lead, in fact, is the only toxic contaminant that has accumulated in humans to average concentrations that approach the threshold for potential clinical poisoning (Southwood 1983; Hoffman *et al*. 1995).

MERCURY

Anthropogenic activities have greatly increased mercury concentrations in some parts of the environment including aquatic ecosystems (Government of Canada 1991). In Canada, the main anthropogenic sources of mercury during the 1960s were chlor-alkali plants that produced bleach and pulp mills that used mercurial slimicides (Fimreite 1971; Noble 1990). These industries were often located on the shores of large rivers and lakes such as in the Great Lakes – St. Lawrence River system, or along the coasts. In 1970, legislation was passed to reduce mercury discharges from sources such as smelters, municipal wastewater treatment plants, and pulp mills. An additional use of mercury, in seed dressings, was banned in 1973 (Noble 1990).

Today, anthropogenic sources of mercury in the environment include leaching from areas of past industrial activity, smelting, coal-burning power plants, the use of some paints and the release of mercury from soil and plants following flooding behind hydroelectric dams (Bodaly *et al*. 1984; Noble 1990). Much of the mercury released by these processes is in a form that poses little threat to aquatic organisms. Bacteria, however, can transform mercury into an organic form of mercury called methylmercury which is toxic.

Methylmercury readily biomagnifies through aquatic food chains (Scheuhammer 1987). It is chemically stable and accumulates in various tissues including muscle (Noble 1990). While methylmercury readily accumulates in some aquatic organisms, animals such as otters, seals and whales, can demethylate it back to the less harmful inorganic mercury (Hoffman *et al*. 1995).

Mercury accumulates rapidly in fish. In Canadian fresh waters it has been found in high concentrations in lake trout, pike and walleye (NRCC 1979; CCREM 1991). Its capacity to biomagnify in aquatic food webs was revealed the presence of high concentrations in fish-eating birds sampled in the late 1960s in areas near chlor-alkali plants and pulp mills using slimicides (Firmreite 1971; Noble 1990). Mercury can inhibit photosynthesis and growth in phytoplankton and is known to cause nervous system disorders, spinal cord degeneration and brain lesions in birds (Scheuhammer 1987; Noble 1990). It can also accumulate in the kidney, liver and brain of fish-eating mammals, including humans, resulting in neurological damage, kidney damage, severe weight loss and even death (Government of Canada 1991; Environment Canada/US EPA 1995).

ORGANOTIN COMPOUNDS

Major uses of organotin compounds in Canada include their use as a heat stabilizer in polyvinyl chloride (PVC) products; catalysts in resins, paints and coatings; and as additives to anti-fouling paints used to protect aquaculture nets and ship hulls. The latter use, however, was restricted in 1989 to vessels greater than 25 m in length (Government of Canada 1993). The release of organotin compounds into the aquatic environment is, in general, the result of direct input into the air or water from production, processing, use or disposal, and indirectly through biotic and abiotic transformations (Thompson *et al*. 1985; Evans and Karpel 1985). The most significant entry into the aquatic environment is from their use, especially that of tributyltin (TBT), in biocidal applications where high concentrations have been attributed to the use of biocidal paints (Thompson *et al*. 1985).

Little information is available on the forms and fate of organotin compounds in the aquatic environment (CCREM 1991). Their persistence is known, however, to depend significantly on ecosystem characteristics, including light, temperature, oxygen concentration and the kinds and concentrations of other organisms that may be present (DFO 1996). In general, they also do not volatilize from water (Maguire *et al*. 1983; Government of Canada 1993). Accordingly, given the climatic conditions present in Canada they may persist for periods ranging from years to decades.

Although methylation and de-methylation of organotin compounds occurs in the aquatic environment via both abiotic and biotic mechanisms they may, nonetheless be bioaccumulated by some organisms (CCREM 1991). Studies indicate that these compounds, especially TBT, cause shell chambering and thickening in bivalves, such as oysters, and induce abnormalities in the reproductive organs of female whelks (snails) which may lead to reduced fertility, possible sterility and eventual population declines (DFO 1996).

References

BODALY, R.A., R.E. Hecky, and R.J.P. Fudge. 1984. Increases in Fish Mercury Levels in Lakes Flooded by the Churchill River Diversion, Northern Manitoba. Can. J. Fish. Aquat. Sci. 41: 682-691.

BROOKS, G.T. 1974. Chlorinated Insecticides. Vol. 1. Technology and Applications. CRC Press, Inc. Cleveland, Ohio.

CANADIAN Council of Resource and Environment Ministers (CCREM). 1991. Can. Wat. Qual. Guide. Inland Wat. Direct., Environ. Can., Ont.

CARSON, R. 1962. Silent Spring. Houghton-Mifflin Co. Boston.

COCHRANE, W.P., and R. Greenhalgh. 1976. Chemical Composition of Technical Chlordane. J. Off. Analyt. Chemists. 59: 696-702.

DEPARTMENT of Fisheries and Oceans (DFO). 1996. Institute of Ocean Sciences Fact Sheet: Snail's Pace Recovery From Restricted Toxic Paint. Inst. Ocean Sci. Sidney, B. C.

DUNN, B.P. 1991. Carcinogen adducts as an indicator for the public health risks of consuming carcinogen-exposed fish and shell-fish. Env. Health Pers. 90: 111-116.

EISLER, R., and J. Jacknow. 1985. Toxaphene Hazards to Fish, Wildlife and Invertebrates: A Synoptic Review. U.S. Fish Wild. Serv. Biol. Rept No. 85.

ENVIRONMENT Canada. 1985. Canada Gazette Announcements - Environmental Contaminants Act. Environ. Prot. Serv., Ottawa, Ont.

ENVIRONMENT Canada. 1991. Health of Our Oceans: A Status Report on Canadian Marine Environmental Quality. Peter G. Wells, and Susan J. Rolston [eds.] Min. Sup. Serv., Ottawa, Ont.1991.

ENVIRONMENT Canada, and U.S. Environmental Protection Agency. 1995. State of the Great Lakes 1995.

ENVIRONMENT Canada, Department of Fisheries and Oceans, and Health and Welfare Canada. 1991. Toxic Chemicals in the Great Lakes and Associated Effects: Synopsis. Min. of Sup. Serv. Can., Ottawa, Ont.

EVANS, C.J., and S. Karpel. 1985. Organotin Compounds In Modern Technology. Elsevier SCI. Publ., Amsterdam, The Netherlands.

FIMREITE, N. 1971. Mercury uses in Canada and their possible hazards as sources of mercury. Environ. Poll. 1: 119-131.

FLORENCE, T.M. 1982. The speciation of trace elements in water. Talanta 29: 345-364.

FRANK, R., M. Van Hove Holdrinet, and W.A. Rapley. 1975. Residue of Organochlorine Compounds and Mercury in Bird's Eggs from the Niagara Peninsula, Ont. Arch. Environ. Contam. Toxic. 3: 205-218.

GOVERNMENT of Canada. 1991. The State of Canada's Environment. Min. of Sup. Serv. Can., Ottawa, Ont.

GOVERNMENT of Canada, Environment Canada, and Health Canada. 1993. Canadian Environmental Protection Act Priority Substances List Assessment Report: Non-pesticidal Organotin Compounds. Min. Sup. Serv. Can., Ottawa, Ont.

HODSON, P.V., B.R. Blunt, and D.M. Whittle. 1983. Suitability of a Biochemical Method for Assessing the Exposure of Feral Fish to Lead, p. 389-405. *In* W.E. Bishop, R.D. Cardwell, and B.B. Heidolph [eds.] Aquatic Toxicology and Hazard Assessment: Sixth Symposium, ASTM STP 802. Amer. Soc. Test. Mat., Philadelphia., Penn.

HOFFMAN, D.J., B.A. Rattner, G.A. Burton, Jr., and J. Cairns, Jr. 1995. Handbook of Ecotoxicology. Ann Arbor, Mich. CRC Press, Inc.

INTERNATIONAL Joint Commission (IJC). 1991. Cleaning up our Great Lakes: A Report from the Water Quality Board to the International Joint Commission on Toxic Substances in the Great Lakes Basin Ecosystem.

JAQUES, A.P. 1985. National Inventory of Sources and Releases of Lead (1982). Environ. Prot. Serv., Environ. Can., Ottawa, Ont.

KAISER, K.L.E. 1978. The rise and fall of mirex. Environ. Sci. Technol. 12: 520-528.

KINCADE, M.L., and H.E. Erdman. 1975. The influence of hardness components (Ca^{2+} and Mg^{2+}) in water on the uptake and concentration of cadmium in a simulated freshwater ecosystem. Environ. Res. 10:308-317.

LELAND, H.V., S.S. Shukla, and N.F. Shimp. 1974. Factors affecting distribution of lead and other trace elements in sediments of southern Lake Michigan. *In* P.C. Singer [ed.] Trace Metals and Metal-Organic Interactions in Natural Waters. Ann Arbor Sci. Publi. Inc., Ann Arbor, Mich.

MAGUIRE, R.J., J.H. Carey, and E.J. Hale. 1983. Degredation of tri-n-butyltin species in water. J. Agric. Food Chem. 31: 1060-1065.

METCALF, R.L., J.P. Kapoor, P.-Y. Lu, C.K. Schuth, and P. Sherman. 1973. Model ecosystem studies of the environmental fate of sic organochlorine pesticides. Environ. Health Perspect. 4: 35-44.

NEELY, W.B. 1980. A method for selecting the most appropriate environmental experiments on a new chemical, p. 287-296. *In* R. Haque [ed.] Dynamics, Exposure and Hazard Assessment of Toxic Chemicals. Ann Arbor Sci. Publ. Inc., Ann Arbor, Mich. 287-296 p.

NOBLE, David G. 1990. Contaminants in Canadian Seabirds. State Environ. Rep. No. 90-2. Min. Sup. Serv., Ottawa, Ont.

NRCC. 1979. Effects of Mercury in the Canadian Environment. Assoc. Comm. Scient. Crit. Environ. Qual., Natl. Res. Coun. Can., Ottawa, Ont. NRCC No. 16739.

NRCC 1983. Polycyclic Aromatic Hydrocarbons in the Aquatic Environment: Formation, Sources, Fate and Effects on Aquatic Biota. Assoc. Comm. Scient. Crit. Environ. Qual., Nat. Res. Coun. Can., Ottawa, Ont. NRCC No. 18981.

NRIAGU, J.O. 1990. Global metal pollution: Poisoning the biosphere? Environ. 32 (7): 7-11, 28-33.

NRIAGU, J.O., and J.M. Pacyna. 1988. Quantitative assessment of worldwide contamination of air, water and soils by trace metals. Nature 333: 134-139.

REMACLE, J., C. Houba, and J. Ninane. 1982. Cadmium fate in bacterial microcosms. Wat. Air Soil Poll. 18: 455-465.

ROMBOUGH, P.J., and E.T. Garside. 1982. Cadmium toxicity and accumulation in eggs and alevins of Atlantic salmon (*Salmo salar*). Can. J. Zool. 60: 2006-2014.

SCHEUHAMMER, A.M. 1987. The Chronic toxicity of Aluminum, Cadmium, Mercury and Lead in Birds: A Review. Environ. Poll. 46: 263-295.

SERGEANT, D.B., M. Munawar, P. V. Hodson, D. T. Bennie, and S.Y. Huestis. 1993. Mirex in the North American Great Lakes: New Detections and Their Confirmation. J. Great Lakes Res. 19(1): 145-157.

SMITH, J.H., W.R. Mabey, N. Bohonos, B.R. Holt, S.S. Lee, T.-W. Chow, D.C. Bomberger, and T. Mill. 1978. Environmental Pathways of Selected Chemicals in Freshwater Systems. Part II. Laboratory studies. Off. Res. Dev., U.S. Environ. Prot. Agency, Athens, Georgia. EPA-600/7-78-074.

SOUTHWOOD, T.R.E. 1983. Lead in the Environment. Cmnd. 8852. HMSO, Royal Comm. Environ. Poll., London, England.

STICKEL, L.F., W.H. Stickel, R.D. McArthur and D.L. Hughes. 1979. Chlordane in Birds: A Study of Lethal Residues and Loss Rates. In W.B. Deichman (org.), Proceedings of 10th International Conference on Toxicology and Occupational Medicine. North Holland, N. Y.: Elsevier. 387-396 p.

THOMPSON, J.A.J., M.G. Sheffer, R.C. Pierce, Y.K. Chau, J.J. Cooney, W.R. Cullen, and R.J. Maguire. 1985. Organotin Compounds in the Aquatic Environment: Scientific Criteria for Assessing their Effects on Environmental Quality. Asso. Comm. Scient. Crit. Environ. Qual., Natl. Res. Coun. Can., Ottawa, Ont. NRCC No. 22494.

U.S. EPA. 1976. Quality Criteria for Water. Off. Wat. Plan. Stand., U. S. Environ. Prot. Agency, Washington, D.C. EPA-440/9-76-023.

U.S. EPA. 1979. "Cadmium" in Water-Related Environmental Fate of 129 Priority Pollutants. Vol. 1. Introduction, Technical Background, Metals and Organics, Pesticides, Polychlorinated Biphenyls. Off. Wat. Plan. Stand., U.S. Environ. Prot. Agency, Washington, D.C. EPA-440/4-79-029a.

U.S. EPA. 1980. Ambient Water Quality Criteria for Toxaphene. Off. Wat. Reg. Stand., Crit. Stand. Div., U. S. Environ. Prot. Agency, Washington, D.C. EPA-440/5-80-076.

VERSCHUEREN, K. 1983. Handbook on Environmental Data on Organic Chemicals. Van Nostrand Reinhold Co., N. Y.

Additional reading

ENVIRONMENT Canada, and U.S. Environmental Protection Agency. 1995. S.O.L.E.C. 1994 State of the Lakes Ecosystem Conference Background Paper: Toxic Contaminants.

GOVERNMENT of Canada, Environment Canada, and Health Canada. 1990. Canadian Environmental Protection Act Priority Substances List Assessment Report No 1: Polychlorinated Dibenzodioxins and Polychlorinated Dibenzofurans. Min. Sup. Serv. Can., Ottawa, Ont.

GOVERNMENT of Canada, Environment Canada, and Health Canada. 1994. Canadian Environmental Protection Act Priority Substances List Assessment Report: Polycyclic Aromatic Hydrocarbons. Min. Sup. Serv. Can., Ottawa, Ont.

Personal Communications

KICENIUK, J. Department of Fisheries and Oceans, St. John's, Newfoundland. 1995.

List of
SPECIES

Common and Scientific Names of Biota of Importance to Contaminant Transfer and Accumulation

ACADIAN REDFISH	*SEBASTES FASCIATUS*
ALEWIFE	*ALOSA PSEUDOHARENGUS*
AMERICAN EEL	*ANGUILLA ROSTRATA*
AMERICAN LOBSTER	*HOMARUS AMERICANUS*
AMERICAN OYSTER	*CRASSOSTREA VIRGINICA*
AMERICAN PLAICE	*HIPPOGLOSSOIDES PLATESSOIDES*
AMERICAN SAND LANCE	*AMMODYTES AMERICANUS*
AMERICAN SHAD	*ALOSA SAPIDISSIMA*
ARCTIC CHAR	*SALVELINUS ALPINUS*
ARCTIC COD	*BOREOGADUS SAIDA*
ARCTIC GRAYLING	*THYMALLUS ARCTICUS*
ATLANTIC COD	*GADUS MORHUA*
ATLANTIC HALIBUT	*HIPPOGLOSSUS HIPPOGLOSSUS*
ATLANTIC HERRING	*CLUPEA HARENGUS HARENGUS*
ATLANTIC MACKEREL	*SCOMBER SCOMBRUS*
ATLANTIC PORBEAGLE SHARK	*LAMNA NASUS*
ATLANTIC PUFFIN	*FRATERCULA ARCTICA*
ATLANTIC SALMON	*SALMO SALAR*
ATLANTIC STURGEON	*ACIPENSER OXYRHYNCHUS*
ATLANTIC TOMCOD	*MICROGADUS TOMCOD*
ATLANTIC WHITE-BEAKED DOLPHIN	*LAGENORHYNCHUS ALBIROSTRIS*
BALD EAGLE	*HALIAEETUS LEUCOCEPHALUS*
BEARDED SEAL	*ERIGNATHUS BARBATUS*
BELUGA	*DELPHINAPTERUS LEUCAS*
BLACK GUILLEMOT	*CEPPHUS GRYLLE*
BLACK-LEGGED KITTIWAKE	*RISSA TRIDACTYLA*
BLUE MUSSEL	*MYTILUS EDULIS*
BLUEBACK HERRING	*ALOSA AESTIVALIS*
BLUEFIN TUNA	*THUNNUS THYNNUS*
BOWHEAD WHALE	*BALAENA MYSTICETUS*

BROOK SILVERSIDE	*LABIDESTHES SICCULUS*
BROWN BULLHEAD	*ICTALURUS NEBULOSUS*
BULL TROUT	*SALVELINUS MALMA*
BURBOT (EELPOUT)	*LOTA LOTA*
CANADA GOOSE	*BRANTA CANADENSIS*
CAPELIN	*MALLOTUS VILLOSUS*
CARIBOU	*RANGIFER TARANDUS GRANTI*
CASPIAN TERN	*HYDROPROGNE CASPIA*
CHINOOK SALMON	*ONCORHYNCHUS TSHAWYTSCHA*
CHUM	*ONCORHYNCHUS KETA*
CISCO	*COREGONUS ARTEDII*
COHO SALMON	*ONCORHYNCHUS KISUTCH*
COMMON DOLPHIN	*DELPHINUS DELPHIS*
COMMON EIDER	*SOMATERIA MOLLISSIMA*
COMMON LOON	*GAVIA IMMER*
COMMON MURRE	*URIA AALGE*
COMMON TERN	*STERNA HIRUNDO*
COPPER ROCKFISH	*SEBASTES CAURINUS*
CRAYFISH	*ORCONECTES SP.*
DALL'S PORPOISE	*PHOCOENOIDES DALLI*
DOUBLE-CRESTED CORMORANT	*PHALACROCORAX AURITUS*
DUNGENESS CRAB	*CANCER MAGISTER*
ENGLISH SOLE	*PAROPHRYS VETULUS*
EUPHAUSID	*EUPHAUSIA SPP.*
EURASIAN RUFFE	*GYMNOCEPHALUS CERNUUS*
GOLDEYE	*HIODON ALOSOIDES*
GREEN SEA URCHIN	*STRONGYLOCENTROTUS DROEBACHIENSIS*
GREENLAND HALIBUT (TURBOT)	*REINHARDTIUS HIPPOGLOSSOIDES*
GREY SEAL	*HALICHOERUS GRYPUS*
HADDOCK	*MELANOGRAMMUS AEGLEFINUS*
HAKE	*UROPHYCIS TENUIS*
HARBOUR PORPOISE	*PHOCOENA PHOCOENA*
HARBOUR SEAL	*PHOCA VITULINA CONCOLOR*
HARP SEAL	*PHOCA GROENLANDICA*
HERRING GULL	*LARUS ARGENTATUS*
HOODED SEAL	*CYSTOPHORA CRISTATA*

HOOKER'S SEA LION	*PHOCARCTOS HOOKERI*
HUMPBACK WHALE	*MEGAPTERA NOVAEANGLIAE*
KILLER WHALE	*ORCINUS ORCA*
KOKANEE SALMON	*ONCORHYNCHUS NERKA*
LAKE TROUT	*SALVELINUS NAMAYCUSH*
LAKE WHITEFISH	*COREGONUS CLUPEAFORMIS*
LEACH'S STORM PETREL	*OCEANODROMA LEUCORHOA*
LOGPERCH	*PERCINA CAPRODES*
LONG-FINNED SQUID	*LOLIGO PEALEI*
LONGNOSE SUCKER	*CATOSTOMUS CATOSTOMUS*
MOUNTAIN WHITEFISH	*PROSOPIUM WILLIAMSONI*
MUD DOG WHELK	*NASSARIUS OBSOLETUS*
MUSKELLUNGE	*ESOX MASQUINONGY*
MYSID SHRIMP	*MYSIS RELICTA*
NARWHAL	*MONODON MONOCEROS*
NORTHERN FULMAR	*FULMARUS GLACIALIS*
NORTHERN GANNET	*MORUS BASSANUS*
NORTHERN PIKE	*ESOX LUCIUS*
NORTHERN SHRIMP	*PANDALUS BOREALIS*
PACIFIC GEODUCK CLAM	*PANOPEA ABRUPTA*
PACIFIC COD	*GADUS MACROCEPHALUS*
PACIFIC HALIBUT	*HIPPOGLOSSUS STENOLEPIS*
PACIFIC OYSTER	*CRASSOSTREA GIGAS*
PILOT WHALE	*GLOBICEPHALA MELAENA*
PLAICE	*HIPPOGLOSSOIDES PLATESSOIDES*
POLAR BEAR	*URSUS MARITIMUS*
POLLOCK	*POLLACHIUS VIRENS*
PRAWN	*PANDALUS PLATYCEROS*
RAINBOW SMELT	*OSMERUS MORDAX*
RAINBOW TROUT	*ONCORHYNCHUS MYKISS*
RIGHT WHALE	*EUBALAENA GLACIALIS*
RINGED SEAL	*PHOCA HISPIDA*
RIVER HERRING	*ALOSA AESTIVALIS*
RIVER OTTER	*LUTRA CANADENSIS*
ROCK GRENADIER	*CORYPHAENOIDES RUPESTRIS*
ROUND GOBY	*NEOGOBIUS MELANOSTOMUS*

ROUND WHITEFISH	*PROSOPIUM CYLINDRACEUM*
SABLEFISH	*ANOPLOPOMA FIMBRIA*
SAUGER	*STIZOSTEDION CANADENSE*
SEA OTTER	*ENHYDRA LUTRIS*
SEA SCALLOP	*PLACOPECTEN MAGELLANICUS*
SHORT CLAM	*MYA TRUNCATA*
SHORT-FINNED SQUID	*ILLEX ILLECEBROSUS*
SILVER HAKE	*MERLUCCIUS BILINEARIS*
SMALLMOUTH BASS	*MICROPTERUS DOLOMIEUI*
SNOW CRAB	*CHIONOECETES OPILIO*
SOCKEYE SALMON	*ONCORHYNCHUS NERKA*
SOFT-SHELL CLAM	*MYA ARENARIA*
SPINY DOGFISH	*SQUALUS ACANTHIAS*
SPOTTAIL SHINER	*NOTROPIS HUDSONIUS*
SPRUCE BUDWORM	*CHORISTONEURA FUMIFERANA*
STEELHEAD	*ONCORHYNCHUS MYKISS*
STELLER SEA LION	*EUMETOPIAS JUBATUS*
STICKLEBACK	*APELTES QUADRACUS*
STRIPED BASS	*ROCCUS SAXATILIS*
SWORDFISH	*XIPHIAS GLADIUS*
THORNY SKATE	*RAJA RADIATA*
WALLEYE (YELLOW PICKEREL)	*STIZOSTEDION VITREUM VITREUM*
WALRUS	*ODOBENUS ROSMARUS*
WHITE PERCH	*APLODINOTUS GRUNNIENS*
WHITE STURGEON	*ACIPENSER TRANSMONTANUS*
WHITE SUCKER	*CATOSTOMUS COMMERSONI*
WINTER FLOUNDER	*PSEUDOPLEURONECTES AMERICANUS*
WRINKLED WHELK	*THAIS LAMELLOSA*
YELLOW PERCH	*PERCA FLAVESCENS*
YELLOWTAIL FLOUNDER	*LIMANDA FERRUGINEA*
ZEBRA MUSSEL	*DREISSENA POLYMORPHA*

Units of Measurement

UNITS OF WEIGHT

1 kilogram (kg)	=	1000 grams (g)
1 gram (g)	=	10^{-3} kg
1 milligram (mg)	=	10^{-3} g
1 microgram (μg)	=	10^{-6} g
1 nanogram (ng)	=	10^{-9} g
1 picogram (pg)	=	10^{-12} g

UNITS OF VOLUME

1 litre (L)

1 cubic metre (m^3)

FRACTIONAL MULTIPLIERS (COEFFICIENTS)

m	= milli	=	10^{-3}
μ	= micro	=	10^{-6}
n	= nano	=	10^{-9}

WEIGHT TO WEIGHT RELATIONSHIPS

1 part per million (ppm)	=	1 mg/kg or 1 μg/g	=	1 μg.g^{-1}
1 part per billion (ppb)	=	1 μg/kg or 1 ng/g	=	1 ng.g^{-1}
1 part per trillion (ppt)	=	1 ng/kg or 1 pg/g	=	1 pg.g^{-1}

CONCENTRATIONS IN LIQUID

1 part per million (ppm)	=	1 mg/kg or 1 μg/g	=	1 μg.L^{-1}
1 part per billion (ppb)	=	1 μg/kg or 1 ng/g	=	1 ng.L^{-1}
1 part per trillion (ppt)	=	1 ng/kg or 1 pg/g	=	1 pg.L^{-1}

WEIGHT TO VOLUME RELATIONSHIPS

1 part per million (ppm)	=	1 mg/m^3	=	1 mg.m^{-3}
1 part per billion (ppb)	=	1 μg/m^3	=	1 μg.m^{-3}
1 part per trillion (ppt)	=	1 ng/m^3	=	1 ng.m^{-3}

In general, the International System (SI) has been used. For considerations of spacing and enhanced readability, the "dot" indicating multiplication has been placed on the line and the term "yr" has been used for year. L is used for the symbol for litre, as per Canadian practice. Where possible, numerical expressions are reduced to two indicative figures for simplicity. Where possible, varying local usage, especially for capitalization, has been retained for generic geographic features. The references and reference lists were supplied by authors. As per practice of the Department of Fisheries and Oceans, the term "Aboriginal" refers to the First Nations of Canada. In tables, DL indicates a measurement at the detection limit. NA, a dash, or a space indicates not available or not applicable.

Index